Flexible Polymer Chains in Elongational Flow

Springer

Berlin
Heidelberg
New York
Barcelona
Hong Kong
London
Milan
Paris
Singapore
Tokyo

Tuan Q. Nguyen · Hans-Henning Kausch (Eds.)

Flexible Polymer Chains in Elongational Flow

Theory and Experiment

With 254 Figures and 12 Tables

 Springer

Dr. Tuan Quoc Nguyen
Prof. Hans-Henning Kausch

Polymer Laboratory
Swiss Federal Institute of Technology
MX-D
CH-1015 Lausanne
Switzerland

ISBN 3-540-65181-0 Springer-Verlag Berlin Heidelberg New York

Library of Congress Cataloging-in-Publication Data
Flexible polymer chains in elongational flow : theory and experiment
/ Tuan Q. Nguyen, Hans-Henning Kausch (eds.)
p. cm.
Includes bibliographical references and index.
ISBN 3-540-65181-0 (hc. : alk. paper)
1. Polymer solutions. 2. Rheology. I. Nguyen, Q. Tuan, 1949–
. II. Kausch, H.H.
QD381.9.S65F54   1999
547′.7045413--dc21   99–17764

© Springer-Verlag Berlin Heidelberg 1999
Printed in Germany

The use of general descriptive names, registered names, trademarks, etc. in this publication does not imply, even in the absence of a specific statement, that such names are exempt from the relevant protective laws and regulations and free for general use.

Coverdesign: Design & Produktion GmbH, Heidelberg
Typesetting: Fotosatz-Service Köhler GmbH, Würzburg

SPIN: 10507054   2/3020-5 4 3 2 1 0 – printed on acid-free paper

Preface

The behavior of polymer solutions in simple shear flows has been the subject of considerable research in the past. On the other hand, reports on polymers in elongational flow have appeared comparatively recently in the literature. Elongational flow with an inherent low vorticity is known to be more effective in extending polymer chains than simple shear flow and thus is more interesting from the point of view of basic (molecular chain dynamics at high deformation) and applied polymer science (rheology, fiber extrusion, drag reduction, flow through porous media). Undoubtly, one landmark in the field of polymer dynamics in elongational flow was the notion of critical strain-rate for chain extension, initially put forward by A. Peterlin (1966) and later refined into the "coil-stretching" transition by P.G. de Gennes and H. Hinch (1974). In the two decades which followed, significant progress in the understanding of chain conformation in "strong" flow has been accomplished through a combination of advances in instrumentation, computation techniques and theoretical studies. As a result of the multidisciplinary nature of the field, information on polymer chains in "strong" flow is accessible only from reviews and research papers scattered in disparate scientific journals. An important objective of this book is to remedy that situation by providing the reader with up-to-date knowledge in a single volume. The editors therefore invited leading specialists to provide both fundamental and applied information on the multiple facets of chain deformation in elongational flow. An important criterion in the selection of subjects was to achieve a balance between theory and experiment. In addition, topics are restricted to the dilute solution behavior of isolated flexible polymer chains in elongational flow. For this reason, the technically important aspect of flow-induced structure formation, more generally encountered in semi-dilute and concentrated polymer systems, is omitted (this topic has been treated in a book edited by K. Sondergaard & J. Lyngaae-Jorgensen, 1995).

From the important advances reported in this book, from theoretical modling to direct single chain visualization, it seems now that we are close to understanding how a real polymer chain may uncoil itself in an elongational flow field. Details on chain conformation, though, are still unsettled and it is our sincere wish that perusal of this book may instigate directions and challenges for future studies.

April 1999   T.Q. Nguyen
                                                              H.-H. Kausch

Contents

Authors

Shlomo Alexander
(1) Department of Chemical Physics, Weizmann Institute of Science,
Rehovot 76100, Isreal
(2) Department of Physics, Bar-Ilan University, Ramat-Gan 52900, Isreal

Oleg V. Borisov
(1) Johannes Gutenberg Universität, Institut für Physik,
D-55099 Mainz, Germany
(2) Institute of Macromolecular Compounds of the Russian Academy of
Sciences, 199004 St Petersburg, Russia

Françoise Brochard-Wyart
Laboratoire de Physicochimie des Surfaces et Interfaces, Universite Pierre et
Marie Curie, 11 rue Pierre et Marie Curie, F-75231 Paris, Cedex 05, France

Axel Buguin
Laboratoire P.S.I. Institut Curie, 11 rue Pierre et Marie Curie 75231 Paris,
Cedex 05, France

Steve P. Carrington
University of Bristol, H.H. Wills Physics Laboratory, Royal Fort, Tyndall Avenue,
Bristol BS8 1TL, UK

Steven Chu
Department of Physics, Stanford University, Stanford, CA 94305, USA

A. A. Darinskii
Institute of Macromolecular Compounds of the Russian Academy of Sciences,
199004 St Petersburg, Russia

Pierre-Gilles de Gennes
Collège de France, 11 place M. Berthelot, 75231 Paris, Cedex 05, France

Hans-Henning Kausch
Polymer Laboratoy, Swiss Federal Institute of Technology, MX-D
CH-1015 Lausanne, Switzerland

Ronald G. Larson
Bell Laboratories, Lucent Technologies, Room 7F-212, 700 Mountain Ave.,
Murray Hill, NJ 07974-0636, USA

Manuel Laso
Dept of Chemical Engineering, ETSII, José Gutiérrez Abascal, 2,
E-28006 Madrid, Spain

Alejandro J. Müller
Grupo de Polímeros USB, Departmento de Termodinámica y Fenómenos de
Transferencia, y Departmento de Ciencia de los Materiales, Universidad Simón
Bolívar, Aptdo. 89000, Caracas 1080-A, Venezuela

Tuan Q. Nguyen
Polymer Laboratoy, Swiss Federal Institute of Technology, MX-D
CH-1015 Lausanne, Switzerland

Jeff A. Odell
University of Bristol, H. H. Wills Physics Laboratory, Royal Fort,
Tyndall Avenue, Bristol BS8 1TL, UK

Hans Christian Öttinger
ETH Zürich, Department of Materials, Institute of Polymers,
CH-8092 Zürich, Switzerland

Thomas T. Perkins
Department of Physics, Stanford University, Stanford, CA 94305, USA

Marco Picasso
Dept. of Mathematics, Ecole Polytechnique Féderal de Lausanne,
CH-1015 Lausanne, Switzerland

Carlo Pierleoni
(1) Dipartimento di Fisica, Università degli Studi, Via Vetoio, Località Coppito,
I-67100 l'Aquila, Italy
(2) INFM, sezione di Roma I, Università "La Sapienza", 00185 Roma, Italy

Réza Porouchani
Laboratoire de Polymères, Ecole Polytechnique Fédérale de Lausanne,
EPFL-Ecublens, CH-1015 Lausanne, Switzerland

Jean-Paul Ryckaert
Unit of Statistical Physics of Condensed Matter, CP 223, Université Libre de
Bruxelles, Bvd du Triomphe, 1050 Brussels, Belgium

Yitzhak Rabin
Department of Physics, Bar-Ilan University, Ramat-Gan 52900, Isreal

A. E. Sáez
Grupo de Polímeros USB, Departmento de Termodinámica y Fenómenos de Transferencia, y Departmento de Ciencia de los Materiales, Universidad Simón Bolívar, Aptdo. 89000, Caracas 1080-A, Venezuela

Douglas E. Smith
Department of Physics, Stanford University, Stanford, CA 94305, USA

CHAPTER 1

Tortured Chains: An Introduction

P.G. de Gennes

A flexible chain is very sensitive to motions of the surrounding solvent. For a single chain floating in a low molecular weight solvent, the response to weak flows was analyzed long ago by the great pioneers: Zimm, Peterlin, ... [1, 2] and determined experimentally by low frequency acoustic measurements [3]. The first basic feature is a relaxation time T_Z:

$$T_Z \sim \frac{\eta R^3}{kT} \tag{1.1}$$

(η: solvent viscosity, R: coil radius, kT: thermal energy) and the weak flow regime corresponds to shear rates $\dot{\gamma} < 1/T_Z$.

The strong flow processes are more complex: here the chain can be seriously distorted. In simple shear, the flow is the superposition of an elongation and a rotation: the net result is that each chain stretches along the flow lines, then rotates, and contracts, etc. Because of these oscillatory features, the effects of strong *shear* flows are not very dramatic.

The effects of *longitudinal* flows are more spectacular: here there is elongation along one direction (x) and constriction in one (or two) other directions (y, z). The net conclusion is that a coil should elongate during the flow: and when it elongates, it offers more grip to the hydrodynamic forces; thus one expects a runaway when $\dot{\gamma}T_Z > 1$, with a highly stretched polymer in the final state [4]. This "coil stretch transition" has been observed in beautiful experiments by the Bristol group [5, 6], using a four-roller machine to generate the flow, and a birefringence measurement to probe the alignment.

For flexible coils in good solvents, one would expect from Eq. (1.1) a critical shear rate $\dot{\gamma}_c \cong M^{-9/5}$. However, the English data (to the difference of the Lausanne data [7], both described in the present book) give $\dot{\gamma}_c \sim M^{-1.5}$, and this discrepancy is not, as far as I know, understood.

The Bristol group also noticed that the chains can break under flow. For a fully stretched chain of length $L = Na$ (N: number of monomers/coil) the overall pulling force is of order

$$f_{hydro} = \eta \, L^2 \, \dot{\gamma} \tag{1.2}$$

This must be compared to the force for rupture, which is of order

$$f_{bond} = U_b/a \tag{1.3}$$

(U_b: bond energy). In practice, U_b is expected to be controlled by any reactive impurity (oxygen ...) present in the solvent.

If we choose $\dot{\gamma} > \dot{\gamma}_c$ and assume that full stretching has taken place, we are led to expect from Eq. (1.2) that breaking will occur at a shear rate $\dot{\gamma}_f \cong M^{-2}$. This agrees with the English data. On the other hand, the Lausanne data (measuring the alignment through the anisotropic fluorescence of a labeled group), taken in a very dilute limit, give different results: the alignment appears to be in complete for $\gamma > \sim \dot{\gamma}_c$, and they find $\dot{\gamma}_f \sim M^{-1}$. This result would be compatible with a model based on hairpins, where the chain appears as a sequence of $N^{1/2}$ hairpins, each of length $\sim N^{1/2}$, and each fully stretched by the flow.

One remarkable observation of Keller and Odell was that the polymer usually breaks *at the mid point*: this would naturally be expected for a fully stretched object. These questions are discussed at length in [6, 7].

Very long chains (typically 20 microns with a phage DNA) can be made visible by insertion of a few fluorescent dyes, and their shapes can be seen under a microscope. The first experiments along these lines were carried out long ago by M. Yanagida et al. [8]. An interesting interpretation was proposed at that time by H. Kuhn [9] – using ideal chains and ignoring hydrodynamic interactions. A more recent scaling approach, incorporating non-ideality, hydrodynamic interactions, and strong stretching effects, is described by F. Brochard in the present book [10].

The most precise experiments, using optical observations of the shapes, have been performed by S. Chu and coworkers at Stanford [11]. They operate in very dilute solutions, choose one coil, watch it distort, and count the duration of the torture (t_{res}).

Here comes the surprise: two chains, which are exactly identical, and have suffered under the same time t_{res}, may display completely different behaviors. Some elongate simply like a dumb-bell. Some others are folded like a hairpin, along the stretching direction: they unfold more slowly. There are many types, the most resilient species being a globular coil (which may in fact be knotted on itself).

My (tentative) picture for this is the following: the various "types" give us a (distorted) image of what the chain looked like, just before it was subjected to torture. For instance, a coil with a slightly protruding tail may become what Chu et al. call a "half dumb-bell".

This does not necessarily contradict the Bristol observation about chains breaking at the mid-point: it is mainly the first type (elongated dumb-bell) which leads to a strong hydrodynamic force.

These observations have initiated a number of simulations of chains under strong flows, which are described in the present book.

The Stanford group has also performed many experiments on distorted chains in other geometries [12]: for instance, with one chain end attached to a bead, they can push the bead (using optical tweezers) and set the system in constant velocity inside the solvent which is globally at rest. This can generate interesting coil shapes ("stems" and "flowers") which have been analyzed in terms of scaling laws [11, 12].

All this work in dilute solution must be complemented by studies on entangled systems, where the dynamics is very different. In the Bristol experiments [5], some semi-dilute solutions were studied, and they showed a remarkable "flare" of birefringence at a certain value of $\dot{\gamma}$. We do not have much theory available on the dynamics of these systems. But Alexander and Rabin make an interesting remark [13]: the osmotic pressure should be significantly altered by shear.

For melts, many non-linear effects can come into play: some are essentially geometrical, related to tube deformations, etc. Some are related to possible changes of the microscopic friction coefficients under strong shear. One dramatic limit could be the formation of "tight knots" [14].

A further complication of interest, could occur with polydisperse melts: namely *segregation by stretching* [15]. Suppose that the melt has been longitudinally stretched for a long time and then relaxed during a time t, which is shorter than the reptation time of the heavy chains, but longer than the reptation time of the short chains: then the latter returns to an isotropic state, while the former stay stretched, with a certain level of alignment $S = 1/2 < 3\cos^2 \theta - 1 >$. Because of the anisotropy of electric polarisabilities inside each monomer, the two species now appear as different, and tend to segregate, with a Flory parameter χ proportional to S^2? This effect has not been seriously considered by the community of experimentalists. But it may still be there, and play a certain role in various industrial processes.

To conclude: we begin to have a good understanding of what happens to a single coil when it is deformed in longitudinal shear flows. The many coil problem is still mysterious, but the powerful experimental techniques which are now available may change the picture very fast. The current situation is very well presented in this book: it should be of significant value.

References

1. Zimm B (1953) J. Chem Phys 21:1273
2. For a global view of these process, see: de Gennes PG (1990) in: Introduction to Polymer Dynamics. Cambridge U Press
3. Ferry JD (1970) Viscoelastic properties of polymers. Wiley (NY), 2nd ed
4. de Gennes PG (1974) J Chem Phys 60:5030
5. Keller A, Odell J (1985) J Colloid Polymer Sci 263:181
6. Odell J, Carrington A, p 137
7. Nguyen TQ et al., p 185
8. Yanagida M, Hiraoka Y, Katsura I (1982) Cold Spring Harbor, Symposia on Quantitative Biology, 47:177
9. Kuhn H (1984) Chimia 38:6
10. Brochard F, p 41
11. Larson RG, Perkins T, Smith P, Chu S, p 259
12. Perkins TT et al., p 283
13. Rabin Y, Alexander S, p 67
14. de Gennes PG (1984) Macromolecules, 17:703
15. Brochard F, de Gennes PG (1988) CR Acad Sci (Paris) 306:699

CHAPTER 2

Polymer Solutions in Flow: A Non-Equilibrium Molecular Dynamics Approach

Jean-Paul Ryckaert and Carlo Pierleoni

2.1
Introduction

The deformation and the orientation of polymers in flows is a widely studied problem which is of high technological and theoretical interest.

Over recent years, a significant amount of new information has been gathered on this subject by various experimental techniques which probe various polymeric systems undergoing shear or elongational flow.

In dilute solutions under shear, the internal structure of a flexible chain was revealed by small angle neutron scattering (SANS) [1] while the global orientation and deformation of high molecular weight polymers was studied by light scattering (LS) for different solvent qualities [2, 3]. Birefringence measurements on polystyrene in elongational flow [4] lead to new predictions on the molecular weight dependence of the critical elongational rate for molecular coil orientation. Semi-dilute solutions have been recently studied in a shear flow by SANS [5, 6] for both good solvent and theta conditions. Homopolymer [7, 8] and block-copolymer melts [9, 10] are being studied by SANS or LS scattering techniques in order to investigate the structural changes which take place at the molecular level when the material is being deformed: these last experiments should obviously help in interpreting the microscopic foundations of the specific rheological behavior of polymer melts.

On the theoretical side, Rouse or Zimm theories have been extended to treat polymer solutions under flow using either perturbation or renormalization group approaches [11–13]. Semi-dilute solutions in shear or elongational flows have been studied on the basis of hydrodynamic models which interrelate monomer concentration, velocity, and stress fields [14, 15]. Polymer blends in shear flow have been treated by the random phase approximation [16].

The present contribution is concerned with the additional route offered by a molecular modeling of these non-equilibrium situations. Simulation at the molecular level of polymers out of equilibrium is a relatively recent extension of both non-equilibrium molecular dynamics (NEMD) techniques developed on small molecules since the 1980s [17, 18] and molecular dynamics (MD) simulations of polymers in continuous space, the latter requiring a computing power which only became available in the last decade [19–21]. Besides the present study of polymers in solution in flows, NEMD studies of polymer melts have now also appeared [22, 23].

In this chapter, we will restrict ourselves to chains in solution at high (almost infinite) dilution. After an initial study of the equilibrium situation, we will investigate the structural and dynamical single chain properties when the solvent is subjected to shear or to elongational flow. A crucial choice must be made at the start regarding whether the solvent will be treated as a continuous medium or as a set of discrete particles. Considering the solvent as an incompressible fluid, a source of stochastic and viscous forces on the polymer beads, leads to the Brownian dynamics (BD) approach. The obvious gain of BD is that it avoids an explicit consideration of the solvent degrees of freedom. Moreover, infinite dilution results do not require an extrapolation of data obtained at finite concentration but directly follow from considering a single polymer chain in a homogeneous velocity field. If finite concentration results are required, BD can then also be used on a many chains system filling a box with periodic boundary conditions (PBC). The basic model of BD is a chain with or without excluded volume (EV) effects and with or without hydrodynamic interactions (HI). The latter are computed in the long time limit under the hypothesis that the solvent reacts on a time scale much faster than the typical range of relevant chain relaxation times. The BD approach was adopted [24–26] to study the rheology of polymer solutions.

The alternative approach of MD is more basic since a microscopic model of solvent, polymer bead, and intra/inter chain interactions must be specified. EV thermodynamic effects, effective friction forces, and the coupling of random forces (HI), solvent viscosity, polymer relaxation, rheological behavior, etc. find their origin in a unique microscopic model. With present day computers, polymers up to $N \approx 50$ beads can be used to model polymer solutions at concentrations well below the overlap concentration. This last situation is obtained as soon as the side L of the MD box satisfies $L > 3R_g$ (R_g is the chain radius of gyration). However, even at those concentrations, the hydrodynamic coupling between a chain and its images in neighboring cells is present as a spurious effect which is detectable on low k dynamical properties. Care must thus be taken to control this unavoidable subtle system size effect.

In the following, we want to show that simulations of polymers in solution at high dilution are indeed feasible and representative of the experimental situation in many cases, namely at equilibrium, in shear flow, and even to some extent in elongational flow! The equilibrium case is an extremely useful preliminary step to the more delicate treatment of a non-equilibrium situation.

In short, simulations of chains in good solvent at equilibrium [27–30] reproduce the high k Zimm universal relaxation behavior of the chain structure factor, confirming the importance of the HI implicitly present through the solvent molecular dynamics. The same HI between the polymer and its own images do lead, however, to a decrease of the center of mass diffusion coefficient which can be interpreted through a Kirkwood-like approach of a periodic system [29]. What is not affected however is the scaling behavior of D vs N, namely $D \propto N^{-0.55}$ when various chain lengths are intercompared at a unique R_g/L ratio, which means that the "concentration" dependence of D appears only as a prefactor $h(R_g/L)$ where h is a decreasing function of the argument.

On the whole, it is nowadays widely accepted that in a 3D continuous space, chains with $N > 20$ practically obey scaling laws. This point is of fundamental importance since it validates the simulation approach as representative of the real polymer behavior at equilibrium. The main message of the present contribution is that this situation seems to remain valid in non-equilibrium situations (outside the linear regime), at least up to some limit where the external forces start to be active at the local scale. In this chapter, we suggest that some form of scaling persists for chains of different lengths under shear provided they are considered at the same reduced shear rate $\beta = \dot{\gamma}\tau_N$ where $\dot{\gamma}$ is the shear rate and τ_N is the equilibrium longest relaxation time of the chain. Our idea is tested on chains up to $N = 50$ at β around unity. In this intermediate β regime and in good solvent conditions, we observe an anisotropic form of static and dynamic scaling at fixed β. More specifically, in a reference frame, depending only on β, which is associated with the principle axes of the deformed gyration tensor, chain dimensions in the three directions scale with N according to a Flory exponent which increases (or decreases) with β for the extensional (compressional) direction.

Dynamical properties of the chain in stationary shear flow are explored in the same spirit by looking at their evolution with polymer size for chains considered at the same reduced shear rate. We find that the diffusion tensor is almost diagonal in the gyration tensor principle axes reference frame. We also find a form of dynamical scaling when k is oriented along one principal axis of the gyration tensor. Each direction leads to a unique normalized intermediate dynamical structure factor (for all chain sizes and for k values in the high k regime) provided the time is properly renormalized by a power of k.

In our contribution, we have adopted the following scheme. Section 2.2 presents a summary of the microscopic model: Newton's equations are used for solvent and bead particles at equilibrium and are extended to set up an homogeneous flow and, in the latter case, we consider flows with zero trace velocity gradient tensor so that the MD cell adjusts its shape according to the flow but the fluid density remains constant. In principle, these techniques are valid for both shear and elongational flows. In the same section, we also mention the single chain properties used to probe the flow influence on the structure and dynamics of chains. In Sect. 2.3 we briefly discuss our microscopic model and analyze the equilibrium situation for our good solvent model. Section 2.4 is devoted to our shear flow simulations for various polymer sizes and various reduced shear rates. We first discuss the structural properties of the chains (orientation, global, and internal deformation) and relate them to experimental data and theoretical predictions. The good agreement with experimental data implicitly suggests that our (relatively short) chains are nevertheless representative of the longer chains used in real experiments. The implicit scaling behavior is then analyzed in detail. A dynamical scaling analysis of the chain dynamical structure factor for a chain undergoing shear flow is added to this section. In Sect. 2.5 we give a short account of the behavior of polymers in extensional flow for simulations based on the same model of polymer solution. Section 2.6 closes our contribution with some conclusions and discussion of future developments.

2.2
Molecular Dynamics of Dilute Solutions of Chains in Homogeneous Flow

We consider a system made of N_s point-mass particles (the solvent) and one linear chain molecule built as N point-mass particles connected through rigid bonds of length d. Solvent particles and polymer beads have the same mass m. The inter(intra)molecular potential model is built as a sum of identical two-body interactions for bead-bead, bead-solvent and solvent-solvent pairs. We adopt a Lennard-Jones potential truncated and shifted to zero at its minimum, i.e.,

$$v(r) = 4\varepsilon \left[\left(\frac{\sigma}{r} \right)^{12} - \left(\frac{\sigma}{r} \right)^6 + \frac{1}{4} \right] \tag{2.1}$$

for $r < 2^{\frac{1}{6}}\sigma$ while $v(r) = 0$ for $r > 2^{\frac{1}{6}}\sigma$. The rigid bond length is $d = 1.075\sigma$: it is sufficiently small to avoid the possibility for two polymers (or two parts of a unique polymer) to intersect each other. Throughout the chapter we will adopt natural Lennard-Jones units specified by setting $\varepsilon = 1$, $\sigma = 1$ and the mass of a particle $m = 1$.

This system is enclosed in a parallelepipedic box defined by three column vectors L_1, L_2, L_3 arranged in a matrix H:

$$H = \begin{pmatrix} L_{1x} & L_{2x} & L_{3x} \\ L_{1y} & L_{2y} & L_{3y} \\ L_{1z} & L_{2z} & L_{3z} \end{pmatrix} \tag{2.2}$$

PBC are applied in order to have an effectively infinite system. This is obtained by replicating the central box by translational operations defined by any linear combination of the three side vectors L_1, L_2, L_3 with integer coefficients.

Homogeneous fluid flows are specified by a constant velocity gradient tensor. For the shear flow (SF) and the elongational flow (EF) that we will be considering, one has respectively

$$K_{SF} = \begin{pmatrix} 0 & 0 & 0 \\ \dot{\gamma} & 0 & 0 \\ 0 & 0 & 0 \end{pmatrix} \tag{2.3}$$

and

$$K_{EF} = \begin{pmatrix} -\frac{1}{2}\dot{\varepsilon} & 0 & 0 \\ 0 & -\frac{1}{2}\dot{\varepsilon} & 0 \\ 0 & 0 & \dot{\varepsilon} \end{pmatrix} \tag{2.4}$$

During the dynamics, the box shape (which also defines the periodic boundary conditions) always evolves according to the equation of motion

$$\dot{H} = K^T H \tag{2.5}$$

The shear and elongational homogeneous flows are illustrated in Fig. 2.1 which shows the typical shape modification of a cubic piece of fluid undergoing the flow for some time. The individual particle or bead dynamics follows the so-called SLLOD equations of motion [31]: we adopt here the version where the hydrodynamic field is set up at time $t = 0$. The heat generated by dissipative forces in the system is removed by a Nosé-Hoover thermostat [32, 33] which couples to the system through a friction term in the acceleration which is proportional to the thermal part of the velocity. The friction parameter is itself a dynamical variable which reacts to the instantaneous deviation between the total thermal kinetic energy and its expectation at the thermostat temperature. For a solvent particle i of mass m_i and located at r_i, the equation of motions reads

$$\ddot{r}_i = \frac{F_i}{m_i} + K^T r_i \delta(t) + \Theta(t) K^T K^T r_i + v_N \xi(t)(\dot{r}_i - \Theta(t) K^T r_i) \tag{2.6}$$

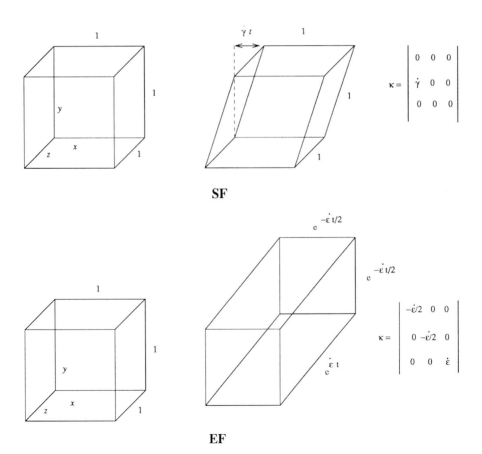

SF

EF

Fig. 2.1. Sketch of the shear and elongational flow deformation of the simulation box

where F_i is the total force due to interactions with the other particles. The $\Theta(t)$ function appearing in the two last terms on the r.h.s. is the unit step function which indicates that the flow is set up at $t = 0$. The $\delta(t)$ function accordingly implies that the particular flow $u(r) = K^T r$ is set up instantaneously. The third term is a "macroscopic" acceleration term which is required to maintain the flow homogeneous. This term is zero in shear flow as in this case fluid velocity is constant along flow lines. The last term on the r.h.s. is the friction force term of the thermostat.

The equation of motion of a polymer bead are similar to those of the solvent. However, here the thermal velocity of a bead is chosen to be the difference between the bead velocity (in laboratory axes) and the flow field velocity taken at the polymer center of mass location. Note also that constraint forces due to first neighbor interactions are now included implicitly in F [34]. One has for the ith bead of the polymer with center of mass R_C

$$\ddot{R}_i = \frac{F_i}{m_i} + K^T R_C \delta(t) + \Theta(t) K^T K^T R_C + v_N \xi(t) (\dot{R}_i - \Theta(t) K^T R_C) \qquad (2.7)$$

The friction variable $\xi(t)$ appearing in the friction force terms follows the thermal kinetic energy. It obeys the equation

$$\dot{\xi} = v_N \left(\frac{2K^{tot}}{g k_B T} - 1 \right) \qquad (2.8)$$

where K^{tot} is the total thermal kinetic energy and g the total number of degrees of freedom in the system. v_N is an inertial parameter governing the response time of the thermostat.

2.3
The Equilibrium Case

In this section, we report a few results concerning molecular dynamics simulations of our polymer + solvent mixture at equilibrium. Our aims are twofold. First, a study at equilibrium allows us to test the extent to which the main static and dynamic single chain properties are verified for the "short" chains/"small" systems considered in the simulations. Moreover, the equilibrium state will provide us with a reference situation with respect to which specific effects due to the presence of a flow can be more easily detected.

2.3.1
Systems Studied

We study several chain sizes, but consider only a single state point characterized by a reduced temperature $k_B T / \varepsilon = 1.5$ and a total number density $\varrho \sigma^3 = 0.8$ where $\varrho = [N + N_s]/L^3$ and L the box side. For each system size, we work at a

Table 2.1. System sizes considered in the simulations. N is the number of beads of the polymer and N_s is the number of solvent point particles

N	N_s
9	207
20	980
30	2167
50	4046

roughly similar reduced polymer concentration $c/c^* \approx 0.02$ where c^* is the overlap concentration (number density) $c^* = R_g^{-3}$. The systems used in simulations are thus characteristic of the dilute limit: they are made explicit in Table 2.1. This athermal model of chain in solvent corresponds to good solvent conditions.

2.3.2
The Static Structure Factor

For static aspects, the main quantity of interest is the single chain structure factor as it reports the chain structure at all length scales. It is defined as

$$S(k) = \frac{1}{N} \left\langle \sum_{i=1}^{N} \sum_{j=1}^{N} \exp\left(i\mathbf{k} \cdot (\mathbf{R}_i - \mathbf{R}_j)\right) \right\rangle \tag{2.9}$$

Its behavior at equilibrium is well known: at low k, $S(k) = N[1 - k^2 R_g^2/3 + O(k^4)]$ and at high k, $S(k) \propto k^{-\frac{1}{\nu}}$ where $\nu = 0.59$ for the good solvent case (valid in the range going from $k \approx 3/R_g$ up to $k \approx 2\pi/b$ where b is the Kuhn segment).

Figure 2.2 shows $S(k)/N$ for different chain lengths. The high k behavior tends to become universal: it yields an exponent of $\nu = 0.584$ which confirms that our model corresponds to good solvent conditions. As could be expected from the rather low c/c^* ratio investigated, no system size effect on the structure is detectable (this was checked by comparing the structure factor of the nine beads chain for different box sizes) but identical overall density covered, over the range $0.1 \leq \dfrac{R_g}{L} \leq 0.3$.

The scaling law obeyed by the radius of gyration $R_g = 0.467(2)N^{0.57(1)}$ agrees with the $S(k)$ power law [28].

2.3.3
Polymer Longest Relaxation Time

The characteristic time of primary interest is the global relaxation time of the polymer of N beads which can be estimated [35] from

$$\tau_N \approx \frac{R_g^2}{D} \approx [\eta] \frac{\eta_s}{k_B T} \tag{2.10}$$

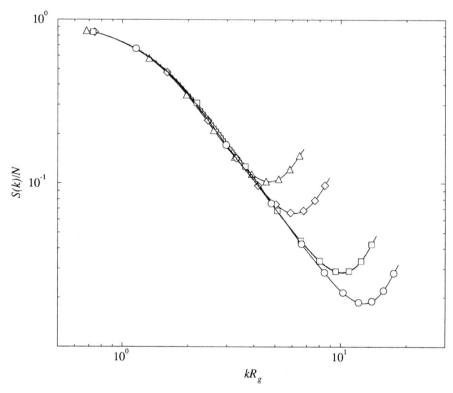

Fig. 2.2. Normalized scattering function vs kR_g at equilibrium for $N = 6$ (triangles), $N = 9$ (diamonds), $N = 20$ (squares) and $N = 30$ (circles)

where D is the center of mass diffusion coefficient, η_s is the pure solvent shear viscosity ($\eta_s = 1.71(1)$ at our thermodynamic state [36]) and $[\eta]$ the intrinsic viscosity defined as

$$[\eta] = \lim_{c \to 0} \frac{\eta - \eta_s}{c\eta_s} \tag{2.11}$$

where η is the shear viscosity of the polymer solution at chain number density c. Note that the second equality of Eq. (2.10) provides the usual route to an experimental estimate of τ_N. To optimize the comparison with experiments, the estimate of the relaxation time in simulations can also be based on the intrinsic viscosity, though indirectly, through the Einstein and Stokes formulas connecting respectively the intrinsic viscosity to the hydrodynamic radius R_H and the latter quantity to the diffusion coefficient D: one thus combines

$$\tau_N = [\eta] \frac{\eta_s}{k_B T} \tag{2.12}$$

with

$$[\eta] = \frac{10}{3} \pi R_H^3 \tag{2.13}$$

and

$$R_H = \frac{k_B T}{6\pi\eta_s D_\infty} \tag{2.14}$$

where D_∞ is the diffusion coefficient estimated at infinite dilution (see below). If R_H scales like R_g, one expects that $\tau_N \sim N^{3\nu}$: in fact, as is well known [35], the scaling exponent for dynamical properties is slightly lower than the static one and we justify below the scaling $\tau_N \sim N^{3\nu'}$ with $\nu' = 0.52$ that we adopted throughout this study. In our units one has $\tau_{50} = 342$, which represents 68,000 MD time steps of size $\Delta t = 0.005$. This is the place therefore to mention that, unless stated otherwise, equilibrium and non-equilibrium simulations are performed on trajectories corresponding to at least $75\tau_N$ to achieve sufficient statistics.

2.3.4
Dynamical Structure Factor

The dynamical structure factor is the central dynamical quantity: it is given by

$$S(k, t) = \frac{1}{N} \left\langle \sum_{i=1}^{N} \sum_{j=1}^{N} \exp\left(ik \cdot (R_i(t) - R_j(0))\right) \right\rangle \tag{2.15}$$

At low k, $S(k, t) \propto \exp(-Dk^2t)$, which allows an estimate of D. The values obtained in the simulation are found to be systematically lower than their expectation value based on Kirkwood's theory [35]. Duenweg and Kremer [29] suggested that, for periodic systems used in simulations, the Kirkwood theory must be modified to take into account the hydrodynamic interactions originating from image polymers. Our MD results support quite well the modified Kirkwood expression [28] which is calculated from the set of conformations generated during the MD run itself. This reinforces the Kirkwood theory at infinite dilution: D_∞ can then be estimated using the same set of MD generated conformations (remembering that structural quantities have been found to be representative of the infinite dilution limit).

Regarding now the N dependence of D, the following scaling has been proposed [29]:

$$D = h\left(\frac{R_g}{L}\right) N^{-\nu'} \tag{2.16}$$

where ν' is the dynamic scaling exponent and where the prefactor is in fact a correction term function of the reduced concentration c/c^*. This picture can be verified by our results [27, 28] covering many chain lengths at the same $\frac{R_g}{L}$:

they appear to follow the scaling law of Eq. (2.16) and provide (1) $D_\infty = 0.120$ (2) $N^{-0.527(5)}$ which confirms that $v' < v$.

At high k, the normalized intermediate scattering function is known to be a universal function of a rescaled time, namely

$$\frac{S(k,t)}{S(k)} = I(tk^x) \tag{2.17}$$

where I is a universal function and x a model dependent exponent [35]. One finds that the Zimm prediction $x = 2 + \dfrac{v'}{v}$ is verified as long as the curves $\dfrac{S(k,t)}{S(k)}$ are exploited up to a maximum (absolute) time $\tau_k(L)$, which is function of the box size L, beyond which interferences with HI originating from image chains would mix up with the intramolecular ones [28]. A detailed analysis gives $x = 2.9$ and thus $v' = 0.53$ in agreement with the scaling of the diffusion coefficient.

To conclude this preliminary section, our MD model reproduces the main features of a chain in good solvent, namely we find the expected structural and dynamical features of excluded volume and hydrodynamic effects both at the global and at the local scales.

2.4
Polymers in Shear Flow

When studying polymers at high dilution in flow, the physical parameter of major importance is the reduced shear rate $\beta = \dot{\gamma}\tau_N$ (also known as the Weissenberg number) or the reduced elongational rate $\beta' = \dot{\varepsilon}\tau_N$ where τ_N is the relaxation time of the polymer discussed in the previous section. Strong deformations of polymers are expected for reduced strain rates β or $\beta' \geq 1$. Most experiments cover a range $1 < \beta < 10$ which gives a reasonable signal/noise ratio but is still sufficiently small to avoid polymer breakage.

In this section, we focus on shear flow situations which are, just as for real experiments, easier to perform than the elongational case. It is interesting to comment briefly on why it is so in the present computer simulation context! In a shear flow case, the fluid element contained in the primary MD cell, originally cubic with side L, deforms (is sheared) in time as Fig. 2.1 illustrates. PBC are based on three time dependent translational symmetry vectors which correspond to the sides of the deformed primary MD cell. However, this periodicity can be viewed in different ways, with the Lees-Edwards [37] sliding brick picture being the most popular one for shear flows. In the Lees-Edwards scheme, the primary cell is redefined as a cube of side L and the periodic image (cubic) boxes are found to be displaced along the flow direction (x axis) by an amount proportial to the elapsed time modulo $\dot{\gamma}^{-1}$. This picture shows that a shear flow can be sustained for an arbitrarily long time.

In the uniaxial elongational flow however (see Fig. 2.1), an initial orthogonal MD cell expands (in one direction) or contracts (in two directions) exponentially in time, e. g., by a factor $\exp\{-\dot{\varepsilon}t/2\}$ in the directions of compression. If we want to observe the polymer deformation in stationary conditions (as materialized in real experiment by a situation where the polymer would stay at the stagnation point for a few relaxation times), we need to maintain the flow for say $T_{max} = 3\tau_N$. If L_{min} is the "minimal" tolerable side of the orthogonal MD box appropriate to deal with the problem at hand (here one has $L_{min} \approx 3R_g$), then the optimum starting point for performing an uniaxial elongational flow by MD simulation would be to start with a flat orthogonal box with dimensions (expressed in L_{min} units) $(\exp\{3\beta'/2\}, \exp\{3\beta'/2\}, 1)$ to finish after a time of $(3\tau_N)$ with an elongated box of dimensions $(1, 1, \exp\{3\beta'\})$. The difficulty in dealing with elongational flows in this MD context can be traced therefore to the increase in the number of solvent molecules by a factor $\exp\{3\beta'\}$ with respect to equilibrium! As the computer time increases linearly with the number of particles, we see that experiments in elongational flow require at CPU budget which is $\exp\{3\beta'\}$ larger than for a corresponding shear experiment. Going significantly over $\beta' = 1$ is thus impossible. In fact, we are facing the profound distinction between weak (shear) and strong (elongational) flows!

2.4.1
Phenomenological Framework for the Shear Flow Case

The simple shear flow is characterized by the velocity gradient given in Eq. (2.3), i.e., with flow velocity $u(r) = \dot{\gamma}y\mathbf{1}_x$ along x. Such a flow field can be regarded as the combination of a purely deformational flow, $u_1 = \frac{1}{2}\dot{\gamma}y\mathbf{1}_x + \frac{1}{2}\dot{\gamma}\mathbf{1}_y$, and a purely rotational flow, $u_2 = \frac{1}{2}\dot{\gamma}y\mathbf{1}_x - \frac{1}{2}\dot{\gamma}x\mathbf{1}_y$, both taking place in the $x - y$ plane. A flexible chain, subject to the combination of these effects, will deform and orient itself with respect to the flow lines. This will be reflected by the anisotropy of tensorial quantities such as the gyration tensor G, the order parameter tensor of individual segments O, and the end-to-end tensor R. These tensors are defined in terms of the bead coordinates R_i as

$$G = \frac{1}{N} \sum_{i=1}^{N} \langle (R_i - R_C) \cdot (R_i - R_C) \rangle \tag{2.18}$$

$$O = \frac{1}{N-1} \sum_{i=1}^{N-1} \frac{\langle (R_{i+1} - R_i) \cdot (R_{i+1} - R_i) \rangle}{d^2} - \frac{1}{3} \mathbf{1} \tag{2.19}$$

$$R = \langle (R_N - R_1) \cdot (R_N - R_1) \rangle \tag{2.20}$$

where R_C is the center of mass of the polymer, d is the bond length, $\mathbf{1}$ the tensor unity, and $\langle ... \rangle$ represents a (non)-equilibrium statistical average.

At equilibrium the system is isotropic and any of these three tensorial quantities, A, for example is proportional to the unit tensor $\mathbf{1}$ (O vanishes). In the shear flow geometry, symmetry considerations require that $A_{xz} = A_{zx} = A_{yz} = A_{zy} = 0$ at any shear rate. Hence there are at most four independent quantities to be monitored, namely the three diagonal elements $A_{\alpha\alpha}$ ($\alpha = x, y, z$) and the off-diagonal element $A_{xy} = A_{yx}$. The orientational angle χ_A, defined through the relation

$$\cot 2\chi_A = \frac{(A_{xx} - A_{yy})}{2A_{xy}} \tag{2.21}$$

measures the rotation around z of the principal axes (x', y', z') of the tensor with respect to the flow (laboratory) frame (x, y, z). In shear flow, A_{xy} starts linearly with $\dot{\gamma}$ while the first contribution to $(A_{xx} - A_{yy})$ is of order $\dot{\gamma}^2$. Therefore the linear (Newtonian) regime is characterized by the value $\chi_A = \frac{\pi}{4}$. Outside the linear regime χ_A decreases to zero for increasing shear rate. In agreement with the limiting behavior expected close to equilibrium, this behavior is generally written as

$$\cot 2\chi_A = \frac{\beta}{m_A(\beta)} \tag{2.22}$$

which essentially defines $m_A(\beta)$, known as the orientational resistance [2, 11] for the property under study, with a non-zero limiting value $m_A^0 = \lim_{\beta \to 0} m_A(\beta)$.

2.4.1.1
Gyration Tensor

Light and neutron scattering experiments performed on solutions flowing in cylindrical shear cells give the most direct access to global structure. Such information is extracted from the low k region where the non-equilibrium statistical average of the structure factor, Eq. (2.9), can be expanded in powers of k:

$$S(k) = N\left[1 - kGk + O(k^4)\right] \tag{2.23}$$

By monitoring the intensity of scattered light in the flow plane as a function of the scattering angle (the incident beam being orthogonal to the plane), Springer and collaborators [2, 38] have determined for several polymer/solvent systems the principal axes of the gyration tensor $(1, 2, 3)$. Such axes form an orthogonal reference frame, rotated with respect to the laboratory (flow) frame by an angle χ_G around the vertical axis z. In this reference frame, the gyration tensor G is diagonal

$$G_{NE} = \begin{pmatrix} G_1 & 0 & 0 \\ 0 & G_2 & 0 \\ 0 & 0 & G_3 \end{pmatrix} \tag{2.24}$$

and G_1, G_2 and G_3 are its eigenvalues. Qualitatively, as $\dot{\gamma}$ increases, χ_G is found to decrease from 45° to 0° and G_1 increases while G_2 decreases with respect to the equilibrium situation where the gyration tensor is $G = \dfrac{R_g^2}{3} \mathbf{1}$. To monitor these deformation features, the best method would be to extract G_1, G_2, G_3 from a linear combination of three noncoplanar mean-square correlation lengths obtained by a simple scattering function of Zimm, monitored at constant detector positions, through wavelength variation [2]. Figure 2.3 shows a superposition of polymer conformations obtained during the simulations under shear: the progressive orientation and deformation of a 30-bead chain is illustrated for two values of β.

2.4.1.2
Birefringence

The O tensor is related, under suitable assumptions [35, 39], to the chain intrinsic birefringence. The stress-optical law further implies that the deviatoric

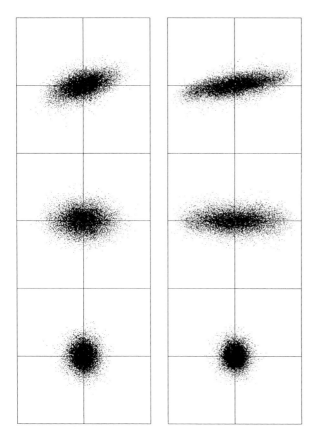

Fig. 2.3. Shape and orientation of a 30-bead chain under shear illustrated by projecting about 500 chain configurations on the three coordinate planes (xy, top; xz, middle; yz, bottom). Two values of the shear rate are shown, namely $\dot{\gamma} = 0.02126$ on the *left* side and $\dot{\gamma} = 0.0671$ on the right side

stress and the birefringence tensors are proportional to each other. In its own principal axes system, we write

$$
O_{NE} = \begin{pmatrix} O_1 & 0 & 0 \\ 0 & O_2 & 0 \\ 0 & 0 & O_3 \end{pmatrix}
$$

(2.25)

At equilibrium, the tensor defined by Eq. (2.19) is zero. Under flow conditions, the order parameter tensor provides the amount of birefringence $(O_1 - O_2)$ and the "so-called" extinction angle χ_O.

2.4.2
Structure of Polymers in Shear Flow

Various system sizes, chain lengths, and reduced shear rates are treated in our non-equilibrium experiments. An exhaustive list of these results can be found elsewhere [40]. Here, our main aim is to analyze some typical results in the light of recent experimental and theoretical predictions.

An important preliminary point is of order. Throughout this study, we spanned the range of shear rates $\dot{\gamma}$ between 0.002 and 1 in Lennard-Jones reduced units. The onset of non-Newtonian behavior in the pure solvent case, detected as a shear thinning, arises around $\dot{\gamma} = 0.5$ [36]. This implies that in our model the Newtonian behavior of the solvent is guaranteed despite the short chains considered.

2.4.2.1
Orientational Resistance

The traditional theoretical analysis of extinction angle evolution with shear flow intensity rests on the pioneering works of Zimm [41] and Peterlin [42] who studied the non-equilibrium birefringence tensor (Eq. 2.19) for Gaussian chains (no excluded volume) in absence (Rouse) or with preaveraged (Zimm) hydrodynamic interactions. These theories suggest that the reduced shear rate β is the "universal" variable which is pertinent to the problem. More specifically, these theories lead to β independent orientational resistances (see Eq. 2.22), respectively $m_O = 2.5$ and $m_O = 4.88$ for Rouse and Zimm dynamics. (Note that the Zimm result needs to be slightly corrected to $m_O = 4.832$ [11]). When light or neutron scattering experiments were performed on dilute solutions [1, 43], it was assumed in the analysis that the gyration and birefringence tensors were oriented similarly. This point has recently been analyzed theoretically in great detail by Bossart and Öttinger [11]. These authors showed that, at the same level of approximation, the orientational resistance of the gyration tensor is significantly lower (more easily aligned along the flow lines) than the one predicted by the extinction angle. A more refined theory, including an approximate treatment of HI fluctuations (still without excluded volume), leads to β-dependent orientational resistances, which is compatible with experimental observations [2, 3, 38, 44]. Values of the orientational resistance predicted by the improved theory at zero shear rate are $m_O^0 = 3.57$ and $m_G^0 = 2.00$.

We show in Fig. 2.4 our simulation results for the orientational resistance of the birefringence and the gyration tensors for various chain lengths and reduced shear rates [45]. The first point to note is that, under shear flow at the same reduced strain rate, namely at $\beta = 3.2$, chains of different length N are effectively (within error bars) oriented at the same angle χ_O or χ_G with respect to the flow. Next, we observe $m_O \geq m_G$ over the β range explored. This "experimental" observation [45] turns out to be in rather good agreement with the recent theoretical prediction for the zero shear rate values (shown by arrows in Fig. 2.4) [11]. It should be stressed however that theory deals with chains at the θ point while our model corresponds to good solvent conditions. There seems to be for the moment no consensus on the role of solvent quality on the orientational resistance: while Bossart and Ottinger [11] suggest a very weak dependence on excluded volume effects, experiments suggest that m_G^0 decreases with solvent

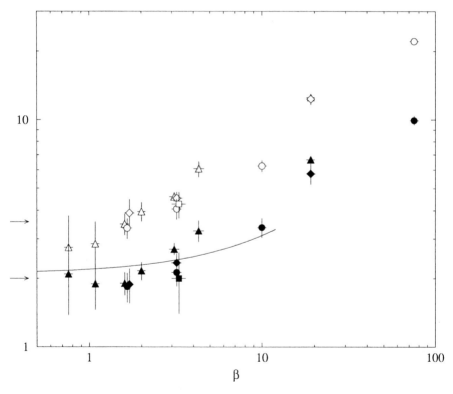

Fig. 2.4. Orientational resistance for the order parameter tensor m_O (open symbols) and for the gyration tensor m_G (filled symbols) vs β for various chain lengths: $N = 9$ (triangles), $N = 20$ (diamonds), $N = 30$ (circles) and $N = 50$ (squares). The continuous curve is a fit to LS data on a θ point chain [2]. Theoretical predictions for θ point chains with HI ($m_O^0 = 3.57$ and $m_G^0 = 2.00$ respectively) are indicated by arrows [11]

power [38] over a range going from ≈ 2.5 to ≈ 1.5. As experimental results are difficult to interpret because of the polydispersity and the need for careful extrapolations to zero concentrations [38], additional simulations by BD or MD on chains at the theta point would be helpful. Let us further note that other tensors describing single chain behavior in shear flow can be studied from our simulations. The end-to-end tensor is found to be oriented like G. The same is true for the center of mass diffusion tensor which is discussed in a later section dealing with chain dynamical properties under steady shear flow.

2.4.2.2
A Scaling Picture at Fixed Reduced Shear Rate

The "universal" dependence of χ_G on the relevant non-equilibrium parameter β suggests an investigation of scaling laws for chains at fixed β and analyzed in the common molecular reference frame. We therefore consider four chain lengths $N = 9, 20, 30$, and 50 at fixed $\beta = 3.2$. We proceed by analogy with the situation in absence of shear flow that we briefly review. Dimension arguments and the existence of a scaling transformation for chains at equilibrium implies that the chain normalized structure factor is a universal function of kR_g, namely $\dfrac{S(k)}{N} = f(kR_g)$, up to $k \approx \dfrac{2\pi}{b}$. For k values corresponding to length scales below the Kuhn statistical segment b, the local structure of the chain appears and scaling is violated: this was illustrated in Fig. 2.2. Furthermore, the well known $k^{-1/\nu}$ power law for $S(k)$ at high k follows from the expected N-independence of the structure factor when it probes spatial correlations over length scales smaller than R_g. The fact that a unique exponent ν governs both global scaling ($R_g \sim N^\nu$) and local scaling ($S(k) \sim k^{1/\nu}$) is a consequence of the self similarity of chain structure at the various length scales.

A similar property is observed for chains of variable lengths in shear flow when they are considered at the same reduced shear rate. For k oriented along one of the principal axes of the gyration tensor ($\alpha = 1, 2, 3$), one finds a regime of k where $S(k_\alpha)/N = f_\alpha(k_\alpha G_\alpha^{1/2})$ in which f_α is a universal function (specific of that direction) for all chain lengths taken at the same reduced shear rate. This is illustrated in Fig. 2.5 for directions 1 and 2. The results for direction 3 are not shown since they do not differ significantly from the equilibrium behavior (see Fig. 2.2). In the high-k region, the N-independence of $S(k_\alpha)$ is indeed observed in directions 1 (and 3) but not in direction 2 as well. We believe that the residual N-dependence is related to the smaller dimension of the chains in the compression direction 2: it can be expected that longer chains are required to observe the properties already noticed for the other directions. It is already clear however that $S(k)$ for each chain length and in each direction follows the "long" chain behavior $S(k_\alpha) \sim k^{-1/\nu_\alpha}$ above $k_\alpha = 1.5/\sqrt{G_\alpha}$. Estimates of ν_1, ν_2, and ν_3 on the basis of $S(k_\alpha)$ power laws are given in detail in [40]. At $\beta = 3.2$, a global analysis of the various chain length investigated lead to estimates $\nu_1 = 0.69(1)$, $\nu_2 = 0.48(1)$, and $\nu_3 = 0.54(2)$. At the same β value, G_1, G_2, and G_3 follow a power law behavior $G_\alpha \propto (N-1)^{2\nu_\alpha}$ as shown in Fig. 2.6. From the

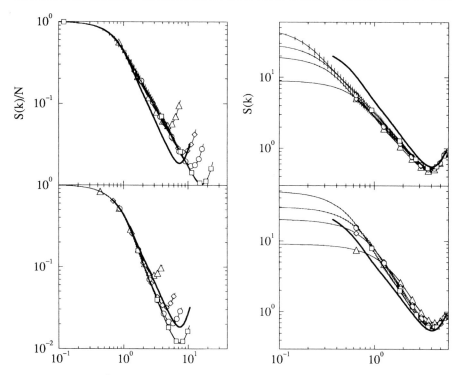

Fig. 2.5. Scattering function at $\beta = 3.2$ for various chain lengths: $N = 9$ (triangles), $N = 20$ (diamonds), $N = 30$ (circles), $N = 50$ (squares). $S(k)/N$ vs $k_\alpha \sqrt{G_\alpha}$ on the *left* and $S(k)$ vs k_α on the right side in the extension direction $\alpha = 1$ (top) and in the compression direction $\alpha = 2$ (bottom). The thick line represents the equilibrium behaviour. The curves in the out-of-plane direction $\alpha = 3$ are not shown here since they are very close to the equilibrium one

slope of this log-log plot, we obtain estimates $v_1 = 0.69(2)$, $v_2 = 0.53(3)$, and $v_3 = 0.58(3)$. The agreement between the high and low k estimates of scaling exponents is rather good, except maybe for the compression direction which requires longer chains as already noted. The observed coherence at high and low k, however, strongly supports an anisotropic scaling picture with β-dependent exponents.

The above scaling picture for chains in shear flow is consistent with the available SANS data for polystyrene in dilute solution in good solvent conditions [1, 46]. Figure 5 of [46] reports measurements of $S(k_x)$ and $S(k_z)$ for a chain of molecular mass $M_w = 2.8\ 10^5\ \text{gmol}^{-1}$ in shear flow at $\beta = 3.17$. At such a β value, $S(k_x)$ and $S(k_1)$ are almost identical [28]. Assuming that the same holds for the experimental case, we infer from the SANS data $v_1 \sim 0.7$ in remarkable agreement with the value obtained from MD. It is unfortunate that experiments on chains of different lengths at a fixed value of β have not been performed so far. A more complete test of our scaling picture obviously requires further experimental investigations.

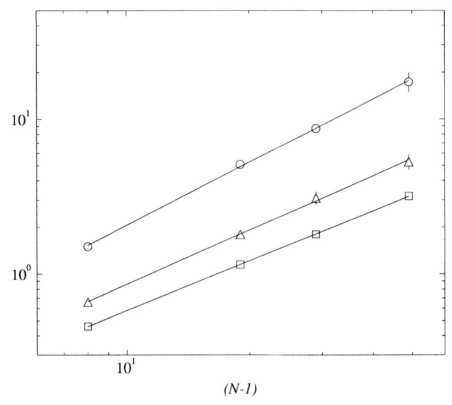

$(N-1)$

Fig. 2.6. $\beta = 3.2$. Eigenvalues of the gyration tensor G_1 (circles), G_2 (squares) and G_3 (triangles) vs $(N-1)$. The straight lines are power law fits to the data. Values of the apparent scaling exponents are indicated in the text

Comparison of our results with high shear rate theories cannot easily be made. The existence of "shear blobs" has been postulated, but never verified experimentally. On the theoretical side, the "shear blob" hypothesis has been considered by Onuki [47]. In this scheme, at large reduced shear rates ($\beta \gg 1$), chains would stretch on a large length scale and orient along the flow direction most of the time while keeping the equilibrium structure on short length scales, with the transition length scale ξ_c being fixed by the absolute shear rate $\dot{\gamma}$. More specifically, chains of N beads, which adopt a coiled structure at equilibrium with radius of gyration $R_g = AN^\nu$, would adopt in a shear flow a linear structure in a coarse grained sense where the basic units are unperturbed chain fragments (blobs) of n monomers with size $\xi_c = An^\nu$, n being fixed by the requirement that the longest internal relaxation time of the blob is of the order of $\approx 1/\dot{\gamma}$. This leads to a number of blobs per chain $N_B = \beta^{\frac{1}{3\nu}}$ and a number of monomers per blob $n = N\beta^{-\frac{1}{3\nu}}$. In this picture the structure factor $S(k)$ should be isotropic and close to the equilibrium curve for $|k| \gg \dfrac{1}{\xi_c}$ while in the flow direction it should exhibit a power law $\propto k^{-1}$ for $|\mathbf{k}| < \dfrac{1}{\xi_c}$. This is indeed observed in situations for which the blob model is known to apply (for instance the case of a chain stretched by a force applied to its ends [48]). Applying the above considerations to our system, we get only a few blobs per chain ($N_B = 2$ at $\beta = 3.2$ and $N_B = 4$ at $\beta = 10$) which immediately indicates that our simulations are outside the β regime where the theory could be valid. We never observe the predicted crossover in the slope of the structure factor for $\mathbf{k}$ parallel to the flow direction while the observed anisotropy in $S(\mathbf{k})$ extends down to the length scale of the single monomer even for $\beta \approx 1$ (see Fig. 2.5). In conclusion, it is difficult to judge about the general validity of the blob model for chains in shear flow on the basis of our data and a definite statement will probably require more flexible and longer chains at much higher β. The same is true for available experimental data which cover the intermediate and not the high β regime.

2.4.2.3
The Evolution of Scaling Exponents with the Reduced Shear Rate

We now investigate the β dependence of the various scaling exponents ν_α ($\alpha = 1, 2, 3$). This can most easily be inferred from the intermediate k regime of $S(k_\alpha)$. In Fig. 2.7 we show $S(\mathbf{k})$ for $\mathbf{k}$ along the three principal axes of the gyration tensor for a 30-bead chain at $\beta = 0$, 1.65, 10, and 75. The figure reveals clearly the general trend of $\nu_\alpha(\beta)$; up to $\beta = 10$ at least, ν_1 increases with β, ν_2 decreases with increasing β, and ν_3 remains practically equal to the equilibrium value. However, for $\beta = 75$, the trends seem to invert: a close look at the curves of $S(k_\alpha)$ suggests the presence of a saturation effect due to the finite length of the chain. Indeed, at such value of β, even the local chain structure at the monomer length scale is strongly deformed in the three directions and therefore the $S(\mathbf{k})$ loses its universal character and becomes strongly N-dependent. Note that $\beta = 75$ for $N = 30$ corresponds to $\dot{\gamma} = 0.5$, which in turn corresponds to the onset of non-Newtonian behavior for the solvent [36]. A detailed discussion of the evolution of the scaling exponents with β can be found elsewhere [40]. In short, the exponents ν_1 and ν_2 remain close to the equilibrium value up to $\beta = 1$ before diverging in opposite directions as previously discussed.

A quantity often discussed in the context of polymer deformation in flows is the so-called deformation ratio δ_{aa} considered along laboratory axes ($a = x, y, z$). These are defined as $\delta_{aa}(\beta) = [G_{aa}(\beta)/G_{aa}(0) - 1]$, the volume change of the chain being related to $\delta_v = \delta_{xx} + \delta_{yy} + \delta_{zz}$. Symmetry reasons demand that δ_{aa} depart from their equilibrium value as $C_{aa}\beta^2$ when β increases. A detailed comparison of our estimate of the constants C_{aa} with theoretical and experimental data is given elsewhere [40]. Let us just mention here that our results for δ_{xx} and δ_{zz} compare rather well with the available SANS data for a chain in good solvent [46, 49]. Along the z axis, a small contraction of the chain is observed above $\beta = 1$, while along x, compatible quadratic laws in β are observed up to $\beta \approx 2$ where both simulations and SANS data indicate a weaker evolution with β. It is important to mention here that δ_{aa} or δ_v are useful concepts in the small β regime.

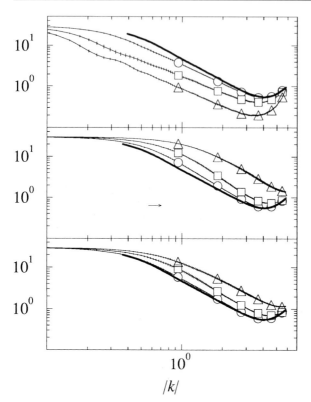

Fig. 2.7. $N = 30$. Evolution of $S(k)$ with β for k along the extension direction (top), the compression direction (middle) and the out-of-plane direction (bottom). Four β values are shown: $\beta = 0$ (thick line), $\beta = 1.6$ (circles), $\beta = 10$ (squares) and $\beta = 75$ (triangles)

At larger β they should become slightly N dependent as, according to our MD results, $G_{aa}(\beta)$ and $G_{aa}(0)$ do not scale with N with the same exponent! Indeed $G_{xx}(\beta)$ and $G_{yy}(\beta)$ are linear combinations of the eigenvalues $G_1(\beta)$ and $G_2(\beta)$ which follow power laws in N with respectively larger and smaller exponents than the equilibrium value. This fact needs to be verified experimentally as the opposite belief, namely that scaling exponents are not affected by the presence of shear flow, is probably widespread (see Eq. 23 in [11]).

2.4.3
Internal Dynamics of Chains Under Steady Shear Flow

In this section, we study the effects of the shear flow on the internal motion of the chain. At equilibrium, chain dynamics are characterized by a well-defined scaling relation connecting length and time scales. As the shear flow is a weak flow, we can expect that not only static but also dynamic properties will not change dramatically from the equilibrium case. We will mainly consider here the effects of the flow on the intermediate scattering function already studied at equilibrium. However, its definition at equilibrium (Eq. 2.15) must be generalized in order to remove the systematic part of the motion due to flow convection effects. For a point particle of coordinate R_i in the laboratory frame, the diffusive displacement is obtained by subtracting the convective part from the total displacement: this convection part is estimated as the time integral of the local velocity field $u(R_i(\tau))$ existing at the particle location at time τ [50]. A chain is an extended object and therefore there is no unique way to define at the microscopic level a many-particle dynamic property such as the dynamic structure factor. The point is that the local velocity field at a bead location is not strictly equal to the velocity field at the center of mass location so that slightly different convection contributions result from the choice of one or the other location. We define the intermediate scattering function under shear flow as

$$S(k, t) = \frac{1}{N} \left\langle \sum_{i=1}^{N} \sum_{j=1}^{N} \exp\{ik \cdot (R_i(t) - R_j(0) - \int_0^t d\tau u(R_C(\tau)))\} \right\rangle_{\dot{\gamma}} \qquad (2.26)$$

where R_C is the chain center of mass coordinate. The use of a center of mass localisation to estimate the local velocity field is appropriate to study center of mass diffusion but is less adequate perhaps for the local dynamics analysis as a flow vorticity contribution remains in the "purely diffusive" part. However in shear flow, the change in the flow field intensity is small over the extension of the chain: the field intensity depends linearly on the y-coordinates and the chain extension in the y-direction is also decreasing linearly with $\dot{\gamma}$ [40].

In Eq. (2.26) $\langle...\rangle_{\dot{\gamma}}$ means a statistical average over the non-equilibrium steady-state ensemble. In the simulation we invoke the ergodic theorem and compute time averages over the non-equilibrium steady-state trajectory rather than non-equilibrium ensemble averages.

Similarly to the equilibrium case, for $(kk: G) \ll 1$, Eq. (2.26) reduces to

$$S(k, t) = N \exp[-(kk: D\, t)] \qquad (2.27)$$

where D is the diffusion tensor defined as

$$D = \lim_{t \to \infty} \frac{\langle [R_C(t) - R_C(0) - \int_0^t dr u(R_C(\tau))] [R_C(t) - R_C(0) - \int_0^t d\tau u(R_C(\tau))]\rangle_{\dot{\gamma}}}{2t} \qquad (2.28)$$

kk is a second rank tensor and $\{:\}$ represents the full contraction of indices. As discussed in [51], this is only one of the possible definitions of the diffusion tensor for chains in shear flow: namely the one related to the Brownian force. If instead we consider the linear response to a driving external force in presence of shear flow, and we relate the diffusion to the mobility by the Einstein relation, we can derive an approximate Kirkwood-like expression for the center of mass diffusion [52]:

$$D^K = \frac{D_o}{N} 1 + \frac{k_B T}{8\pi\eta_s N^2} \sum_i \sum_{j \neq i} \left\langle \frac{1}{|R_{ij}|} \left(1 + \frac{R_{ij} R_{ij}}{|R_{ij}|^2}\right) \right\rangle_{\dot{\gamma}} \qquad (2.29)$$

where η_s is the solvent viscosity, D_0 is the free monomer diffusion coefficient, and R_{ij} the vector joining beads i and j. This relation is only an approximate estimate of the true diffusion in shear flow. However we learned from the equilibrium case that the Kirkwood formula is quite accurate for our model and is an important tool to extrapolate the low k dynamical results to infinite system size. We expect that finite system size effects are even more complex in shear flow than at equilibrium as the chain structure becomes anisotropic. Here we limit our analysis of dynamics under shear at the high k regime where finite size effects are believed to be negligible. In all our experiments, the solvent is in practice at equilibrium and we applied the same hydrodynamic argument of the equilibrium case to estimate the onset of finite size effects at a given length scale [28]. As for the low-k dynamics we exploit the relation in Eq. (2.29) to estimate the infinite system size diffusion tensor.

Again we consider the internal dynamics along the three principal axes of the chain gyration tensor. In Fig. 2.8 we present the normalized intermediate scattering function for an $N = 30$ beads chain at $\beta = 3.2$ for the vector k oriented along the chain principal axis and for $|k| = 1.432$. The corresponding equilibrium behavior is superimposed on the relaxation along the out-of-plane

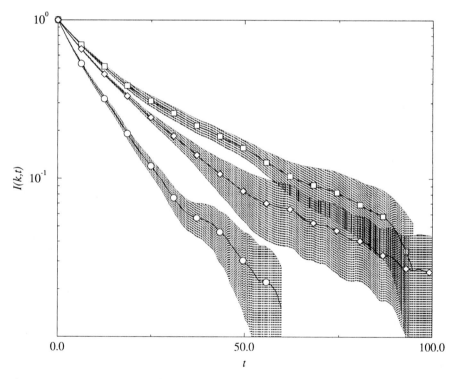

Fig. 2.8. Anisotropic internal relaxation of the chain as represented by the normalized dynamic structure factor $I(k, t)$. Data for $N = 30$, $\beta = 3.2$, $|k| = 1.432$ and for k along the extension direction (circles), the compression direction (squares) and the out-of-plane direction (diamond) are shown. The equilibrium relaxation at the same $|k|$ value is superimposed to the out-of-plane relaxation within error bars

axis, while the relaxation along the extension direction is faster and along the compression direction is slower. This is the typical behavior observed in shear flow.

As for the static case, we can consider two separate problems: (a) the existence of self similar dynamic behavior of chains of different lengths at fixed β and (b) the variation of such behavior for increasing β.

2.4.3.1
Dynamics at Fixed β

As for static properties, in order to consider our results as representative of the behavior of long chains, it is important to show that there exists some form of scaling for the dynamics. To establish this universal behavior, we study the normalized dynamical structure factor $I(k, t) = S(k, t)/S(k)$ for k along the principal axes of the gyration tensor at $\beta = 3.2$ for $N = 9, 20, 30$, and 50. First we note that $I(k_\alpha, t)$, $(\alpha = 1, 2, 3)$ is N-independent within error bars and within the range of N investigated. This observation and the known equilibrium behavior [35] (see Sect. 2.3.4) lead us to write, in the high k regime ($k_\alpha G_\alpha^{1/2} \gg 1$)

$$I(k_\alpha, t) = f_\alpha(t|k|^{x_\alpha}) \tag{2.30}$$

where f_α should be universal functions. In order to verify this conjecture and to extract estimates of the exponents x_α, we numerically inverted the function $I(k, t)$ for k along the three principal directions. The results of such procedure are shown in Fig. 2.9 where, for each direction, we report the function $t(k_\alpha, I)$ for two values of I, namely $I = 0.1$ and $I = 0.6$. As for the equilibrium case, we neglect data for times longer than the box time limit $\tau_k(L)$ beyond which interferences with image chains could be present [28]. Note that results for two chain lengths ($N = 30$ and $N = 50$) are superimposed, proving the required N-independence of $I(k, t)$. As at equilibrium, we observe well defined power law behavior of t vs $|k|$. We observe some residual dependence of x_α on the value of I but we have learned that this could be induced by the PBC or by too short chains [27] and we are not in a position here to discuss this effect. However, if we take typical values for such exponents, namely $x_1 = 2.6, x_2 = x_3 = 2.9$, and we plot $I(tk^{x_\alpha})$ for different k-values and for different chain lengths we obtain the sought after universal behavior as shown in the right hand side of Fig. 2.9. In this figure we report data for $k = 2.016$ and $k = 4.06$ which are well within the region of the power law behavior $t_\alpha \approx k_\alpha^{-x_\alpha}$. Also data for $N = 30$ and $N = 50$ are shown. Comparison with the equilibrium behavior leads us to the following conclusions: (1) the apparent dynamical exponent in the extension direction decreases from the equilibrium value $x_{eq} = 2.9 \pm 0.1$ which is instead preserved in the compression and out-of-plane directions; (2) the universal dynamics along the out-of-plane direction is unaffected by the presence of the flow while it is faster along the extension direction and sometimes slower along the compression direction. All the three universal relaxations are well fitted by a stretched exponential form like

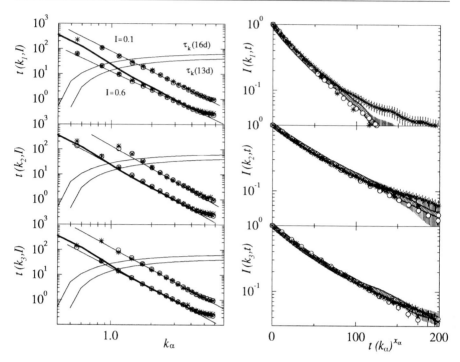

Fig. 2.9. $\beta = 3.2$. Left side: parametric plot of $t(k_\alpha, I)$ for $I = 0.1$ and $I = 0.6$ and for k along the extension direction (*top*), the compression direction (*middle*) and the out-of-plane direction (*bottom*). Data for two chain lengths are reported: $N = 30$ (circles) and $N = 50$ (stars). The typical box time $\tau_k(L)$ for $L/d = 13$ and 16 are represented by the thin continuous curves [28]. The straight lines are power law fits $t \sim |k_\alpha|^{-x_\alpha}$ to the $N = 30$ data. In all plots the thick line represents the equilibrium behavior for $I = 0.6$ and $N = 30$. Right side: universal relaxation $I(tk_\alpha^{x_\alpha})$ vs $tk_\alpha^{x_\alpha}$ for $N = 30$ and 50 (symbols are the same as on the left side). For each chain length, data for $|k| = 2.016$ and $|k| = 4.06$ are plotted

$$f_\alpha(t|k|^{x_\alpha}) = f_\alpha(y_\alpha) = \exp\left[-\left(\frac{y_\alpha}{m_\alpha}\right)^\gamma\right] \tag{2.31}$$

with a direction independent stretching exponent $\gamma = 0.88$ and with $m_1 = 30$, $m_2 = 53.5$, $m_3 = 46.4$. This suggests that the universal relaxation is unique and what changes with the direction is the argument: $f_\alpha(t|k|^{x_\alpha}) = \phi(t|k|^{x_\alpha}/m_\alpha)$. Note that: (1) the coefficients m_α are N independent; (2) the out-of-plane relaxation is unaffected by the flow which also represents the isotropic equilibrium behavior.

To complete the analysis of dynamical scaling in SF we should study the low k dynamics ($k_\alpha G_\alpha^{1/2} \ll 1$) which is directly related to the behavior of the diffusion tensor D. However we know that this is the regime where finite size effects may change dramatically the chain dynamics, so that a straight scaling analysis is not possible. A much less ambitious program, the one we pursue here, is to assume that the diffusion tensor is well approximated by its Kirkwood form given by Eq. (2.29) and to study the N dependence of the Kirkwood diffusion at fixed β. It is also reasonable to expect that, as in the equilibrium case, the dynamic scaling exponent will not depend on the size of the system under study. Therefore we limit ourselves to compute the Kirkwood diffusion for the isolated chain without considering the hydrodynamic interactions from image chains. We first note that the diffusion tensor is diagonal, within error bars, in the reference frame defined by the principal axes of the gyration tensor. This validates the above analysis of high k dynamics. Diffusion along each principal direction follows an apparent scaling with N as shown in Fig. 2.10. The best power law fits to the data provided $D_1 = 0.129\,(1)\ N^{0.546\,(5)}$, $D_2 = 0.122\,(2)\ N^{0.560\,(5)}$ and $D_3 = 0.124\,(2)\ N^{0.556\,(5)}$ to be compared with the equilibrium fit [28] $D = 0.120\,(2)\ N^{0.527\,(1)}$. Note that the apparent dynamic exponents are all slightly larger than at equilibrium.

In analogy with the equilibrium case, one is tempted to relate high-k and low-k dynamic exponents by $v'_\alpha = (x_\alpha - 2)\,v_\alpha$. However at equilibrium this rela-

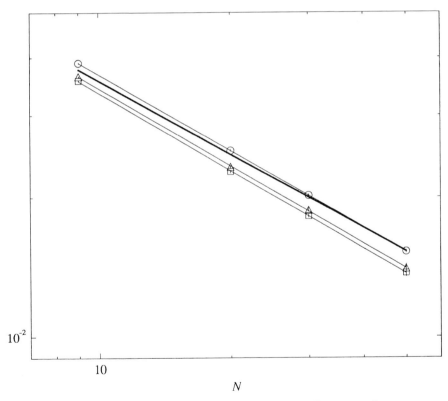

Fig. 2.10. $\beta = 3.2$. Eigenvalues of the Kirkwood diffusion tensor D_1^K (circles), D_2^K (squares) and D_3^K (triangles) vs N. The straight lines are power law fits to the data with the apparent exponents v'_α as indicated in the text. The thick line is the equilibrium fit (see text)

tion was derived from symmetry arguments which are probably not directly applicable when the system becomes anisotropic as in the case under shear. In fact, estimates of v'_α from these relations are systematically 15% lower than the previous estimates of low-k dynamic exponents. Investigation of this discrepancy deserves a careful study of finite size effects.

2.4.3.2
β-Dependence of Dynamical Properties

We first note that diffusion and gyration tensors orientations agree (within error bars) over the whole β range explored. This again justifies the choice of an analysis of the dynamics in the molecular reference frame. As before, we investigate the high-k regime of the normalized intermediate scattering function and the zero-k Kirkwood diffusion.

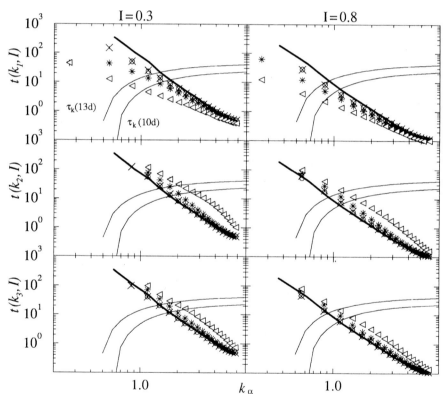

Fig. 2.11. $N = 30$. Parametric plot of $t(k, I)$ for $I = 0.3$ (left side) and $I = 0.8$ (right side) and for k along the extension direction (*top*), the compression direction (*middle*) and the out-of-plane direction (*bottom*). On each plot data for five values of β are reported: $\beta = 0$ (thick line), $\beta = 1.6$ (crosses), $\beta = 3.2$ (circles), $\beta = 10$ (stars) and $\beta = 75$ (triangles). Data at $\beta = 1.6$ are for $N = 20$. Scaling must be analyzed below the continuous curves giving the box times $\tau_k L$ for $L/d = 13, 10$

In Fig. 2.11 we show, separately for k vectors along the three principal axes and for two values of I ($I = 0.3$ and $I = 0.8$), how the time to reach a particular I value depends on the k modulus. Data are for $N = 20$ and 30 and for five values of the reduced shear rate, namely $\beta = 0, 1.6, 3.2, 10$, and 75. We immediately note that in the extension direction a power law behavior is observed at any value of β covering a large range of k-values. The value of the exponent x_1 decreases from the equilibrium value 2.9, to the value 1.7 observed for $\beta = 75$. Intermediate values of x_1 are reported in Table 2.2. In the other two directions, power law behaviors are observed for $\beta \le 10$ while at $\beta = 75$ data show two different regimes with different apparent power laws. As discussed previously [40], we believe that such behavior is not universal as a 30-bead chain is too short to be studied at such a high β value. In Table 2.2 we report values of the high-k dynamic exponent x_α estimated from a power law fits of data shown in Fig. 2.11 whenever possible.

Dynamical scaling demands that $I(k, t)$ computed for different chains and for different k modules collapse onto a single curve when time is renormalized by the proper x power of k. In Fig. 2.12, we show the intermediate scattering function for two values of $|k|$ and for $\beta = 1.6, 10$, and 75. On the left hand side we plot data vs $|k|$ while on the right hand side we plot data vs $t |k|^{x_\alpha}$ using the x estimate obtained from the Fig. 2.11 analysis. In the extension direction the collapse is very good. In the other two directions, the same is true for $\beta = 1.6$ and 10. At $\beta = 75$, in the compression and the out-of-plane direction, there is no universal curve. It is also very interesting to note the quite strong anisotropy induced by the flow and enhanced by increasing the flow strength. In Table 2.3 we report, for increasing β, the values of the parameters extracted from fits by a stretched exponential function of the universal relaxations.

Finally we want to see how the flow changes the Kirkwood diffusion in the principal directions. In Fig. 2.13 we plot D_α/D vs β for a 30-bead chain. We ob-

Table 2.2. High-k apparent dynamics exponents x_α vs β. At $\beta = 75$ only the estimate for x_1 is reported since no unique slopes $t \approx k^{-x_\alpha}$ are observed in the two other directions (see text)

β	x_1	x_2	x_3
1.6	2.6	2.9	2.9
3.2	2.6	2.9	2.9
10.0	2.25	3.5	3.5
75	1.7	–	–

Table 2.3. Values of the parameters of the stretched exponential fit to the universal relaxation $I(tk^{x_\alpha})$ for increasing β

β	m_1	γ_1	m_2	γ_2	m_3	γ_3
1.6	30.6	0.88	50.3	0.88	48.0	0.87
3.2	29.8	0.88	53.5	0.89	46.4	0.88
10	16.7	0.98	136	0.77	108	0.77
75	5.8	1.2	189	0.92	131	0.90

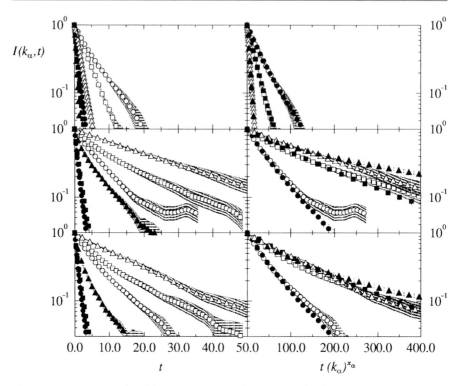

Fig. 2.12. $N = 30$. Normalized dynamic structure factor vs t (left side) and vs $tk_\alpha^{x_\alpha}$ (right side). Data for $\beta = 1.6$ (circles), $\beta = 10$ (squares) and $\beta = 75$ (triangles) are presented. For each β value we report the internal relaxation at two values of $k = |\mathbf{k}|$, $k = 2.016$ (open symbols) and $k = 4.06$ (closed symbols)

serve that the presence of the flow inhibits the diffusion in all directions, with the effect being very small in the extension direction and more pronounced in the other two directions. This is not surprising as one could expect that the diffusion in a given direction (defined in the molecular reference frame) is determined mainly by the inverse of the chain cross section in a plane orthogonal to the same direction. Therefore, as the chain dimension is much increased along the extension direction while it is little changed in the two others, the diffusion should decrease mainly in the compression and out-of-plane directions. This phenomenon has been previously discussed by Zylka and Ottinger [52] for a Zimm-like model using Renormalization Group techniques. Their predictions are qualitatively in agreement with our findings; quantitatively they predict a much stronger β dependence than the one observed in our simulations. However, we must note that our scaling analysis of the previous subsection suggests that the scaling exponents for D_α are slightly different from the equilibrium exponent. Therefore such ratios should present a residual (very weak) N dependence which appears to be very difficult to discuss on the basis of results for the short chains studied in the simulations.

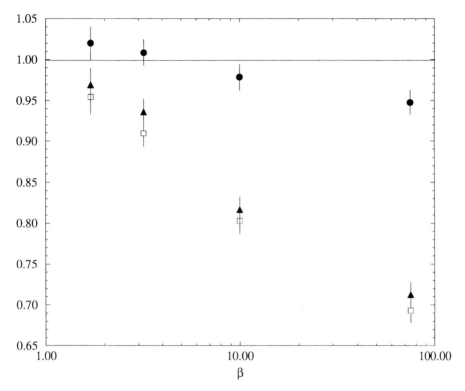

Fig. 2.13. β dependence of the eigenvalues of the Kirkwood diffusion tensor normalized to the equilibrium value for $N = 30$. Extension direction (filled circles), compression direction (open squares) and out-of-plane direction (filled triangles). Note that at $\beta = 1.6$, data for $N = 30$ are missing and have been substituted by the $N = 20$ result

2.5
Transient Behavior of a Nine-Bead Chain in Elongational Flow

As a last topic, we present an exploratory study of polymer solutions under elongational flow obtained by NEMD techniques. As discussed before, such a study is severely limited by the exponential shrinking of the periodic simulation box. As there is no way to maintain an elongational flow indefinitely with the present NEMD technique, we need to use the dynamical version of this technique which consists of following in time the transient behavior of the system originally at equilibrium and then suddenly subjected to the specific time independent homogeneous flow. In favorable cases, a stationary state is achieved before box size problems start manifesting [18]. The statistical average of any observable is taken over the initial equilibrium ensemble [53].

The particular elongational flow we are considering has cylindrical symmetry around the z-axis and therefore all tensorial quantities related to the chain structure, like the gyration tensor (Eq. 2.18), the order parameter tensor (Eq. 2.19) and the end-to-end tensor (Eq. 2.20), remain diagonal. Moreover two of their eigenvalues are equal by symmetry (along x and y axes) and each tensorial observable is fully specified by at most two scalar quantities, e.g., the trace of the tensor and the quantity $[A_{zz} - (A_{xx} + A_{yy})/2]$. The former quantity starts at order $\dot{\varepsilon}^2$ while the latter starts linearly with $\dot{\varepsilon}$. In the limit of vanishing flows we have by symmetry considerations

$$\lim_{\dot{\varepsilon} \to 0} \frac{A_{zz} - (A_{xx} + A_{yy})/2}{3\dot{\varepsilon}} = \lim_{\dot{\gamma} \to 0} \frac{A_{xy}}{\dot{\gamma}} \qquad (2.32)$$

which defines the *Newtonian* limit for the quantity A independently of the specific flow applied and related to time correlations of equilibrium fluctuations. The Trouton rule for the viscosity [53, 55] is a particular case of Eq. (2.32). If we want to compare the behavior of the system under shear and elongational flows at finite strain rates (beyond the linear regime), we need to define an effective flow strength. This is usually given by the second scalar invariant Γ_2 of the symmetrized velocity gradient tensor [53] (see Eq. 2.3 and 2.4):

$$\Gamma_2 = T_r[(K + K^T)^2] \qquad (2.33)$$

which is related to the heat dissipation rate per unit volume within the flow. In elongational flow $\Gamma_2 = 6\dot{\varepsilon}^2$ while in shear flow $\Gamma_2 = 2\dot{\gamma}^2$.

We limit our study to a nine bead chain, at two values of the elongational rate $\dot{\varepsilon} = 0.04083$ and $\dot{\varepsilon} = 0.08167$ corresponding to $\Gamma_2 = 0.01$ and 0.04 respectively. As discussed in Sect. 2.4, it is convenient to choose a non-cubic initial box. We fixed $L_x(0) = L_y(0) = 20d$ and $L_z(0) = 6d$ where d is the bond length. At the chosen state point ($\rho = 0.8$, $T = 1.5$) this corresponds to 2400 particles, chain beads included. The time length of the non-equilibrium trajectories was chosen to reduce the box size in the compression directions down to $5d$. This gives a maximum time of $t_{max} = 2(ln4)\dot{\varepsilon}^{-1}$. In each experiment, statistical averages are taken over 50 independent initial conditions.

In Fig. 2.14 we plot, for both experiments, the transient behavior of the quantities $[O_{zz} - (O_{xx} + O_{yy})/2]/3\dot{\varepsilon}$ proportional to the amount of birefringence (upper plot) and $[G_{zz} - (G_{xx} + G_{yy})/2]/3\dot{\varepsilon}$ (lower plot). Such behaviors should be compared to the transient of the quantities $\Delta O/\dot{\gamma} = [(O_{xx} - O_{yy})^2 + 4 O_{xy}^2]^{1/2}/\dot{\gamma}$ and $\Delta G/\dot{\gamma} = [(G_{xx} - G_{yy})^2 + 4 G_{xy}^2]^{1/2}/\dot{\gamma}$ obtained in shear flow at the same value of the second scalar invariants. In shear flow, we have only stationary values of the observables. Data reported in [40] provide $\Delta O/\dot{\gamma} = 7.5$, $\Delta G/\dot{\gamma} = 9.3$ at $\Gamma_2 = 0.010$ and $\Delta O/\dot{\gamma} = 6.76$ and $\Delta G/\dot{\gamma} = 7.48$ at $\Gamma_2 = 0.0386$ which are represented by horizontal lines in Fig. 2.14. Despite the large noise, we note that the first transient is quite independent of the elongational rate for both quantities. For the order parameter tensor an apparent plateau is observed at the smallest elongational rate ($\Gamma_2 = 0.01$); it appears around $t \approx 20$ and its value seems to be slightly below the shear flow result. At the highest elongational rate the segments are too short to observe a stationary state. For the gyration tensor, which is known to relax more slowly than the order parameter tensor, a plateau value is more dif-

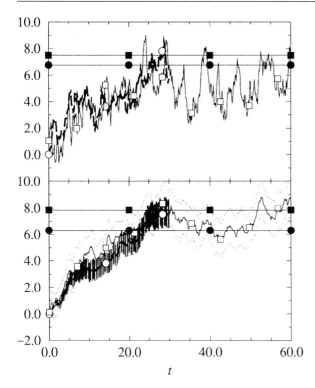

Fig. 2.14. Transient behavior of the amount of birefringence $\{[O_{zz} - (O_{xx} + O_{yy})/2]/3\dot{\varepsilon}\}$ (top) and the quantity $\{[G_{zz} - (G_{xx} + G_{yy})/2]/3\dot{\varepsilon}\}$ (bottom) in EF for $\dot{\varepsilon} = 0.08167$ (open circles) and $\dot{\varepsilon} = 0.04083$ (open squares). In each plot, horizontals represent the steady state value in S.F. of the corresponding quantity $(O_1 - O_2)/2\dot{\gamma}$ (top) and $(G_1 - G_2)/2\dot{\gamma}$ (bottom) at the same values of the second scalar invariant, namely $\Gamma_2 = 0.01$ (closed squares) and $\Gamma_2 = 0.04$ (closed circles)

ficult to infer even at the smallest elongational rate. The order of magnitude of the effect is, however, close to the shear flow values. Finally in Fig. 2.15 we plot the transient behavior of the deformation ratio $\delta(t) = Tr[G(\Gamma_2, t)]/Tr[G(0)]$ divided by the second scalar invariant. No plateau is observed for data even at the smallest value of the elongational rate. This indicates that the plateau observed in the order parameter behavior was only apparent and therefore such technique can only be exploited to study the first transient behavior of chains in EF. In Fig. 2.15 we also report the transient behavior observed in shear flow at $\Gamma_2 = 0.08$. In this case we see the onset of the steady state around $t = 20$. It is interesting however to note that the first transient seems to be independent of Γ_2 and also of the specific applied flow.

2.6
Conclusions and Perspectives

In this contribution we have shown that, besides experimental and theoretical approaches, the simulation route provides its own clues to improve our understanding of complex liquids. In the particular case where a dilute polymer solution is subjected to various homogeneous flows, the numerical experiments we have discussed can be seen in some sense as a rather "cheap" repe-

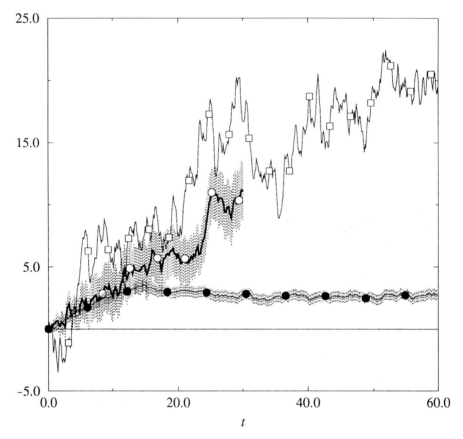

Fig. 2.15. Transient behavior of the trace of the gyration tensor normalized to the second scalar invariant in EF at $\Gamma_2 = 0.01$ (open squares) and $\Gamma_2 = 0.04$ (open circles) and in SF at $\Gamma_2 = 0.08$ (closed circles)

tition of real experiments. The data accumulated by simulation, on the internal structure of polymers in the dilute regime subject to shear flow, today outnumber those obtained so far by small angle neutron scattering using a Couette cell, the appropriate experimental technique to access such properties. Indeed, we have obtained data for scattering vectors oriented in all directions (and not two particular directions only as in the usual SANS/ Couette cell geometry).

Many system sizes have been considered: this allowed us to establish new self similarities in the anisotropic structure of polymers in shear flow at intermediate shear rates. Although, chain size limitations prevent us from getting a clear picture of the structure of polymers for k parallel to the compressional direction, an anisotropic form of scaling laws emerges rather clearly provided chains are compared at the same fixed reduced shear rate. About the global average structure of polymers in flows, we noted that the principal radii of the average gyration tensor still follow a power law in terms of the number of monomers in the chain but the exponent, which first keeps its Flory equilibrium value up to $\beta = 1$, is found to depart slowly from this value with increasing β. This implies that the commonly introduced deformation ratios at fixed β become N-dependent at high β, a point to keep in mind when interpreting experimental data. This scaling picture also has implications on the internal structure of the polymer and on dynamical fluctuations at high and low k.

The next step in the study of polymers in flows is to return to new experimental investigations in order to verify the existence of the self similarities observed in simulations. Such an experiment involving a SANS study of various polymer sizes at the same reduced shear rate (intermediate shear rate regime) is planned at the Laue-Langevin Institute in collaboration with P. Lindner. At this occasion, attempts will also be made to search for structural effects at large reduced shear rate, a situation which is expensive to handle by simulation given the practical limitations on chain size.

Time correlations of monomer concentration fluctuations (intermediate structure factor) in a non-equilibrium stationary state such as shear flow have been analyzed for the first time in the present chapter. For a fixed k module, we find that the relaxation is faster or slower respectively in the elongational or in the compressional direction. Actually, this could be a simple manifestation of the fact that monomer dynamics taking place on a given time scale is probed respectively at shorter or larger length scales with respect to equilibrium depending on whether the scattering vector is oriented in the compressional or elongational direction of the ellipsoid. These dynamical effects will be hard, if not impossible, to test experimentally as they concern structural fluctuations around the systematic motion imposed by the flow.

The transient behavior of a relatively short chain subject to homogeneous uniaxial elongational flow has been studied by NEMD. It is found that the order (birefringence) tensor reaches a plateau much faster than the gyration tensor. Similarities between elongational and shear flow situations considered at "equivalent" strain rates (i.e., at the same second scalar invariant) have been established. Nevertheless, it is clear that these kinds of simulations cannot be extended to long chains and long times (to probe the stationary case) because of the prohibitive computing time. The polymer relaxation time increases as $\tau_N \approx N^{3\nu}$ but the amount of solvent needed to maintain the elongational flow for a total time τ_N increases with N and the reduced elongational rate β' like $\propto N^{3\nu} \exp(3\beta')$. The CPU time is thus $\propto N^{6\nu} \exp(3\beta')$ which will make calculations quickly prohibitive for "long" chains and/or "high" elongational rates. Actually, elongational flows are also complex to control experimentally. NEMD simulations schemes could be devised to mimic "model flows" like the two dimensional elongational flow obtained with the four rolls mill apparatus (see [56] and references therein). Then there is some hope that reasonable system sizes could be employed and the coil-stretch transition of dilute polymers could be studied by a microscopic simulation on the basis of a flexible polymer model.

Acknowledgments. We thank G. Destrée for technical help and CECAM (Centre Européen de Calcul Atomique et Moléculaire) for financial help within the EEC contract ERB-CHRX-CT-930351.

List of Symbols and Abbreviations

b	Kuhn segment
BD	Brownian dynamics
c^*	overlap concentration (number density)
d	rigid bond lenth between adjacent beads of the polymer
D	polymer center of mass diffusion coefficient
D_0	free monomer diffusion coefficient
D_∞	polymer center of mass diffusion coefficient (Kirkwood's formula)
D_α	diffusion coefficient in non-equilibrium stationary state along principal axes directions of the gyration tensor
EF	elongational flow
EV	excluded volume
G	average gyration tensor
G_1, G_2, G_3	eigenvalues of G
H	MD cell matrix
HI	hydrodynamic interactions
$I(k, t)$	normalized dynamical structure factor
k_B	Boltzmann constant
K	shear rate tensor
L	side of the cubic box enclosing the MD system
L_1, L_2, L_3	MD cell defining vectors
L_x, L_y, L_z	sides of the rectangular parallelepipedic box followed in elongational flow
LS	light scattering
m	mass of both solvent particles and polymer beads
m_G^0	orientational resistance of the average gyration tensor in the zero shear rate limit
m_G	orientational resistance of the average gyration tensor
m_O^0	orientational resistance of the average order parameter tensor in the zero shear rate limit
m_O	orientational resistance of the average order parameter tensor
MD	molecular dynamics
M_w	molecular weight in g/mol
N	number of beads per chain
n	number of monomers per blob
N_B	number of shear blobs per chain
NEMD	non-equilibrium molecular dynamics
N_s	number of solvent point-mass particles
O	average order parameter tensor of polymer segments
O_1, O_2, O_3	eigenvalues of O

PBC	periodic boundary conditions
$\mathbf{R}$	average polymer end-to-end tensor
$\mathbf{R}_C$	polymer center of mass coordinate
R_H	hydrodynamic radius
$\mathbf{R}_i$	cartesian coordinate of the i-th polymer bead
R_g	chain radius of gyration
SANS	small angle neutron scattering
SF	shear flow
$S(k)$	single chain structure factor
$\mathbf{u}(\mathbf{r})$	local velocity field
x	scaling exponent $x = 2 + \dfrac{v'}{v}$
β	reduced shear rate
β'	reduced elongational rate
$\dot{\gamma}$	absolute shear rate
Γ_2	second scalar invariant of the symmetrized velocity gradient tensor
δ_{aa}	deformation ratio considered along laboratory axes ($a = x, y, z$)
$\dot{\varepsilon}$	absolute elongational rate
ε	Lennard-Jones energy parameter
$[\eta]$	intrinsic shear viscosity
η_s	solvent shear viscosity
v	Flory scaling exponent
v'	dynamical scaling exponent
v_1, v_2 and v_3	scaling exponents along the principal axes directions
ξ_c	shear blob size
ϱ	bead + solvent number density
σ	Lennard-Jones hard core parameter
τ_k	time limit to probe HI of intramolecular origin
τ_N	longest relaxation time of the chain
χ_G	orientation angle of the principal axis of the gyration tensor with respect to flow lines in shear flow
χ_O	extinction angle (orientation angle of the principal axis of the order parameter tensor with respect to flow lines in shear flow)

References

1. Lindner P (1991) Neutron, X-Ray and light scattering. Elsevier
2. Link A, Springer J (1993) Macromolecules 26:464
3. Zizenis M, Springer J (1994) Polymer 35(15):3156
4. Nguyen TQ, Yu G, Kausch H-H (1995) Macromolecules 28:4851
5. Boué F, Lindner P (1994) Europhysics Lett 25(6):421
6. Boué F, Lindner P (1995) Flow-induced structure in polymers, ACS Symposium series. ACS Washington D.C. 1005
7. Muller R, Picot C, Zang YH, Froelich D (1990) Macromolecules 23:2577
8. Muller R, Pesce JJ, Picot C (1993) Macromolecules 26:4356
9. Koppi KA, Tirell M, Bates FS (1993) Phys Rev Lett 70:1449

10. Nakatani AI, Morrison FA, Douglas JF, Mays JW, Jackson CL, Muthukumar M, Han CC (1996) J Chem Phys 104(4):1589
11. Bossart J, Ottinger HC (1995) Macromolecules 28:5852
12. Bruns W, Carl W (1993) Macromolecules 26:557
13. Carl W, Bruns W (1994) Macromol Theory Simul 3:295
14. Helfand E, Fredrickson GH (1989) Phys Rev Lett 62:2468
15. van Egmont JW, Fuller GG (1993) Macromolecules 26:7182
16. Pistoor N, Binder K (1988) Colloid & Polymer Science 266:132
17. Brown D, Clarke JHR (1983) Chem Phys Lett 98:579
18. Hounkonnou MN, Pierleoni C, Ryckaert JP (1992) J Chem Phys 97:9335
19. Kremer K, Grest GS (1990) J Chem Phys 92:5057
20. Brown D, Clarke JHR, Okuda M, Yamazaki T (1996) J Chem Phys 104(5):2078
21. Smith SW, Hall CK, Freeman BD (1996) J Chem Phys 104(14):5616
22. Kröger M, Loose W, Hess S (1993) J of Rheology 37(6):1057
23. Gao J, Weiner JH (1995) J Chem Phys 103(4):1614
24. Lopez Cascales JJ, Navarro S, Garcia de la Torre J (1992) Macromolecules 25:3574
25. Knudsen KD, Elgsaeter A, Lopez Cascales JJ, Garcia de la Torre J (1993) Macromolecules 26:3851
26. Zylka W (1991) J Chem Phys 94(6):4628
27. Pierleoni C, Ryckaert JP (1991) Phys Rev Lett 66(23):2992
28. Pierleoni C, Ryckaert JP (1992) J Chem Phys 96(11):8539
29. Duenweg B, Kremer K (1991) Phys Rev Lett 66:2996
30. Duenweg B, Kremer K (1993) J Chem Phys 99:6938
31. Ladd AJC (1984) Mol Phys 53:459
32. Nosé S (1991) Prog Theor Phys Suppl 103:1
33. Evans DJ, Holian BL (1985) J Chem Phys 83(8):4069
34. Pierleoni C, Ryckaert JP (1992) Molecular Physics 75(3):731
35. Doi M, Edwards SF (1986) The Theory of Polymer Dynamics. Clarendon Press, Oxford
36. Pierleoni C, Ryckaert JP (1991) Phys Rev A 44(8):5314
37. Lee AW, Edwards SF (1972) J Phys C 5:1921
38. Zizenis M, Springer J (1995) Polymer 36(18):3459
39. Flory P (1969) Statistical Mechanics of Chain Molecules. Interscience N.Y.
40. Pierleoni C, Ryckaert JP (1995) Macromolecules 28:5097
41. Zimm BH (1956) J Chem Phys 24:269
42. Peterlin A (1976) Ann Rev Fluid Mech, page 35
43. Cottrell FR, Merrill EW, Smith KA (1969) J of Polymer Science: Part A-2, 7:1415
44. Janeschitz-Kriegl H (1969) Adv Polym Sci 6:170
45. Pierleoni C, Ryckaert JP (1993) Phys Rev Lett 71(11):1724
46. Lindner P, Oberthur RC (1989) Physica B 156 & 157:410
47. Onuki A (1985) J of Phys Soc Japan 54(10):3656
48. de Gennes PG (1979) Scaling Concepts in Polymer Physics. Cornell University Press, Ithaca, N.Y.
49. Lindner P, Oberthur RC (1988) Colloid Polym Sci 266:886
50. Cummings PT, Wang BY, Evans DJ, Fraser KJ (1991) J Chem Phys 94(3):2149
51. Ottinger HC (1987) J Chem Phys 87:6185
52. Zylka W, Ottinger HC (1991) Macromolecules 24:484
53. Ciccotti G, Pierleoni C, Ryckaert JP (1991) Theoretical Foundation and Rheological Application of Nonequilibrium Molecular Dynamics. In: Maréschal M, Holian BL, editors, Microscopic Simulation of Complex Hydrodynamic Phenomena. Plenum Press, New York
54. Bird RB, Armstrong RC, Hassager O (1987) Dynamics of Polymeric Liquids, volume 1. Wiley Interscience, New York
55. Larson RG (1988) Constitutive Equations for Polymer Melts and Solutions. Butterworth, Boston
56. Odell JA, Keller A, Paul W, Atkins EDT (1985) Macromolecules 18(7):1443

CHAPTER 3

Tethered Polymer Chains under Strong Flows: Stems and Flowers

F. Brochard-Wyart and A. Buguin

3.1
Introduction

A long DNA molecule can be seen directly under an optical microscope if it is suitably treated with fluorescent dyes [1]. This allows one to see how a long flexible chain (N monomers) responds to external forces f. The conformation of the uniformly stretched chain has been discussed by Pincus [2]. Under a force f, acting on the two free ends, the chain configuration is modified only above a scale length $\xi = kT/f$. The chain (Fig. 3.1a) can be pictured as a string of blobs of size ξ containing g monomers with $\xi = g^\nu a$ ($\nu = 1/2$ in θ solvent, $\nu = 3/5$ in good solvent). The end to end distance is $L = \dfrac{N}{g}\xi$, i.e.

$$L = Na\,\frac{fa}{kT} \quad (\nu = 1/2, \text{ideal}) \tag{3.1}$$

$$L = Na\left(\frac{fa}{kT}\right)^{2/3} \quad (\nu = 3/5, \text{swollen}) \tag{3.2}$$

For DNA, which is a semi-rigid chain, ideal behavior has been observed for chain's length up to 20 microns when f is small. At higher f ($\cong 100$ piconewtons) Cluzel et al. [3] found a structural phase transition, corresponding to a plateau in the curve f vs elongation. Going even further, one reaches the break point [4].

Another method is based on optical tweezers [5]. Here one end of the DNA chain is attached to a glass bead of high polarisability, which is attracted by a region of large (optical) electric field. This was used by the Stanford group: in one series of experiments they observed the conformations of the chain, pulled at one end by a constant force f. The friction on monomers are cumulative, and the tension along the chain is not uniform: it increases from the free end to the tethered end. At low f, the shape is a "trumpet" (Fig. 3.1b): near the bead all the drag forces add up, giving a relatively large tension; in this region, the trumpet is thin. At the bottom end, the tension is lower and the chain is more contorted. At higher f, the portion near the bead is under so high a tension that it becomes completely aligned: we have called this the "stem" [6]. Behind the stem, the free end is still expanded: we call it the "flower" (Fig. 3.1c). The border between stem and flower is relatively sharp. At large f, the flower shrinks to zero.

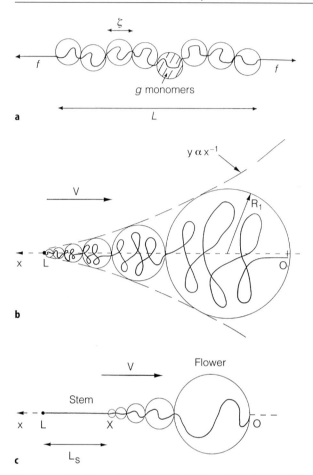

Fig. 3.1a–c. Deformation of a single chain: **a** free chain under traction f pictured as a string of independent blobs; **b** deformation of a tethered chain under uniform flow: trumpet regime; **c** deformation of a tethered chain under uniform flow: stem and flower regime

In Sect. 3.2 we discuss the unwinding of one chain under uniform and shear flows (i) for swollen chains in good solvent, and (ii) for ideal chains, which corresponds to the case of DNA molecules.

In Sect. 3.3 we consider one N chain immersed in a semi dilute solution of free longer P chains (P monomers) chemically identical. The Stanford group also looked at a labeled DNA chain not in water, but in a solution of other (unlabeled) chains [5]. This leads to entanglements: the test chain is trapped into a "tube", because it cannot intersect its neighbors. By pulling the test chain along a curved path, such as the letter C, they could show that all the chain follows the trajectory of the "head" because it stays in the same tube. The idea of the tube was invented long ago by Edwards [7] and exploited by de Gennes in the dynamics of entangled polymers (Reptation model) [8]. However, the Stanford experiment gave the first visual proof of the existence of these tubes. Keeping this situation of one test chain entangled with other chains, Wirtz [9] measured both the force applied at one end (via a magnetic bead) and the resulting steady state velocity V. This is the first direct measurement of the friction on one single chain. In a certain range of velocity they found that the friction coefficient $\xi = f/V$ is constant, and also that the conformation is progressively changing. At higher velocity, the chain undergoes a disentanglement transition, and enters in a "marginal state", where the friction force becomes constant in a broad interval in velocities. At high velocity, the force starts to increase again. We discuss these different regimes of the stretching under flow of one tethered chain immersed in a semi-dilute solution.

We study in Sect. 3.4 the unwinding of globular macromolecules under flow: the tethered flexible chain is now immersed in a poor solvent. This case, which has not yet been studied experimentally, may be important for the denaturation of proteins under strong flows.

In Sect. 3.5 we extend our discussion to polymer in a good solvent, but confined in a slit. Visualization of individual nucleic acid moving in microlithographic arrays of ordered posts has shown that the dynamics of confined chains is slowed down because the hydrodynamic interactions are screened in a slit.

The Stanford group and the Curie group have also monitored the relaxation of one chain when the flow is abruptly suppressed. We discuss here the relaxation for chains in pure solvent, in semi-dilute solution, and in a poor solvent. Our discussion is restricted to the level of scaling laws (all numerical coefficients are ignored).

3.2
Chains Immersed in a Pure Solvent

3.2.1
Steady State in Uniform Flows

3.2.1.1
Good Solvent ($v = 3/5$)

We discuss the progressive unwinding vs increasing flow velocity V.

3.2.1.1.1
The Unperturbed State ($V < V_{c1}$)

We start with a coil of size the Flory radius $R_F = N^{3/5}a$. At very low velocities we have a Stokes friction, with the force $f \cong \eta R_F V$, η being the solvent viscosity.

When the force f is much smaller than kT/R_F, the distortions in the chains are small (of the order of fR_F/kT). Things change when we reach a velocity

$$V_{c1} = \frac{kT}{\eta R_F^2} = \frac{R_F}{\tau_Z} \tag{3.3}$$

where $\tau_Z = \eta R_F^3/kT$ is the Zimm relaxation time for weak perturbations.

3.2.1.1.2
The Trumpet ($V_{c1} < V < V_{c2}$)

Under stronger flow, the polymer is stretched, but not in a cigar shape: the tension increases from the free end toward the grafted end. From the Pincus rule, the size of the blobs decreases and the conformation can be described as a trumpet [6]. Consider the n-th monomer (counted from the free end) with a distance $x(n)$ to the free end. For a (roughly) linear object of length x (and diameter smaller than x), the viscous force is $f_v \cong \eta x V$. The local deformation dx/dn is related to the force f_v by the derivative of the Pincus law [2] giving the stretching of polymers in good solvent conditions (Eq. 3.2):

$$\frac{dx}{dn} = a \left(\frac{f_v a}{kT} \right)^{2/3} \tag{3.4}$$

This leads to

$$x(n) = \left(\frac{na}{3} \right)^3 \left(\frac{\eta V a}{kT} \right)^2 \tag{3.5}$$

The total extension is

$$L = x(N) = R_F \left(\frac{V\tau_Z}{R_F} \right)^2 \cong N^3 V^2 \tag{3.6}$$

The profile $y(x)$ of the trumpet is given by the Pincus relation [2]: $y(x) = \frac{kT}{f_v(x)}$ $\cong \frac{kT}{\eta V_x}$.

One can picture the chain as a string of blobs (Fig. 3.1b). The largest blob (near the free end) has a size R_1, $\left(\eta V R_1 = \frac{kT}{R_1} \right)$. The smallest blob (at the tethered end) has a size $R_N = \frac{kT}{\eta V L} = R_F \left(\frac{R_F}{V\tau_Z} \right)^3$.

The trumpet regime holds only if R_N is larger than a monomer size; this corresponds to

$$V < V_{c2} = N^{1/5} V_{c1} \tag{3.7}$$

Our trumpet regime is thus valid only in the range of velocity $V_{c1} < V < V_{c2}$, which is extremely narrow ($V_{c2} \cong 4V_{c1}$ for $N = 10^3$).

3.2.1.1.3
The Stem and Flower Regime ($V > V_{c2}$)

Above V_{c2}, the chain is elongated into a stem of diameter a and is terminated by a little "flower" (Fig. 3.1 c). The number n^* of monomers in the "flower" of length X_f is given by Eq. (3.5), with $\eta V X_f = kT/a$. This leads to

$$n^* = N\frac{V_{c2}}{V}, \quad X_f = (n^*/3)\, a \qquad\qquad (3.8)$$

We purposely keep the coefficient in Eq. (3.8) to show that $X_f \cong n^*a$, but that the stretching is not complete. The length of the stem is $L_S \cong (N - n^*)\, a$. The full extension of the chain is $L = L_S + X_f$, i.e., $L = \left(1 - \dfrac{2}{3}\dfrac{V_{c2}}{V}\right) Na$, where again the coefficient 2/3 is only indicative.

   This regime itself is limited. At $V_{c3} = NV_{c2}$, $n^* = 1$ and the chain is fully elongated. But V_{c3} is several orders of magnitude larger than V_{c2}. The main regime is thus the "stem and flower" regime.

3.2.1.1.4
Chain Rupture

The force acting on the grafted site is the total viscous force $f_v = \eta VL$. In the limit $N \gg n^*$, this leads to $f_v \cong \dfrac{N}{n^*}\dfrac{kT}{a}$. This force becomes rapidly comparable to the force necessary to break a chemical bond U/a (where U is a chemical binding energy).

3.2.1.2
Ideal θ Solvent ($v = 1/2$)

In θ solvents, the grafted chains are ideal ($R_0 = N^{1/2}a$) and the deformation is enhanced. For $V > V_{c1} = \dfrac{kT}{\eta R_0^2}$, Eq. (3.4) is replaced by $dx/dn = a\,(fa/kT)$. For grafted chains in a uniform flow, we find a stretching increasing exponentially with velocity,

$$L = R_0 \left(\frac{R_0}{V\tau_Z}\right)^{1/2} \exp\left(\frac{V\tau_Z}{R_0}\right) \qquad\qquad (3.9)$$

$$\tau_Z = \frac{\eta R_0^3}{kT}$$

3.2.2
Tethered Chains in Shear Flow (Fig. 3.2)

3.2.2.1
Good Solvent

This is relatively easy to set up with a grafted plate exposed to a Couette or a Poiseuille flow. The solid surface on which a few long, flexible chains are grafted is immersed in a shear flow $V_x(y) = s y$ (Fig. 3.2).

3.2.2.1.1
"Trumpet" Regime

The friction on a blob of size y is $\eta s y^2$. The total friction force is $f = \sum \eta s y_i^2 = \int \eta s y^2 (dx/y)$. The size $y(x)$ is deduced from the Pincus rule:

$$f(x) = \int_0^x \eta s y(x')\, dx' = kT/y \tag{3.10}$$

Equation (3.10) leads to $\eta s y(x) = kTy'/y^2$, i.e., a profile of a "horn"

$$y(x) = \sqrt{\frac{kT}{\eta s x}} \tag{3.11}$$

The first blob has a size given by Eq. (3.11) ($y = x = R_1$):

$$R_1 = \left(\frac{kT}{\eta s}\right)^{1/3} \tag{3.12}$$

The strong deformation regime starts for $R_1 < R_F$, i.e., $s_{c1}\tau_Z > 1$.

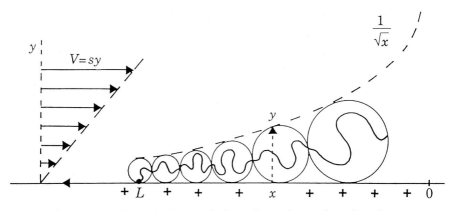

Fig. 3.2. Deformation of a single chain grafted on a flat surface under a shear flow

Equation (3.4) with $f(x) = \dfrac{kT}{y(x)} = (\eta s k T x)^{1/2}$ leads to

$$\frac{dx}{dn} = a\left(\frac{a}{y}\right)^{2/3} = a^{5/3}\left(\frac{\eta s x}{kT}\right)^{1/3} \tag{3.13}$$

Integrating over the total chain length, we arrive at

$$L = N^{3/2} a^{5/2}\left(\frac{\eta s}{kT}\right)^{1/2} = R_F(s\tau_Z)^{1/2} \tag{3.14}$$

3.2.2.1.2
"Stem and Flower" Regime

It starts when the size of the smallest blob becomes a, the monomer size. From Eq. (3.11–3.14) one finds $s_{c2}\tau_Z = (R_F/a)^{4/3} = N^{4/5}$. For $s > s_{c2}$, the chain is composed of a flower (containing $N_f = N\dfrac{s_{c2}}{s}$ monomers) and a rod, of length $L_f = (N - n_f)\,a$.

It can be concluded that the trumpet regime starts at $s\tau_Z = 1$ and extends up to $s\tau_Z = N^{4/5}$. It can be observed on few decades in s and it will be interesting to check the power law for $L(s)$.

3.2.2.2
Ideal Chains

For ideal chains, such as DNA grafted on a wall, the deformation is enhanced. The profile of the trumpet is still given by Eq. (3.11). The local deformation is now proportional to the local tension $\dfrac{dx}{dn} = a\left(\dfrac{fa}{kT}\right)$. With f given by Eq. (3.10):

$$\frac{dx}{dn} = a^2\left(\frac{\eta s x}{kT}\right)^{1/2} \tag{3.15}$$

This leads to an extension

$$L = R_0(s\tau_Z) \tag{3.16}$$

The trumpet regime where $L \sim s$ is valid in the range $1 < s\tau_Z < N^{1/2}$. At higher shear rates, the chain enters in the stem-flower regime, where no simple scaling law is expected.

3.2.3
Relaxation Processes: Stretch to Coil

3.2.3.1
Good Solvent ($v = 3/5$)

Let us consider an initial state of uniform flow with $V > V_{c2}$, i.e., in the stem plus flower regime. At time $t = 0$, we stop the flow abruptly. The chain then relaxes towards the Flory coil conformation. But the stem, being fully stretched, cannot relax internally: it is progressively destroyed at one end. (Fig. 3.3)

A related problem was considered long ago by P.G. de Gennes [10]. This was the helix-coil transition of a short polypeptide, where the relaxation can proceed only from the free ends: then there is a driving force f_D (related to the free energy difference between helix and coil) pulling the coil portion, and balanced by the Stokes friction on this portion. De Gennes studied the limit of *small* driving forces, where the coil was relaxed.

For our problem, the driving force is of entropic origin: the monomers of the stem would like to recover some entropy, and this corresponds to a *large* driving force $f_D \cong \dfrac{kT}{a}$.

During relaxation, the flower expands, containing an increasing number of monomers n_t. The stem shrinks by a length $x_t \sim a\, n_t$, which is also the length of the flower. Let us assume that at any instant t, the flower has the shape (discussed before) corresponding to a velocity $\dfrac{dx_t}{dt}$. This velocity must be equal to V_{c2}, defined in Eq. (3.7), provided that, in the definition of V_{c2}, we replace N by n_t. This choice will assure that the pulling force on the flower is indeed $\dfrac{kT}{a}$. Thus we are led to

$$\frac{dx_t}{dt} = V_{c2}(n_t) = n_t^{1/5} \frac{kT}{\eta a^2 n_t^{6/5}} = \frac{kT}{\eta a x_t} \tag{3.17}$$

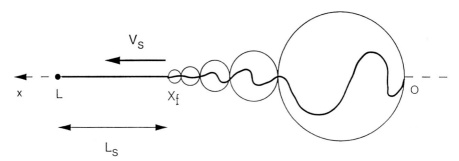

Fig. 3.3. Relaxation of a strongly stretched polymer chain in the stem-flower model. The stem retracts at velocity V_S and pulls the growing flower, of extension X_f, increasing as the square root of time

For instance, if the initial size ($n_{t=0} = n^*$) of the flower is negligible ($n^* \ll N$) we may write from Eq. (3.10):

$$x_t^2(t) = \frac{kT}{\eta a} t \tag{3.18}$$

and relaxation is complete after a time

$$\tau_r \cong \frac{N^2 \eta a^3}{kT} = N^{1/5} \tau_Z \tag{3.19}$$

Thus the recovery time $\tau_r \sim N^2$ is similar to a Rouse time. This is not entirely surprising, since back flow effects become nearly negligible (within logarithmic factors) for strongly stretched chains. The stem retracts as $L_S = L_0 - \left(\frac{kT}{\eta a} t \right)^{1/2}$, and one notices that the full length does not obey simple power laws.

One question remains: is it correct to discuss the transient flower as a steady state flower with a variable velocity $(dx/dt) \sim t^{-1/2}$? Here our earlier discussion [11] of collective modes in a trumpet is useful. The slowest collective mode of the whole trumpet τ_t is given by the simple scaling relationship $V\tau_t = n_t a$, i.e., $\tau_t = n_t^2 \tau_0$, where $\tau_0^{-1} = \frac{kT}{\eta a^3}$ is a monomer relaxation time. We see from Eq. (3.18) that $t \sim \tau_t$, and the steady state approximation is valid.

3.2.3.2
Ideal θ Solvent ($v = 1/2$)

This case is important in practice because it has been studied experimentally with DNA [5, 12].

Consider a tethered polymer chain fully elongated, i.e., $L = Na$. In θ solvents, the grafted chain is ideal ($R_0 = N^{1/2}a$) and Eq. (3.5) is replaced by $x(n) = \sqrt{\frac{kT}{\eta V}}$ $\exp \left(\frac{\eta n a^2}{kT} V \right)$. If n_f is the number of monomers in the expanding flower, its size X_f is given by $X_f = \sqrt{\frac{kT}{\eta V}} \exp \left(\frac{\eta n_f a^2}{kT} V \right)$. Moreover the total viscous force on this flower is equal to the driving force $\eta V X_f = \frac{kT}{a}$.

With $V = a \dfrac{dn_f}{dt}$, this leads to

$$2n_f \frac{dn_f}{d\tilde{t}} = - \log \left(\frac{dn_f}{d\tilde{t}} \right) \tag{3.20}$$

where $\tilde{t}$ is the time normalized by $\tau_0 = \dfrac{\eta a^3}{kT}$, the monomer relaxation time.

A second order approximation of the solution of Eq. (3.20) gives

$$n_f(t) \approx \tilde{t}^{1/2} \log^{1/2}(\tilde{t}^{1/2}) \qquad\qquad (3.21)$$

We should notice that a first order approximation gives the same solution as in the case of good solvents. In practice the logarithmic factor could be neglected. The end-to-end distance does not follow simple scaling law, but $L(0) - L(t) \approx n_f a \approx t^{-1/2}$ is predicted to fit universal scaling features. The "stem and flower" model for this relaxation of a stretched chain is in agreement with experimental data of Manneville et al. [12]. They measure the relaxation on a fluorescent free DNA molecule after stretching by Poiseuille flow in a capillary vessel (Fig. 3.4). The scaling exponent they obtained for the decay law was 0.51 ± 0.05.

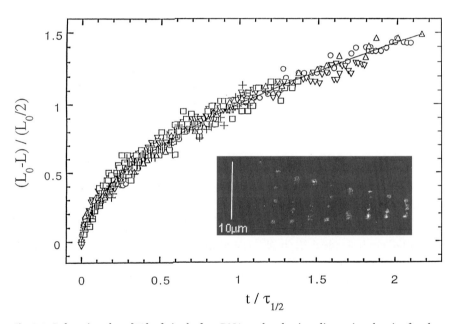

Fig. 3.4. Relaxation data [12] of single free DNA molecules in adimensional units for three values of initial length L_o [($\triangledown$) 8.1, (+) 12.6, and ($\square$) 16.1 µm] and three viscosities [η = ($\triangle$) 37, ($\bigcirc$) 70, and ($\square$) 170 cP]. $\tau_{1/2}$ is the half relaxation time given by $L(\tau_{1/2}) = L_o/2$ (S. Manneville et al.)

3.3
Tethered Chains Immersed in a Polymer Solution

We discuss here the case where the tethered chain (N monomers) is immersed in a solution of chains, chemically identical, but longer (P monomers, $P > N$) (Fig. 3.5). This corresponds to the Wirtz [9] experiment, where a fluorescently labeled DNA molecule is end-tethered to a small magnetic bead and pulled by a calibrated magnetic force through a solution of unlabeled DNA molecules (Fig. 3.6). This also occurs if the N chain is grafted on a flat wall and the P solution flows tangentially to the wall.

The semi-dilute solution is pictured as a transient network of mesh size ξ. One P chain can be represented as an ideal chain of blobs of size ξ and the unperturbed size is $R_{0N} = (N/g)^{1/2}\xi$. Inside one blob, one expects the configuration of an isolated chain and the number of monomers is g, with $g^\nu a = \xi$ and $\nu = 1/2$ (ideal) or $\nu = 3/5$ (swollen).

3.3.1
Friction on the Test Chain ($V \geq 0$)

Assume that the N chain has to move on a distance ξ equal to the diameter of an Edwards tube in a semi-dilute solution. To allow for this motion of the N chain, each P chain entangled with N must move along its own tube $\left(\text{of length } L_t = \dfrac{P}{g} \xi\right)$ by a length L_t. Thus the sliding velocity of the P chain is not V, but is much larger:

$$V_S = V \frac{L_t}{\xi} = V \frac{P}{g} \tag{3.22}$$

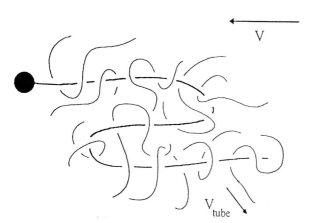

Fig. 3.5. In the relative motion of the solution with respect to the test chain, ambient chains entangled with tethered chain have to reptate along their own tube to disengage from the test chain

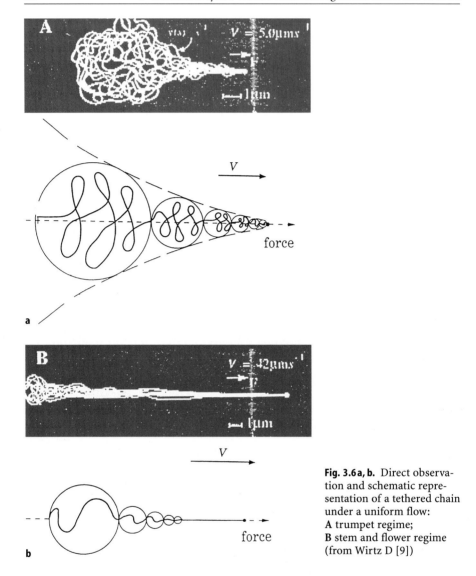

Fig. 3.6a, b. Direct observation and schematic representation of a tethered chain under a uniform flow: **A** trumpet regime; **B** stem and flower regime (from Wirtz D [9])

The dissipation $T\mathring{S}$ due to the motion of the test chain corresponds to X chains moving at velocity V_S in the ambient solution:

$$T\mathring{S} = X\eta_0 \left(\frac{P}{g}\right)^3 \xi V^2 = fV \tag{3.23}$$

Following [13], we assume that the number X of running ambient chains is proportional to the length of the test chain $X = N/g$.

Comparing the two expression for $T\mathring{S}$, we get

$$f = \frac{N}{g} \xi \eta_P V \tag{3.24}$$

where $\eta_P = \eta_0 \left(\dfrac{P}{g}\right)^3$ is the viscosity of the solution, and η_0 a monomer friction coefficient.

3.3.2
Deformation Under Flow

3.3.2.1
Coil Regime

As long as $f = \dfrac{N}{g} \xi \eta_P V < \dfrac{kT}{R_{0N}}$, where $R_{0N} = \left(\dfrac{N}{g}\right)^{1/2} \xi$, the chain configuration is a coil. This correspond to $V < V_{c1} = \dfrac{R_{0N}}{\tau_R}$, where τ_R is the coil relaxation time:

$$\tau_R = \left(\frac{N}{g}\right)^2 \tau_\xi$$

$$\tau_\xi = \frac{\eta_P \xi^3}{kT} \tag{3.25}$$

3.3.2.2
Trumpet Regime

For $V > V_{c1}$, the chain enters in the trumpet regime. In the monobloc approximation, the chain is pictured as a cigar of diameter D given by $\dfrac{N}{g} \eta_P \xi V = \dfrac{kT}{D}$, and the extension is

$$L = \frac{N}{g_D} D = N \frac{a^2}{D} = Na^2 \frac{N}{g} \eta_P \frac{\xi V}{kT}$$

In fact, the friction force increases along the chain, and the conformation is a trumpet:

$$\frac{dx}{d\tilde{n}} = \frac{\xi^2 f}{kT} = \xi^2 \tilde{n} \eta_P \frac{\xi V}{kT} \tag{3.26}$$

where $\tilde{n} = n/g$ is the index of the blob ξ and x the extension (from the free end). Integration of Eq. (3.26) leads to a full extension:

$$L = \left(\frac{N}{g}\right)^2 \eta_P \frac{\xi^3}{kT} V \tag{3.27}$$

The profile of the trumpet is given by

$$\frac{n}{g}\eta_P\xi V = \frac{kT}{y} \tag{3.28}$$

Equation (3.26) shows that $x \cong \left(\frac{n}{g}\right)^2$ and $y \cong 1/\sqrt{x}$. Notice that the friction coefficient ζ in the trumpet regime is constant $\left(\zeta = \frac{N}{g}\eta_P\xi\right)$. This corresponds to the "Rouse entangled regime" shown in Fig. 3.7.

3.3.2.3
Marginal Regime: Stem and Flower

The trumpet regime ends when the size of the smallest blob becomes equal to ξ. This correspond to a velocity given by $f(V^*) = \frac{kT}{\xi}$:

$$\frac{N}{g}\eta_P\xi V^* = \frac{kT}{\xi}$$

$$\text{i.e., } V^* = V_{cl}\sqrt{N/g} \approx N^{-1}P^{-3} \tag{3.29}$$

Above V^*, the chain enters the "*marginal*" regime. The friction force (Fig. 3.7) becomes constant and equal to $\frac{kT}{\xi}$. This plateau corresponds to a disentanglement transition. As soon as the lateral size D becomes smaller than ξ, the chain is no more entangled to the free chain and the friction becomes very small, and the chain has to swell again. The chain is constrained in the marginal state for a broad range of velocity. The number $X(V)$ of trapped chains decreases as V increases:

$$X(V)\eta_P\xi V = \frac{kT}{\xi}$$

$$\text{i.e., } X(V) = \frac{N}{g}\frac{V^*}{V} \tag{3.30}$$

The chains can be pictured as a stem of disentangled units of size ξ, length $L_S = \left(\frac{N}{g} - X\right)\xi$, and a flower entangled with $X(V)$ "P" chains (Fig. 3.7: marginal regime).

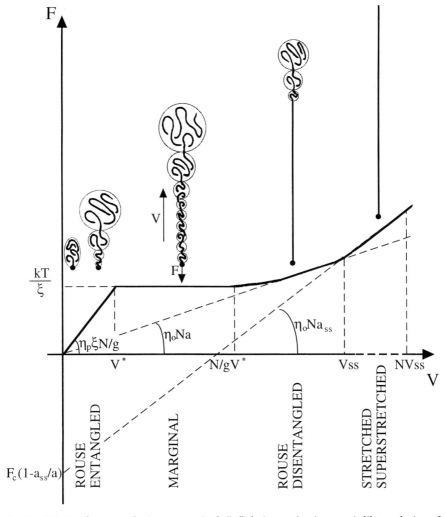

Fig. 3.7. Friction force vs velocity on one single "N" chain moving in a semi-dilute solution of free "P" chains

3.3.2.4
Ideal Stokes Regime

The marginal regime ends when the friction on the disentangled chain becomes sufficient to stretch the chain. This corresponds to a velocity V^{**}:

$$\eta_0 LV^{**} = \frac{kT}{\xi}$$

$$L = \frac{N}{g} \xi$$

i.e., $V^{**} = \dfrac{kT}{\eta_0 R_N^2}$ $\qquad$ (3.31)

Above V^{**}, we find the regime discussed in Sect. 3.2 for a tethered chain in a pure solvent.

For an ideal DNA chain, the extension L increases exponentially:

$$L = R_1 e^{Na^2/R_1^2}$$
$\qquad$ (3.32)
$$R_1^2 = \frac{kT}{\eta_0 V}$$

The chain configuration is a stem of monomers, terminated by a little flower. The friction coefficient increases up to a plateau value $\zeta = \eta_0 Na$.

At larger friction force, the internal configuration of the DNA chain can be modified, as in the Cluzel experiment [3]. The coil-superstretched DNA transition has been observed around a tension of about $F_c \approx 100$ pN. Such transition can easily been induced under strong flows, above a threshold velocity

$$V_{ss} = \frac{F_c}{\eta_0 Na} \text{ (mms}^{-1} \text{ for } Na = 25 \ \mu\text{m}, \ \eta = 10 \text{ cp). For } V > V_{ss}, \text{ the chain is in a}$$

two-state (stretched-superstretched) conformation. The friction force is now

$$F_v = F_c + 7 \ \eta_0 a_{ss} NV \left(1 - \frac{V_{ss}}{V} \right), \text{ where } a_{ss} \cong 1.5 \, a \text{ is the size of the superstretched}$$

monomer.

3.4
Tethered Chains in Poor Solvents

3.4.1
Conformation of a Single Chain in a Poor Solvent

If we suddenly cool a dilute solution of flexible homopolymer chains (containing N monomers) below the θ Flory's temperature, we observe that each chain collapses into a compact globule. Initially, the chain is ideal (its radius is given by a random walk of monomers and scales like $N^{1/2}a$), while at a temperature $T = \theta - \Delta T$, in order to favor its monomer-monomer interactions, the chain collapses into a globule (if the condition $\Delta T > \theta/N^{1/2}$ is satisfied). This collapsed chain can be pictured as a collection of thermal blobs of size ξ (the correlation length), each blob containing $g = (\theta/\Delta T)^2$ monomers. At scales smaller than ξ, the chain is nearly ideal and exhibits Gaussian statistics because the thermal energy dominates; we have $\xi = g^{1/2}a$, where a is the monomer size. At larger scales, between two blobs, there is an attractive potential of the order of $- k\theta$. Higher order development accounting for three blobs interactions would give a repulsive term. Thus, the blobs tend to stick together to build up a globule of radius $r_c = (N/g)^{1/3} \xi$. The radius of the chain scales like N^ν with $\nu = 1/3$. This globule is very similar to a simple liquid droplet with a homogeneous monomer volume fraction $\phi_g = ga^3/\xi^3 = \dfrac{\Delta T}{\theta}$. It is then possible to define a surface tension between the droplet and the pure solvent:

$$\gamma = kT/\xi^2 \tag{3.33}$$

This description, based on a single chain, masks the fact that the collapse behavior is impossible to decouple from a collective process of demixion. As a final equilibrium state, we will obtain a two-phase system (polymer precipitation), with poor and rich polymer volume fraction regions. This is the reason why the dynamic of collapse of a single chain, which has been studied theoretically [14–16], is very difficult to observe experimentally [17]. You need to use *very dilute solutions* in order to avoid the growth of rich polymer volume fraction regions (demixion) rather than the single chain collapse.

In this discussion, we consider that the chain is always flexible whatever the temperature of the final state. We assume, thus, that ΔT is not large to avoid the glassy state for our dense polymer droplet.

3.4.2
Force-Elongation Diagram Under Uniform Tension

The elongation L of one collapsed chain, when external constraint f is applied at both ends of the chain, has been studied by Halperin and Zhulina [18]. This case is drastically different from the good or θ solvent case because the surface tension plays an important role. As long as $f < \dfrac{kT}{\xi}$, the local structure is not modified. The chain behaves like a liquid droplet of monomers of constant volume, $\Omega_c = Na^3 \dfrac{\Theta}{\Delta T}$ and capillary energies are dominant. On the other hand, where $f > \dfrac{kT}{\xi}$, the chain conformation is perturbed at scales smaller than the blob size. The chain behaves now like a regular ideal chain.

As f increases, one expects three regimes.

(i) For small tensions f, the globule is very close to its equilibrium state, and it looks like a liquid droplet. If there is no change in the structure of the globule, its shape becomes ellipsoidal (Fig. 3.8a). The tension is due to an increase of the surface energy; $F = \gamma \Delta A$ with $\Delta A \approx (L - r_c)^2$. The minimization of the free energy gives a linear response:

$$f = \gamma (L - r_c) \tag{3.34}$$

(ii) For large tensions, the chain is stretched into a "liquid" filament of diameter d (Fig. 3.8b), with $\Delta A = Ld$. Using the condition of volume conservation

$\Omega_c = Ld^2$ we obtain

$$f = \gamma(\Omega_c/L)^{1/2} \tag{3.35}$$

In these two first stages there is no change in the internal blobs structure, as $d > \xi$, i.e. $L = (N/g)\xi$ and $f = kT/\xi$.

(iii) For very large tensions, the chain elongates as a string of N/g_p blobs of size $\xi_p = kT/f$ (Fig. 3.8c). At each scale, for this regime, the chain has an ideal behavior. The deformation follows the Kuhn law for ideal chains:

$$L = R_0^2 f/kT \tag{3.36}$$

where $R_0 = N^{1/2}a$ is the radius of the ideal chain.

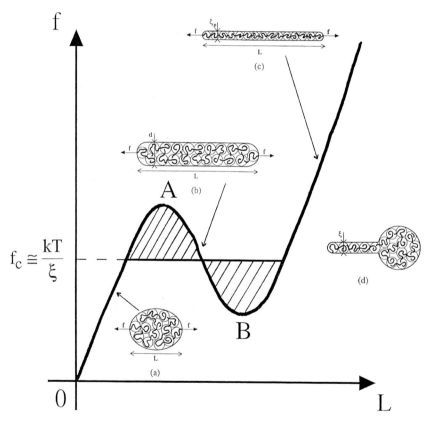

Fig. 3.8a–d. Force f vs deformation L of a globular chain and corresponding conformation: Van der Waals loop in the $f(L)$ diagram: **a** the ellipsoidal shape for small tensions; **b** the cylindrical shape for large tensions which corresponds to the unstable region A–B of the $f(L)$ curve (Rayleigh instability); **c** the string of blobs of size $\xi_p = kT/f$ for very large tensions; **d** in the plateau region $f = f_c$ (given by the Maxwell construction) the conformation of the chain is a globule coexisting with a rod of diameter ξ

The sequence $f \approx L, f \approx L^{-1/2}$, and $f \approx L$ as the deformation increases (Fig. 3.8) describes a Van der Waals loop in the $f(L)$ diagram. This is a manifestation of the Rayleigh instability of liquid cylinders. The Maxwell equal area construction shows, at $f = f_c = kT/\xi$, a first-order stretching transition from a globular form $L = r_c + \xi$ to a stretched form $(L = (N/g)\xi)$, involving a coexistence of a weakly deformed globule and a stretched string of blobs of size ξ (Fig. 3.8 d).

3.4.3
Deformation Under Solvent Flow

The tethered collapsed chain is now exposed to a uniform flow of (poor) solvent. The external force which acts on the globular chain is the Stokes friction. The chain is attached at one end: experimentally, this can be achieved by grafting a magnetic or a glass bead and by keeping it motionless (with respectively a magnetic gradient [19] or optical tweezers [5]) in a uniform flow of solvent (viscosity η) at velocity V. One assumption of our model is that there is no hydrodynamic interaction between the particle and the chain: the bead is small enough not to perturb the solvent flow around the polymer. One can also think of a globular protein attached to a glass bead. We will not describe the mono-block approximation for this system and we will consider only the case where the tension along the chain is not uniform (the friction force increases from the free end to the fixed extremity).

3.4.3.1
Small Deformation: $(V < V_1)$

At low velocity, the drag forces are small and the chain becomes a slightly elongated ellipsoid. By balancing the friction force $f_v \cong \eta r_c V$ and the chain tension (Eq. 3.34) we obtain:

$$\eta r_c V = \gamma (L - r_c) \qquad (3.37)$$

To determine the drag force on the chain we have assumed that the solvent flow does not penetrate the globule, which is a good approximation because the flow is screened on a length scale of the order of ξ. When the drag force becomes of the order of the critical tension $f_c = kT/\xi$, the ellipsoid gets strongly distorted and the assumption of small deformation is not longer valid. This defines the threshold velocity $V_1 = \dfrac{\gamma \xi}{\eta r_c}$.

3.4.3.2
"Stem and Globule": $(V_1 < V < V_2)$

For velocities greater than V_1, because the chain deformation is non-uniform, we should observe a coexistence between a globule (near the free end where the ten-

ROD + TRUMPET   GLOBULE
N-n*   n*

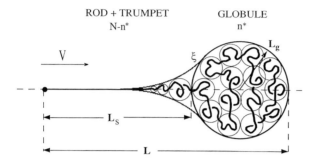

Fig. 3.9. Schematic representation of a collapsed chain deformation in a uniform flow

sion is low) and a stretched part of the chain (near the fixed extremity). Thus the main point is that at the end of the tail, there is still a globule of size L_g with a number $n*$ of monomers (Fig. 3.9). We have $L_g = (n*/g)^{1/3}\xi$. The elastic force acting on this globule remains equal to the threshold value kT/ξ and balances exactly the Stokes friction: $\eta L_g V = kT/\xi$. Near the attachment point we expect a "stem" of $N - n*$ monomers of length L_s. In this portion of the chain the tension is necessarily larger than kT/ξ and it implies that the "stem" diameter is smaller than ξ. Moreover, because at scales smaller than ξ, the chain is ideal, and we can use the local equation between force and elongation, assuming an increase of tension from the fixed point to the globule $\dfrac{kT}{a^2} \dfrac{dx}{dn} = f = \eta x V + \dfrac{kT}{\xi}$ where x is the distance from the globule to the monomer. The total extension L is derived by integrating this equation:

$$L = L_s + L_g = L_g e^{\frac{\eta V a^2}{kT}(N-n*)} \tag{3.38}$$

The exponential increase of the extension vs velocity in Eq. (3.38) shows that when V is just above V_1, it is possible to consider the stem as almost fully elongated, with a length

$$L_s \cong (N - n*)a = N\dot{a}(1 - (V_1/V)^3) \tag{3.39}$$

However, there is still a small portion of the chain between the globule and the rod where the polymer assumes a horn shape. A rather simple calculation [20] allows us to neglect it.

This regime disappears when the globule size becomes equal to the blob size ξ. This corresponds to a solvent velocity V_2 when $L_g = \xi$:

$$V_2 = \frac{kT}{\eta \xi^2} = \frac{\gamma}{\eta} = \frac{r_c}{\xi} V_1 \tag{3.40}$$

3.4.3.3
Above V_2

In this regime the globule fades out, and the chain is ideal and fully elongated ($L = Na$) because V_2 is much higher than $V_3 = \dfrac{kT}{\eta Na^2}$, the velocity required to extend completely an ideal chain.

3.4.3.4
Relaxation

We consider a chain in poor solvent, which is fully elongated at time $t = 0$. The chain can be tethered or free. During the relaxation, the chain (which always relaxes by its free ends) is in the two-state model pictured in Fig. 3.9: the globule ($n^*(t)$ monomers, length $L_g(t)$) absorbs the stem at velocity $V = a\, dn^*/dt$. The Stokes friction on the globule $f_v = \eta L_g(t) V$, which now varies with time, is always equal to the transition tension. By integrating this equality we obtain $\eta n^{*4/3} \dfrac{a\xi}{g^{1/3}} = \dfrac{kT}{\xi} t$. The total relaxation time τ_r corresponds to $n^* = N$, and it leads to

$$\tau_r = \tau_\xi \frac{N^{4/3}}{g^{5/6}} \tag{3.41}$$

where $\tau_\xi = \eta\xi^3/kT$ is the blob relaxation time.
  The length $L(t)$ of the chain decreases as

$$L = Na\,(1 - (t/\tau_r)^{3/4}) \tag{3.42}$$

A scaling law ($t^{3/4}$) can be extracted by fitting $L(0)$-$L(t)$ vs time. One should notice that this scaling law is very different from $t^{1/2}$ expected [6] and monitored experimentally [12] for the relaxation of chains in a θ solvent.

3.5
Tethered Chains Confined in a Slit

The deformation and relaxation of DNA chains confined in a slit and moving under an electrical field in microlithographic arrays has been observed very recently [21]. We discuss here the unwinding under flow of chains in a good solvent and in a θ solvent. We assume that the chains are repulsed at both interfaces. V is the mean flow in the slit.

3.5.1
Good Solvent

The chain is squeezed between two walls separated by a gap $h < R_F$. It can be pictured as a 2D string of blobs of size h, each blob containing $g = (h/a)^{5/3}$ monomers.

In the absence of external forces the overall string is a self avoiding walk, of size $R_F = \tilde{N}^{3/4}h$ where $\tilde{N} = N/g$ is the number of blobs.

3.5.1.1
Uniform Tension

For a 2D chain under tension, we define disks of size $\xi = kT/f$, each subunit containing g blobs ($g = (\xi/h)^{4/3}$). The chain extension is

$$R = \frac{\tilde{N}}{g}\xi = \tilde{N}h^{4/3}\left(\frac{f}{kT}\right)^{1/3} \tag{3.43}$$

The extension x measured from the free end is $x = \dfrac{\tilde{N}}{g}\xi$ and the deformation of the $\tilde{n}$ blob is

$$\frac{dx}{d\tilde{n}} = h^{4/3}\left(\frac{f}{kT}\right)^{1/3} \tag{3.44}$$

We shall be mainly concerned with this 2D stretching regime, where $x > h$. We note that for larger tensions, one returns to the 3D behavior.

3.5.1.2
Uniform Flow

3.5.1.2.1
Trumpet

A flow $J = hV$ of the solvent is applied between the plates. An important difference with the previous cases is due to the screening of hydrodynamic interactions in a slit [22]: the chain behaves as a Rouse chain of impenetrable blobs of size h. The friction force on the $\tilde{n}$-th blob starting from the free end is simply

$$f(\tilde{n}) = \eta V h\tilde{n} \tag{3.45}$$

The deformation of a 2D chain $\dfrac{dx}{d\tilde{n}}$ is given by Eq. (3.44):

$$\frac{dx}{d\tilde{n}} = h^{4/3}\left(\eta\,\frac{hV}{kT}\,\tilde{n}\right)^{1/3} \tag{3.46}$$

The full extension is given by integration of Eq. (3.46):

$$L = h^{4/3} \left(\eta \frac{hV}{kT} \right)^{1/3} \tilde{N}^{4/3} \approx R_{F^{2D}} \left(\frac{V\tau_R}{R_{F^{2D}}} \right)^{1/3} \tag{3.47}$$

where $\tau_R = \eta h \tilde{N} R_{F^{2D}}/kT$ (Rouse time) is the confined chain relaxation time, much larger than the 3D Zimm time.

The size $y(x)$ of the blobs, deduced from Eqs. (3.45–3.47), decreases from the free end as

$$y = R_1 (R_1/x)^{3/4} \tag{3.48}$$

where $R_1 = h^{4/7} \left(\frac{kT}{\eta hV} \right)^{3/7}$ is the size of the largest blob.

The 2D trumpet regime start for $V = V_{c1} = R_{F^{2D}}/\tau_R(h)$ and ends up at $V_{c2} = V_{c1}\tilde{N}^{3/4} \approx N^{-1}$. Above V_{c2} the chain enters in the 3D regime discussed previously.

3.5.2
Ideal Chains

The extension of an ideal confined chain is $R_0 = N^{1/2}a$. Stretched under a uniform tension, the extension L is still given by Eq. (3.1): $L = Na \left(\frac{fa}{kT} \right)$. The confinement does not alter the gaussian statistics of polymer chains, but modifies the dynamical properties. The chain has a Rouse behavior, because the backflows are screened by h. The Rouse relaxation time τ_{Ri} of a confined ideal chain is given by the balance equation between elastic and viscous forces acting on a stretched chain

$$\frac{N}{g} \eta h \delta \dot{R} = kT \frac{\delta R}{R_0^2} \tag{3.49}$$

where $g = (h/a)^2$, i.e.,

$$\tau_{Ri} = \frac{(N/g)^2 \eta h^3}{kT} = \left(\frac{N}{g} \right)^2 \tau_h \tag{3.50}$$

Under uniform flow, the tension $f(x)$ is proportional to the number n/g of Zimm blobs, counted from the free end. The local deformation is

$$\frac{dx}{dn} = a^2 \frac{f(x)}{kT} = a^2 \left(\frac{n}{g} \right) \eta \frac{hV}{kT} \tag{3.51}$$

The integral leads to the extension

$$x = na^2 \left(\frac{n}{g} \right) \eta \frac{hV}{kT} \approx \left(\frac{n}{g} \right)^2 \eta \frac{h^3 V}{kT} \tag{3.52}$$

Setting $n = N$ leads to the full chain extension

$$L = V\tau_{Ri} \tag{3.53}$$

The 2D trumpet regime starts at $V_{c1} = R_0/\tau_{Ri}$ and ends up at $V_{c2} = (N/g)^{1/2}V_{c1}$. Above V_{c2}, a part of the chain enters in the 3D regime.

It can be seen that $V_{c2}^{2D} = V_{c1}^{3D}$. As soon as $V > V_{c2}^{2D}$ the chain is composed of a 2D flower containing $N^* = NV_{c2}^{2D}/V$ monomers and a 3D stem.

For the relaxation after stretching, the first stage "stem-flower" is identical to the 3D case ($L(O) - L(t) \approx t^{1/2}$). Then the exponential relaxation of the extended flower to the ideal coil is extremely slow: the characteristic time involved is $\tau_{Ri} = (N/g)^2\tau_h$ instead of $\tau_Z = N^{3/2}\tau_O$.

3.6
Concluding Remarks

We have discussed some experiments of nanomanipulations of one single molecule. These also apply to chains grafted on a solid surface, in the mushroom regime, at very low grafting density. To date, one may find many experimental and theoretical papers on the deformation under shear flows of dense polymer brushes. However, in practice the mushroom regime is more important for many applications, such as grafted layers used as adhesion promoters. Experiments on grafted layers under solvent flows are underway. In shear flows, the deformation is more progressive, and the trumpet regime, where simple scaling laws are predicted for the chain extension, might be observed over several decades. One still needs experiments for this case.

Electrophoresis of DNA fragments was commonly performed in gels, where reptation was the source of separation. Separation has also been observed in dilute solution of neutral polymers. The deformation of the neutral chain driven at constant velocity by the moving charged chains discussed here is now responsible for the separation. Polymer, charged or uncharged, can also be separated using hydrodynamic flow in microlithographic arrays. Our discussion on the deformation and relaxation of polymers confined in a slit may well be useful.

The *marginal state* has also been observed when a "hairy" surface, grafted with a few long polymer chains, is exposed to a polymer melt; the melt can slip tangentially to the surface, and in a certain regime the shear stress due to the hair becomes independent of velocity for the same reasons [23, 24]. Thus there is an amusing convergence between two different fields: long DNA molecules under flows and rheological measurements of polymer melts!

References

1. Glazer AN, Ryo HS (1992) Nature 359:859
2. Pincus P (1976) Macromolecules 9:386
3. Cluzel P, Lebrun A, Heller C, Lavery R, Viovy JL, Chatenay D, Caron F (1996) Science 271:792
4. Bensimon D, Simon A, Croquette V, Bensimon A (1995) Phys Rev Lett 74:4754
5. (a) Perkins TT, Smith DE, Chu S (1994) Science 264:819; (b) Perkins TT, Quake SR, Smith DE, Chu S (1994) Science 264:822
6. Brochard-Wyart F (1993) Europhysics Letters 23:105 (1995) 30:387
7. Edwards SF (1976) Molecular Fluids. In: Balian R and Weill G (eds) (Gordon and Breach, NY)
8. de Gennes PG (1985) "Scaling laws in polymer physics", Cornell University Press
9. Wirtz D (1995) Phys Rev Letters 75:12–2436
10. de Gennes PG (1967) J de Chimie-Physique 87:962
11. Marciano Y, Brochard-Wyart F (1995) Macromolecules 28:985
12. Manneville S, Cluzel P, Viovy JL, Chatenay D, Caron F (1996) Europhys Lett 36:413
13. Ajdari A, Brochard-Wyart F, Gay C, de Gennes PG, Viovy JL (1995) J Phys II France 5:491
14. de Gennes PG (1985) J Phys Lett 46:639
15. Ostrovsky B, Bar Yam Y (1994) Europhys Lett 25:409
16. Buguin A, Brochard F, de Gennes PG (1996) CR Acad Sci Paris 322:741
17. Chu B, Ying Q, Grosberg AYu (1995) Macromolecules 28:180
18. Halperin A, Zhulina EB (1991) Europhys Lett 15:417
19. Smith SB, Finzi L, Bustamante C (1992) Science 258:1122
20. Buguin A, Brochard F (1996) Macromolecules 29:4937
21. Volmuth WD, Austin RH (1992) Nature 358:600
22. Brochard F, de Gennes PG (1977) The journal of Chemical Physics 67:52
23. Migler K, Leger L, Hervet H (1993) Phys Rev Letters 70:287
24. Brochard F, de Gennes PG (1992) Langmuir 8:3033

CHAPTER 4

Osmotic Pressure in Solutions of Stretched Polymers

Yitzhak Rabin and Shlomo Alexander †

4.1
Introduction

Studies of flow-induced deformation of isolated chains in solution [1] confirm the theoretical prediction of the coil-stretch transition in elongational flow [2]. Although there are still several unresolved issues related to the scaling of the critical strain rate with molecular weight [3, 4], it is believed that the basic physics of the coil-stretch transition in a single polymer chain is well-understood.

The next logical step is to study the behavior of semi-dilute solutions of stretched polymers where, as will be shown in the following, new physical effects are expected. Several years ago we predicted that, when a polymer brush is subjected to shear, the shear-induced stretching of the grafted polymers along the direction parallel to the surface of the brush will lead to increased osmotic pressure and the brush will swell [5]. Such an effect was indeed observed by Klein and coworkers [6]. The idea was then applied to model deformation-induced swelling of polymer gels [7]. Tests utilizing gels swollen inside small capillaries proved to be inconclusive, due to the limited range of experimentally accessible deformation ratios [8]. Since the underlying physical picture is quite general and applies to any concentrated system of stretched polymers (whether freely suspended in solution, grafted to a wall or cross-linked to each other), in Sect. 4.2 we consider the case of a semi-dilute solution of polymers and show how the osmotic pressure in such a solution is affected by the stretching of the chains. In Sect. 4.3 we propose an osmotic experiment which could test our predictions and make some tentative comments on their possible significance for the hydrodynamics of polymer solutions in strong shear and elongational flows.

4.2
Semi-Dilute Solution of Stretched Polymers

Consider a semi-dilute solution of overlapping polymer chains in a good solvent, with a given monomer volume fraction ϕ. Each macromolecule consists of N monomers of length a. According to the scaling theory of semi-dilute solutions [9], each polymer can be represented as a Gaussian chain of "blobs" of radius $R_0 = \xi_0 N_{0b}^{1/2}$ where $\xi_0 \simeq a/\phi^{3/4}$ is the radius of a blob and N_{0b} is the number of blobs per chain (Fig. 4.1 a). The latter can be calculated by noticing that the blob radius is related to the number of monomers inside it, n_{0b}, by $\xi_0 = an_{0b}^{3/5}$ and therefore, $N_{0b} = N/n_{0b} = N/(\xi_0/a)^{5/3}$. Since ξ_0 is the correlation length over which excluded-volume interactions between monomers belonging to the same chain are screened by the presence of other chains, the entire solution can be considered as a "melt" of densely packed Gaussian chains of blobs. The osmotic pressure in the solution can be estimated from the free energy density due to excluded volume interactions inside a blob,

$$\Pi_0 \simeq \frac{k_B T}{\xi_0^3} = \frac{k_B T}{a^3} \, \phi^{9/4} \equiv \frac{k_B T}{R_F^3} \, (\phi/\phi^*)^{9/4} \tag{4.1}$$

where k_B is the Boltzmann constant, T is the temperature, $R_F = aN^{3/5}$ is the radius of a polymer in a dilute solution in a good solvent, and $\phi^* = N^{-4/5}$ is the monomer volume fraction at chain overlap (the onset of the semi-dilute regime). The same result is obtained from a mean-field estimate of excluded volume interactions between blobs, $k_B T v_{0b} c_b^2$, where $v_{0b} = \xi_0^3$ is the second virial coefficient of a blob and $c_b = \phi/(a^3 n_b)$ is the concentration of blobs.

The above argument applies to a semi-dilute solution of undeformed chains. We now proceed to calculate the osmotic pressure in a semi-dilute solution of *stretched* chains of end-to-end distance $R_0 \le R \le \xi_0 N_{0b}$ (Fig. 4.1 b). Since our derivation was based only on the assumption of a closed packed system of blobs and made no reference to the dimensions of an individual chain, we conclude that the stretching of Gaussian chains of blobs does not affect the osmotic pressure in the solution and that it is the same as under ambient conditions, $\Pi = \Pi_0$. In this regime, the force needed to stretch any of the chains scales linearly with the deformation, i.e., $f \propto R$, as expected for Gaussian chains.

For deformations in the range $\xi_0 N_{0b} \le R \le aN$, the chains can no longer be considered as Gaussian (Fig. 4.1 c). In order to understand the physics of this regime, we recall the behavior of a single strongly stretched chain in a good solvent. Such a polymer can be considered as a stretched chain of length $R = \xi N_b$, where the size of a Pincus blob [9]

$$\xi = \frac{k_B T}{f} \tag{4.2}$$

is the length scale on which the thermal energy, $k_B T$, is of the same order as the deformation energy, $f\xi$ (f is the force on the ends of the chain). The number of Pincus blobs per chain is $N_b = N/n_b$ and since $n_b = (\xi/a)^{5/3} = (k_B T/fa)^{5/3}$, the force-length relation is nonlinear, $f = (k_B T/a)(R/Na)^{3/2}$. This yields $\xi = a (Na/R)^{3/2}$, $n_b = (Na/R)^{5/2}$, and $N_b = N (R/Na)^{5/2}$.

Returning to the semi-dilute solution of stretched chains we note that, when $\xi_0 N_{0b} < R$, the size of the Pincus blob ξ is smaller than that of the concentration blob ξ_0. In this regime the deformation of individual chains, rather than the presence of other chains, determines the length scale over which excluded volume is "screened". However, since blobs of size ξ do not fill the volume of the system (Fig. 4.1 c), the osmotic pressure cannot be obtained by replacing ξ_0 by ξ

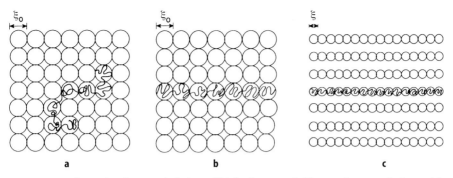

Fig. 4.1a–c. Schematic picture of chains of blobs in a semi-dilute polymer solution, with one of the chains shown by the wiggly line: **a** undeformed chains of concentration blobs, $R_0 = \xi_0 N_{0b}^{1/2}$; **b** stretched chains of concentration blobs, $R = \xi_0 N_{0b}$, (linear regime); **c** stretched chains of Pincus blobs, $R = \xi N_b$, (nonlinear regime)

in Eq. (4.1), and we need to estimate the excluded volume interaction between non-overlapping Pincus blobs of different chains.

Assuming that the stretched chains are parallel to each other, the average inter-chain distance r can be calculated from the relation $\phi = Na^3/Rr^2$:

$$r = a \left(\frac{aN}{R\phi} \right)^{1/2} \tag{4.3}$$

This distance should be compared to the rms radius of transverse fluctuations of a chain,

$$R_\perp = \xi N_b^{1/2} = aN^{1/2} \left(\frac{aN}{R} \right)^{1/4} \tag{4.4}$$

which decreases from $\xi_0 N_{0b}^{1/2}$ to $aN^{1/2}$ with progressive stretching. Using Eqs. (4.3) and (4.4) yields $r/R_\perp = (\phi^*/\phi)^{1/2}(aN^{3/5}/R)^{1/4} \ll 1$, where the inequality holds for stretched chains in the semi-dilute regime. Since the length scale of transverse fluctuations exceeds the average spacing between the chains, we conclude that there are many other chains within the volume probed by the fluctuations of a single chain and that one may use a mean field estimate of the excluded volume interaction between the chains [5]. The interaction energy per blob is estimated as $k_B T$ times the volume fraction of blobs, ξ^2/r^2, and the osmotic pressure is obtained by multiplying this energy by the number of blobs per chain, n_b, and dividing by the volume per chain, Rr^2. This yields

$$\Pi = \frac{k_B T}{a^3} \left(\frac{R}{aN} \right)^{1/2} \phi^2 \tag{4.5}$$

and thus

$$\frac{\Pi}{\Pi_0} = \left(\frac{R}{R_F}\right)^{1/2}\left(\frac{\phi^*}{\phi}\right)^{1/4} > 1 \quad \text{for} \quad \xi_0 > \xi \tag{4.6}$$

where the inequality follows from the observation that the above equation can be recast into the form (recall that in the strong deformation regime, $\xi_0 > \xi$):

$$\Pi/\Pi_0 = (\xi_0/\xi)^{1/3} \tag{4.7}$$

4.3
Discussion

Are the predicted effects experimentally observable? In order to answer this question let us consider a gedanken experiment in which we introduce a rigid semi-permeable partition (permeable to the solvent but not to the polymer) into an open channel which contains a polymer solution at volume fraction ϕ and osmotic pressure $\Pi_0(\phi)$. The partition separates the channel into two regions, A and B (see Fig. 4.2) such that the volume of A is much larger than that of B. If we impose an external field which stretches the polymers on the B side of the partition (the solution on the A side of the partition remains quiescent), the stretching of the macromolecules will result in an increased osmotic pressure in B and solvent will move through the partition from A to B until the difference of the osmotic pressures $\Delta\Pi = \Pi(\phi_B) - \Pi(\phi_A)$ is balanced by gravity. In order to make a rough estimate of the resulting change in concentration we will neglect gravity, i.e., assume that osmotic balance between the two sides of the partition is reached. Using the expressions for Π_0 and Π, Eqs. (4.1) and (4.5), the condition $\Pi(\phi_B) = \Pi(\phi_A)$ can be written as

$$\left(\frac{R}{R_F}\right)^{1/2}\left(\frac{\phi_B}{\phi^*}\right)^2 = \left(\frac{\phi_A}{\phi^*}\right)^{9/4} \tag{4.8}$$

Taking $R/R_F = 10$ and $\phi_A/\phi^* = 10$ we obtain $\phi_B/\phi^* = 7.5$, i.e., a 25% decrease in volume fraction of the stretched chains.

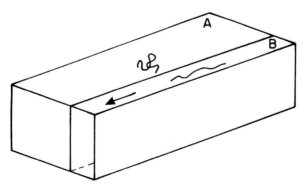

Fig. 4.2. A pipe separated by a semi-permeable partition into regions A (quiescent polymer solution) and B (flowing polymer solution). Schematic drawings of undeformed and of stretched polymers in A and in B respectively, are shown. The *arrow* indicates the direction of flow in channel B

The predicted increase in the osmotic pressure raises the possibility of observing similar phenomena in flowing polymer solutions in which large velocity gradients stretch the polymers. The analysis of such situations is difficult since the physics of entanglements in semi-dilute solutions subjected to strong shear flows is not well-understood at present and other effects, apart from the increase of osmotic pressure, may play an important role. However, some conclusions can be reached on the basis of very general arguments. Note that at Weissenberg numbers exceeding unity, $We \equiv \beta\tau > 1$ (i.e., at shear rates β greater than the inverse reptation time τ^{-1}), an entangled polymer solution must respond as an elastic network, since polymers cannot disentangle on time scales smaller than the reptation time. This would result in dramatic increase of the shear stresses, contrary to experimental observations which report shear thinning in the strong shear regime [10]. We conclude that polymers must become disentangled in regions where $We > 1$ and that further increase of the shear rate will lead to progressive alignment and stretching of the macromolecules along the flow lines. Although the deformation of the polymers may be more complicated than that assumed in the Pincus model (see, e.g., [11]), this will result in increased osmotic pressure in the solution and may lead to flow instabilities and to the development of flow-induced polymer concentration inhomogeneities. Several scenarios are possible.

4.3.1
Instabilities in Steady Homogeneous Flows

In this case, e.g., plane Couette flow, at low shear rates ($We < 1$) polymers are deformed uniformly throughout the flow field. At higher shear rates the solution is expected to become unstable against segregation into regions of increased polymer concentration and small velocity gradients and into domains oriented along the flow axis, in which the polymer concentration is reduced and the local shear rates are high. Support for the above picture comes from recent light scattering and optical microscopy observations of "string" phases (accompanied by rheological anomalies) in strongly sheared semi-dilute polystyrene solutions in DOP [12]. Note, however, that while our mechanism applies only to polymers in good solvents, the above experiments, as well as other observations of shear-induced phase separation [13], were performed under θ solvent conditions. Experiments on sheared polymer solutions in good solvents are needed in order to provide a direct test of our predictions.

4.3.2
Inhomogeneous Flows

Polymers are strongly stretched in regions in which the velocity gradients are high and the increased osmotic pressure in these regions will result in diffusion to regions of smaller velocity gradients where the osmotic pressure is lower. In Poiseuille flow, this will further increase the velocity gradients near the walls and decrease them at the center of the pipe, and may lead to plug flow and apparent wall "slip". A deformation-induced increase in the osmotic pressure may also occur along the symmetry axis in axisymmetric elongational flow. This phenomenon should affect the polymer conformation, pressure head, and flow velocity profile. Although such effects were reported in the literature [14], the connection between experimental observations and our predictions is unclear since they were observed at concentrations lower than c^*, to which the present mechanism does not strictly apply.

Acknowledgment. This research was supported by grants from the Israeli Academy of Sciences and Humanities, the Israeli Ministry of Science and Technology, and the Research Authority of Bar-Ilan University.

References

1. Frank FC, Keller A, Mackley MR (1971) Polymer 12:467
2. de Gennes P-G (1974) J Chem Phys 60:751
3. Keller A, Odell JA (1985) Coll & Polym Sci 263:181
4. Nguyen TQ, Yu G, Kausch H-H (1995) Macromolecules 28:4851
5. Rabin Y, Alexander S (1990) Europhys Lett 13:49
6. Klein J, Perhia D, Warburg S (1991) Nature 352:143
7. Alexander S, Rabin Y (1990) J Phys: Condens Matter 2:SA313
8. Rabin Y, Samulski ET (1992) Macromolecules Commun 25:2985
9. de Gennes P-G (1979) Scaling methods in polymer physics. Cornell University Press, Ithaca
10. Bird RB, Armstrong RC, Hassager O (1977) Dynamics of polymeric liquids, vol 1. Wiley, New York
11. Brochard-Wyart F (1993) Europhys Lett 23:105
12. Kume T, Hashimoto T (1995) In: Nakatani AI, Dadmun MD (eds) Flow-induced structure in polymers. ACS Symposium Series 597:35
13. Migler K, Liu C-H, Pine DJ (1996) Phys Rev Lett 29:1422
14. Chow A, Keller A, Müller AJ, Odell JA (1988) Macromolecules 21:250

CHAPTER 5

Stretching of Polyelectrolytes in Elongational Flow

O. V. Borisov and A. A. Darinskii

5.1
Introduction

Polyelectrolytes are long chain molecules containing a certain fraction of ionizable monomer units. NaPSS or polyacrylic acid are the typical examples of flexible chain polyelectrolytes while DNA presents an example of an intrinsically rigid polyelectrolyte. As these macromolecules are immersed in the polar solvent (e.g., in water), the dissociation of salt or acidic functional groups results in the appearance of charges on some monomers in the chain and of free counterions in the solution.

The main molecular characteristics of the polyelectrolyte are the total molecular weight or the degree of polymerization N, the monomer size a, and the intrinsic rigidity (Kuhn segment length) A of the chain in the absence of charges and the fraction f of charged monomers in the chain.

For intrinsically flexible chains $A \cong a$, while for rigid polyelectrolytes (such as DNA, for example) the intrinsic persistence length (measured under high ionic strength conditions) exceeds by far the monomer size which coincides approximately with the chain thickness.

The last parameter f is determined by the "primary structure", i.e., by the fraction of ionizable monomer units in the chain (the degree of sulfonation in the case of NaPSS) and by external conditions such as the concentration of counterions in the solution.

The conformation of a polyelectrolyte molecule is determined by strong Coulomb repulsion between charged monomer. It is strongly affected by the concentrations of polyelectrolyte and of low molecular weight salt in the solution due to the screening of electrostatic interactions.

The non-electrostatic excluded volume interactions between monomers of the polyelectrolyte chain also play an important role, especially at small f and under the conditions of poor solvent when these interactions have a character of short-range attraction which may result in the collapse of the chain.

The behavior of charged macromolecules in shear flows is of great interest from the point of view of general theory of the rheological properties of polyelectrolyte solutions. At the same time the understanding of the main relationships of this behavior is very important for some technological applications in the oil industry, for reducing hydrodynamic friction, etc.

It is known [1] that the specific rheological properties of solution of flexible polymers are determined by the ability of polymer coils to undergo strong deformations in response to comparatively weak forces applied to polymer in a shear flow. One of the most exciting phenomena related to polymer deformation in the shear flow is the stretching of polymer coil in elongational flow field. It was shown that the coil-stretch transition for neutral polymer molecule occurs as the first order dynamical phase transition at the certain value of the flow rate gradient [1, 2]. At weak shear the chain retains in general an unperturbed coil conformation whereas at strong shear (exceeding the critical value) the dynamically stable conformation corresponds to the completely stretched chain. The critical value of the gradient is determined by the relation between the chain elastic modulus and the friction coefficient and is inversely proportional to the fundamental relaxation time. As polyelectrolyte chain is partially stretched by intramolecular Coulomb repulsion between charged monomers, one can expect at least an additivity of the effects of two stretching forces (the electrostatic and the hydrodynamic one) and, consequently, the difference in the behavior of uncharged and charged polymer molecules in the elongational flows.

The main goal of the theory is to connect the behavior of charged polymer chain in the flow with molecular parameters, N and f, and with solvent characteristics.

The structure of the chapter is the following. In Sect. 5.2 we give the brief review of the experimental data concerning the behavior of polyelectrolyte solutions in an elongational flow. Section 5.3 contains a description of the previous theoretical approaches to the problem. The main content is concentrated in Sects. 5.4 and 5.5. We present there along the lines of [3, 4] the scaling theory of stretching of weakly charged polyelectrolyte chains in an elongational flow in solvents with different strength. Section 5.6 concludes the chapter.

5.2
Experimental Evidence

In this section we consider some experimental data concerning the behavior of polyelectrolyte chains in elongational flow. There is a number of works devoted to the studies of polyelectrolyte solutions in flows involving the extensional motion: turbulent pipe flow [5], orifice flow [6], converging channels [7], porous media flow [8]. We restrict ourselves only to results obtained for well-defined purely elongational flow fields created in opposite jets or slots, or in two- or four-roll mill. There are only a few of the experimental works devoted to the study of the behavior of dilute polyelectrolyte solutions in such flows.

In the first work of Miles et al. [9] the 0.1% solutions of NaPSS in water without salt and in the presence of NaCl and $CaCl_2$ were studied in flows created by opposed jets or cross-slots. The molecular weight was varied from $M_w = 1.06 \times 10^6$ to $M_w = 1.77 \times 10^5$. The degree of sulfonation was about 100%. The ionic strength was changed up to the maximum NaCl concentration of about 0.06 mol/l. In the more recent publication of Dunlap and Leal [10] the influence of the elongational flow on the NaPSS molecules dissolved in a less polar solvent (glycerol) in the range of added NaCl concentrations from zero to 0.4 mol/l was studied. The linear two-dimensional elongational flows were produced by a four-roll mill. Two samples of NaPSS with $M_w = 1.2 \times 10^6$ at concentration 100 ppm and $M_w = 4.16 \times 10^6$ at concentration 89 ppm were used. In both cases the stretching or orientation of chains was monitored by the onset of birefringence (BF). Independent data on the equilibrium size of NaPSS at different salt concentrations were obtained by quasi-elastic light-scattering spectroscopy (QELS) and intrinsic viscosity (IV) measurements.

The previous studies of neutral macromolecules solutions in flow have shown that for rigid and flexible polymers distinctly different behavior is observed. Rod-like molecules tend to align parallel to the flow, and the degree of alignment, as measured by the birefringence, increases continuously with increasing strain rate s (see Fig. 5.1 a). For monodisperse neutral flexible polymers the onset of the birefringence occurs in a narrow range of s within the localized region along the center line or axis of the flow system (Fig. 5.1 b). NaPSS is a strongly charged polyelectrolyte, i.e., for this polymer $f \cong 1$. According to the existing theories [11, 12] the strongly charged polyelectrolyte chain has a rod-like conformation at very dilute concentrations. Therefore one should expect a rod-like behavior of NaPSS in an elongational flow. However, for NaPSS solution in pure water the localized birefringent line was observed at strain rates s above a certain critical value s_c. (Fig. 5.2). This behavior is similar to that of flexible neutral polymers with the only difference being that after the sudden initial step the measured intensity characterizing the chain orientation continues to rise gently with increasing s. For neutral PS [10] flattening off at a maximum corresponding to the stretched out chain was observed.

Hence NaPSS in water reveals pronounced flexibility contrary to the theory predictions. Such behavior is consistent with results of recent computer simulations [13–15] which show that for chemically realizable intrinsically flexible polyelectrolytes the rod-like structure should not be observable at experimentally realizable concentrations. In dilute solution the chain retains certain local flexibility and on the large scale it has a horseshoe-shaped conformation.

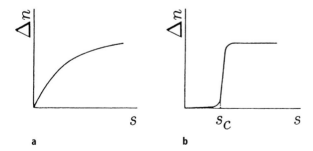

Fig. 5.1 a, b. Schematic representation of the different responses of: **a** rod-like; **b** flexible macromolecules to the elongational flow field. The diagrams represent the birefringence Δn measured in experiments as a function of the flow rate s. Reprinted from [9] with permission from Butterworth-Heinemann Journals, Elsevier Science

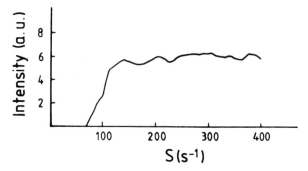

Fig. 5.2. Plot of the measured intensity of the localized birefringent line *versus* flow rate showing a step-like increase in intensity with increasing flow rate for a 0.1 % solution of PSS ($M_w = 1.06 \times 10^6$) in a 0.0005 mol/l NaCl solution, i.e., in region I of Fig. 5.3. Reprinted from [9] with permission from Butterworth-Heinemann Journals, Elsevier Science

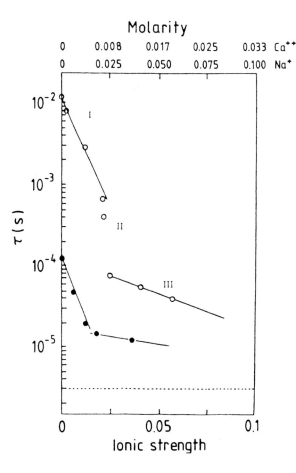

Fig. 5.3. Plots of the log of relaxation time, τ, vs ionic strength of added NaCl (open circles) and CaCl$_2$ (black circles). The corresponding molarity of the salt is also indicated on the top. The broken line indicates the relaxation time expected for a polystyrene molecule of the same degree of polymerization as that of the polystyrene sulphonate used, and in a solvent of the same viscosity. Reprinted from [9] with permission from Butterworth-Heinemann Journals, Elsevier Science

For salt-free solutions of NaPSS in glycerol the BF data have the shape typical for rigid molecules orienting in the field. No distinct "onset shear rate" behavior was observed. The spatial extent of the birefringent region was also specific for rigid molecules. However, macromolecules were not really rigid and showed significant additional stretching due to the flow. The enhanced rigidity of NaPSS in glycerol could be explained by the lower value of the dielectric constant for this solvent in comparison to water.

At intermediate and high salt concentrations for both systems the BF was mostly localized along the streamline of the flow and the shape of BF vs s curves was similar to that for flexible polymers. The results of IV and QLS measurements revealed the pronounced decrease of the molecule size in this region of salt concentration. For NaPSS dissolved in water the characteristic relaxation times τ were determined from the condition $\tau s_c = 1$. Three distinct regions in the dependence of τ on the ionic strength were identified (Fig. 5.3).

In region I a large relaxation time at very low salt concentration decreased steeply with increasing salt concentration. In region II a huge drop in τ was observed. Beyond the drop τ decreased slowly at higher salt concentrations.

The molecular weight dependence of τ was measured only in region I. It appeared to be quadratic (Fig. 5.4). For solutions of NaPSS in glycerol the values of τ were estimated as an inverse value of the strain rate at which the BF reaches 20% of the asymptotic birefringence level. Relaxation times for the salt-free solution in glycerol (normalized with respect to the solvent viscosity) were about 10% of the values of τ for the same polymer in water and remained less over the entire range of salt concentrations studied. As in aqueous solutions the dramatic 10-fold drop of τ for NaPSS in glycerol was observed but at a much lower concentration of salt (0.001–0.004 mol/l compared to 0.025 mol/l in

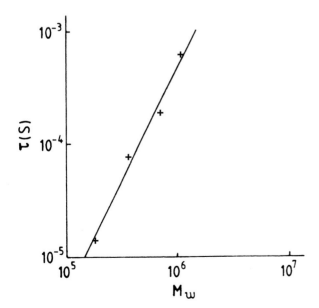

Fig. 5.4. Plot of the relaxation time, τ, vs molecular weight of PSS in log scales for a 0.1% solution of NaPSS in 0.004 mol/l NaCl, i.e., in region I of Fig. 5.3. Reprinted from [9] with permission from Butterworth-Heinemann Journals, Elsevier Science

water). This difference was ascribed to the lower dielectric constant of glycerol. The change of τ with an increase in the ionic strength from no salt to high salt concentrations was of the same order as that of τ_{iv} values obtained by intrinsic viscosity measurements.

For divalent calcium as the counterion, the values of relaxation times were always smaller than those obtained with sodium at the same normality. This difference was ascribed to the greater compactness of the molecule due to the ionic bridges between various parts of the same molecule.

5.3
Early Theoretical Approaches

The first theory for the behavior of dilute solutions of polyelectrolytes in elongational flow was developed by Dunlap and Leal [10]. They used a dumbbell model applied previously to nonionic polymers. The dumbbell consisted of two beads with conformation dependent friction coefficients ζ connected by an entropic spring. The elastic law of the spring was suggested by Warner [16] and takes into account the finite chain extensibility:

$$\mathcal{F}_{el} = \frac{6NkT}{R_{max}^2} \left[\frac{R}{1 - \left(\dfrac{R}{R_{max}}\right)^2} \right] \tag{5.1}$$

where R is the distance between beads. $R_{max} = N a$ is the maximal extension of the chain. For the dependence of the friction coefficients of beads ζ on R the simple de Gennes-Hinch law [18, 19] was used:

$$\zeta = \frac{\zeta_0 R}{N^{1/2} a} \tag{5.2}$$

where ζ_0 is the Stokes' law friction coefficient for the undeformed dumbbell. The dependence at Eq. (5.2) takes into account the change of hydrodynamic properties of a chain by its deformation: from Zimm's non-draining coil at small R to free-draining chain at large R. The Coulomb repulsion effect was modeled as effective charges placed at the beads separated by a distance R and acting through fluid with dielectric constant ε. The values of charges were estimated by using the values of the equilibrium expansion factor $\alpha = \dfrac{R}{R_0}$ at every salt concentration from the balance of elastic and electrostatic forces. R_0 is the dimension of the undeformed dumbbell. The value of α can be measured independently of intrinsic viscosity or quasi-elastic light scattering experiments. The simple Coulomb force law was assumed to be applicable, independent of the concentration of the counterions in the solution. Thus the influence of the counterions due to the shielding of the charges along the polymer backbone was only accounted for by decreasing values of the effective charge.

The authors of [17] also used another approach to take into account the screening of electrostatic interactions through the change of the number N of the statistical segments in the chain without introducing the effective charge. The value of N can be determined from the expression for α:

$$\alpha = (N/N_o)^{12(\nu-1)} \tag{5.3}$$

where N_o is reference value for most flexible polymers in the high-salt limit. Here ν is the value of the exponent in the dependence of the mean square end-to-end distance $\langle r^2 \rangle$ on N:

$$\langle r^2 \rangle = N^{1+\nu} A^2 . \tag{5.4}$$

and A is the effective segment length.

The model relaxation time τ_{db} increased with the decrease in N according to the equation

$$\tau_{db} \propto \alpha^3. \tag{5.5}$$

The onset rate of the coil-stretch transition was determined only by expansion factor α but the shape of the dependence of birefringence on s depended on N and therefore on the choice of ν which was considered as a parameter. For salt free solutions the value of ν equal to $0.2-0.3$ was required in order to fit the model predictions to the experimental data for NaPSS in glycerol. The authors of [17] concluded that the first approach (which assumes electrostatic repulsion via the charged dumbbell) slightly underestimates the shift of BF data at low salt concentrations from that of the high-salt solutions, while the second approach (which assumes local stiffening) slightly overestimates the same shift of the data.

The theory of Dunlap and Leal describes qualitatively the effects of chain expansion or chain stiffening due to the Coulomb interactions on the behavior of polyelectrolyte chain in an elongational flow. The theory was successfully applied to experimental data for solutions of NaPSS. As we have mentioned above this polyelectrolyte is the strongly charged one, i.e., $f \cong 1$. For such polymers the effect of Coulomb interactions is dominant at all scales and the role of the solvent strength is not very important. However, there is a broad class of weakly charged polyelectrolytes, i.e., with $f \ll 1$. Even if all the monomers of the chain are capable of ionization, as is the case for polyacrylic acid, the fraction of charged monomers f may remain much smaller than 1 in solutions with low pH.

For weakly charged polyelectrolytes short-range interactions determine the chain conformation on scales of the order of the distance between charges. This distance could be rather large and the theory should include the solvent strength explicitly. In Dunlap's model the added salt only changes the value of the effective charge or the number of statistical segments. However, modern theories of conformational properties of polyelectrolyte in solution [20, 21] predict more complex behavior of the weakly charged polyelectrolyte by addition of salt. As we will see for weakly charged polyelectrolyte coils in θ- or good solvents the coupling between the electrostatic and hydrodynamic interactions results in more complicated behavior of the chain in the elongational flow: in the case of salt-added solution we predict the existence of two stretching transitions. The first of these transitions is from that of a coil swollen by screened Coulomb repulsions to the partially stretched conformation and the second one is from the partially stretched to the completely stretched state. The physical origin of the stability of the intermediate state is an increase in the elastic modulus of swollen polyelectrolyte chain accompanying an increase in the chain stretching. Below we present a scaling theory of stretching of weakly charged polyelectrolytes in elongational flow which includes all these effects. We start with a brief review of the main ideas concerning the conformation of weakly charged polyelectrolyte in the bulk of the salt-free and salt-added solutions in the absence of flow.

5.4
Equilibrium Conformation of the Polyelectrolyte Chain

5.4.1
Conformation in a Salt-Free Solution

In a salt-free dilute solution of long polyelectrolyte chains the average concentration of counterions is too small to ensure the screening of the Coulomb interaction on the scale of the order of polyelectrolyte chain size. As a result, the interaction between charged monomers of individual polyelectrolyte chain in a dilute salt-free solution has the character of unscreened Coulomb repulsion. This repulsion strongly affects the polyelectrolyte conformation causing partial stretching of the chain whose characteristic size becomes proportional to the degree of polymerization N [22, 23].

The short-range excluded volume interactions between uncharged monomers also affect the chain conformation. These interactions can be described in the virial approximation using the second, va^3, and the third, v_3a^6, virial coefficients. The former depends on the solvent strength (v is positive or negative under good or poor solvent conditions, respectively and vanishes at the θ-point) whereas the latter is supposed to be of the order of unity, $v_3 \cong 1$, independent of the solvent strength.

The equilibrium chain conformation and the chain dimension, $R = R_e$, are determined by the balance between Coulomb repulsion of charged monomers and conformational free energy penalty for the deformation of a neutral polymer coil (or globule). This balance can be expressed as a condition of minimum of the free energy of the chain:

$$F(\alpha) = F_{Coulomb}(\alpha) + F_{elastic}(\alpha, z). \tag{5.6}$$

The electrostatic free energy of the Coulomb repulsion between charged monomers is given by

$$\frac{F_{Coulomb}(\alpha)}{kT} \cong \frac{l_B(fN)^2}{R} \cong \frac{w}{\alpha} \tag{5.7}$$

where we have introduced in a conventional way a reduced chain dimension $\alpha = R/N^{1/2}a$ and a dimensionless interaction parameter

$$w = \frac{l_B(f N)^2}{N^{1/2}a} \tag{5.8}$$

equal to the energy (in kT units) of electrostatic interaction in a Gaussian coil. Here l_B is the Bjerrum length, $l_B = e^2/kT\varepsilon$, which characterizes the strength of the Coulomb interaction and is usually of the order of a.

The elastic free energy, $F_{elastic}(\alpha, z)$, whose derivative determines the elastic force arising in a deformed polymer and preventing the stretching of the chain, can be presented in different forms depending on the solvent strength. The latter determines the character (repulsive or attractive in the case of good and poor solvent, respectively) of excluded volume interactions between uncharged monomers. We shall describe these interactions using another conventional reduced variable

$$z \cong \upsilon N^{1/2}$$

equal to the free energy (in kT units) of excluded volume interactions in the chain when it acquires the conformation of a Gaussian coil.

Another parameter used as a characteristic of the solvent strength is the thermal correlation length, $\xi_t \cong a|\upsilon|^{-1}$, equal to the size of the part of the chain remaining unperturbed by excluded volume interactions and retaining the Gaussian statistics (we call this part of the chain "the thermal blob").

As we shall see below, the reduced size α of the polyelectrolyte chain in a salt-free solution can always be expressed as a function of two dimensionless interaction parameters, w and z.

For a neutral polymer chain, $w = 0$, the range of $z \gg 1$ corresponds to good solvent conditions, when the chain acquires the conformation of a swollen coil, $\alpha \cong z^{1/5}$, whereas at $z < 0$, $|z| \gg 1$ the chain collapses into a spherical globule, $\alpha \cong z^{-1/3}$; in the θ-region $|z| \ll 1$ and the chain has the conformation of a Gaussian coil, $\alpha \cong 1$.

The elastic response arising under the deformation of a swollen or Gaussian polymer coil is determined by the conformational entropy losses due to chain stretching. The corresponding increase in the free energy can be presented in the scaling form [1, 2] as $F_{elastic}(R) \cong kT(R/R_0)^\delta$ where the exponent $\delta = 5/2$ or 2 under good or θ-conditions, respectively. R_0 is an unperturbed dimension of a swollen or of a Gaussian coil and is equal to $R_0 \cong N^{3/5}a\upsilon^{1/5}$ in the former case or to $R_0 \cong N^{1/2}a$ in the latter case. In reduced units the elastic free energy can be expressed as

$$\frac{F_{elastic}(\alpha, z)}{kT} \cong \begin{cases} \alpha^{5/2}z^{-1/2} \\ \alpha^2 \end{cases} \tag{5.9}$$

under the conditions of good and θ-solvent, respectively.

We have omitted in Eq. (5.9) the logarithmic term which is irrelevant at large α; we have also not included a non-linear term which is significant at strong extensions and takes into account the limited extensibility of a chain. The latter term determines the stretched state of the chain and is irrelevant for present consideration.

As we have seen, the deformation behavior of a swollen or of a Gaussian coil under good or θ-solvent conditions is dominated by the configurational entropy losses. This is in pronounced contrast to the deformation behavior of a collapsed polymer chain under poor solvent conditions when the surface free energy of a globule plays an important role. As has been shown in [24, 25] the deformation of a neutral polymer chain under poor solvent conditions is accompanied by intramolecular microphase separation.

If the elongation of the chain is much larger than the unperturbed globular size, then the chain splits into a spherical globular core with the density $\cong |v|$ coexisting with a stretched string of thermal blobs. The globular core is stabilized by the surface tension at the globule-solvent interface, and the corresponding excess free energy per unit area is given by $\gamma/kT \cong \xi_t^{-2}$. The total elastic free energy related to the deformation of the chain under poor solvent conditions includes this surface contribution and the conformational free energy of a stretched string of thermal blobs. If a distance R between chain ends much larger than unperturbed globular size is imposed, then the latter contribution is proportional to the number of thermal blobs in the string, $\sim R/\xi_t$, and dominates over the surface term. Finally, the expression for elastic free energy in reduced units reads

$$\frac{F_{elastic}}{kT} \cong \alpha |z|. \tag{5.10}$$

Equation (5.10) describes the intermediate range of elongations, $R \ll Na^3\xi_t^{-2}$, whereas at stronger stretching the elastic behavior of a Gaussian chain, $F_{elastic} \cong \alpha^2 kT$, is recovered.

In the case of a polyelectrolyte globule the intramolecular Coulomb repulsion of charges in the globular core overcomes the surface tension and the globule splits into two or several bead-globules joined by strings of thermal blobs [26], forming a "necklace" globule. The size of each bead-globule is determined by the balance between the surface tension and the Coulomb repulsion of charged monomers inside the bead. The size of the whole necklace globule is determined by the balance between the Coulomb repulsion of bead-globules and the tension in joining the strings. The corresponding elastic free energy is equal to the total number of thermal blobs in the strings and is still given (in our approximation) by Eq. (5.10).

Note that we utilize here the idea of the necklace polyelectrolyte globule suggested in [26] rather than more traditional cylindrical globule picture [27] used earlier in [4].

Minimization of the free energy defined by Eqs. (5.6), (5.7), (5.9) and (5.10) results in the equilibrium chain dimension determined by the competition between the intramolecular Coulomb repulsion of charged monomer units and an elastic force arising in the deformed polymer coil (under good or θ-solvent conditions) or in a globule (under poor solvent conditions):

$$\alpha_e \cong \left\{ \begin{array}{l} w^{2/7}z^{1/7} \\ w^{1/3} \\ w^{1/2}\,|z|^{-1/2} \end{array} \right\} \tag{5.11}$$

Here and below, the first, the second, and the third lines in the array correspond to good, θ-, and poor solvent conditions, respectively. The corresponding values of the elastic free energy in the minimum are given by

$$F_e/kT \cong \left\{ \begin{array}{l} w^{5/7}z^{-1/7} \\ w^{2/3} \\ w^{1/2}\,|z|^{1/2} \end{array} \right\} \tag{5.12}$$

It is easy to be convinced that equilibrium dimensions of the polyion described by Eq. (5.11) are proportional to N regardless of the solvent strength.

In the framework of the blob picture [1] the polyion partially stretched by the intramolecular Coulomb repulsion under good or θ-solvent conditions can be presented as a succession of N_B "electrostatic" blobs of size ξ [22, 23]:

$$R_e = \alpha_e N^{1/2}a \cong N_B\xi. \tag{5.13}$$

Under good or θ-solvent conditions the chain part inside every blob retains excluded volume or Gaussian statistics, respectively, and the energy of electrostatic interactions in the blob is of the order of kT so that they are equivalent to the Pincus stretching blobs [2] formed by the force of intramolecular electrostatic repulsion. The number of monomer units in the blob, $g = N/N_B$, and the blob size ξ are determined by equations

$$f^2l_Bg_+^2/\xi_+ \cong 1\,, \qquad\qquad \xi_+ \cong g_+^{3/5}av^{1/5} \tag{5.14}$$

or

$$f^2l_Bg_\theta^2/\xi_\theta \cong 1\,, \qquad\qquad \xi_\theta \cong g_\theta^{1/2}a \tag{5.15}$$

under good or θ-solvent conditions, respectively.

Under poor solvent conditions (in a polyelectrolyte globule) the electrostatic blobs are collapsed, and their size exceeds the thermal correlation length, $\xi_- \gg \xi_t$. The monomer density inside the blob is equal to the density in an ordinary neutral globule $\sim |v|$ and the blob size is determined from the balance of the excess surface free energy and the energy of electrostatic repulsion inside the blob:

$$f^2l_Bg_-^2/\xi_- \cong (\xi_-/\xi_t)^2\,, \qquad g_-/\xi_-^3 \cong \xi_t^{-1} \tag{5.16}$$

As follows from Eq. (5.16), $\xi_- \cong \xi_\theta$.

The polyelectrolyte globule as a whole can be envisioned [26] as a succession of collapsed electrostatic blobs ("beads") connected by stretched strings of thermal blobs. The length $l_{str} \cong af^{-1}(a\,|\,v|\,l_B^{-1})^{1/2}$ of each string corresponds to the condition, that the Coulomb repulsive force between neighboring electrostatic blobs is equal to the tension in the string, which is $\sim kT\xi_t^{-1}$. Therefore the size of a polyelectrolyte globule is given by $R \cong l_{str}N/g_-$ that coincides with Eq. (5.11) for poor solvent cases. Note that at $\xi_t \ll \xi_\theta$, i.e., in the globular regime, most monomers are contained in collapsed bead-blobs and only a minority of them form strings; at the same time strings give the main contribution to the length of the globule.

Taking into account Eqs. (5.14)–(5.16), we can rewrite the equilibrium dimension, $R_e = \alpha_e N^{1/2}a$, of a polyion under good or poor solvent conditions as

$$R_e^+ \cong R_e^\theta(\xi_\theta/\xi_t)^{1/7}, \qquad\qquad R_e^- \cong R_e^\theta(\xi_\theta/\xi_t)^{-1/2} \qquad\qquad (5.17)$$

respectively; here $R_e^\theta \cong Na^2/\xi_\theta$ is the size of the polyion under θ-solvent conditions.

In order to visualize all different regimes of the behavior of a polyelectrolyte molecule in a salt-free solution, we present them in the diagram of states in z, w

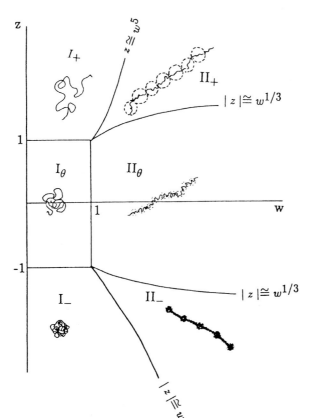

Fig. 5.5. Diagram of states of a polyelectrolyte molecule in a salt-free solution

coordinates in Fig. 5.5. As we have mentioned above, two dimensionless interaction parameters, z and w, completely determine the conformational state of a
polyelectrolyte in a salt-free solution.

Regions I_+, I_θ, I_- of the diagram correspond to the weak intramolecular
Coulomb interaction limit, i.e., to the quasi-neutral behavior. These regions oc-
cupy the range of small values of w or large values of z where electrostatic interactions in the chain are negligible in comparison with short-range excluded
volume interactions of uncharged monomers (in I_+ and I_- regimes) or with a
conformational entropy (in I_θ regime) and as a result are unable to affect the
chain conformation. The chain acquires the conformation of a swollen or of a
Gaussian coil in I_+ and I_θ regimes, respectively, and a collapsed globular conformation in I_- regime. The boundaries between I_θ/I_+ and I_θ/I_- regimes corre-
spond to the condition $R_0 \cong \xi_t$ or $|z| \cong 1$, i.e., the free energy of excluded
volume interactions in the chain is of the order kT. We note that at the former
boundary the smooth crossover for all thermodynamic and structural characteristics occurs, whereas the latter corresponds in the limit of infinitely long
chain to the line of the second-order coil-globule phase transition.

Regions II_+, II_θ, II_- correspond to the conformational states strongly perturbed by intramolecular Coulomb repulsion, i.e., to large values of the para-
meter w. Under good or θ-solvent conditions, i.e., in regions II_+ or II_θ, the chain
conformation can be envisioned as a succession of electrostatic blobs which are
swollen (obey the excluded volume statistics) in the former case, or Gaussian
in the latter case. Region II_- corresponds to the poor solvent conditions, where
the necklace-globule consisting of collapsed electrostatic blobs connected by
stretched strings of thermal blobs is formed.

The boundaries between II_θ/II_+ and II_θ/II_- regimes correspond to smooth
crossover and are determined by the condition $\xi_\theta \cong \xi_t$. Smooth crossover
occurs at II_+/I_+ and II_θ/I_θ boundaries as well. The boundary between neutral, I_-,
and polyelectrolyte, II_-, globule regimes corresponds to the first order phase
transition related to the splitting of a spherical globule into a dumbbell consisting of two collapsed ξ_- blobs joined by a string of thermal blobs. This
boundary is determined by the condition of equality of surface free energy of
the spherical globule to the energy of Coulomb repulsion of charged monomers
in it. We have not marked the lines of successive first order phase transitions related to splitting of a dumbbell globule into three-bead, four-bead states and so
on. These transitions occur inside II_- regime as the value of w increases. We
shall also not discuss the condensation of counterions on the polyelectrolyte
globule which occurs when the charge per unit length of a globule reaches the
value of e/l_B; the latter regime takes place at large $|z|$ and w (see [26] for
details).

5.4.2
Conformation in a Salt-Added Solution

As low molecular weight salt is added into the solution, a new characteristic
length in the system, i.e., the Debye screening length κ^{-1} appears. This screen-

ing length is related to the salt concentration c_s via conventional equation $\kappa^{-1} \cong (l_B c_s)^{-1/2}$, and we suppose the salt ions to be monovalent as well as charged monomers.

Partial screening of intramolecular Coulomb interactions enriches significantly the spectrum of possible conformational states of a polyelectrolyte molecule.

Let us consider the polyelectrolyte conformation in a salt-added solution, supposing that the values of z and w correspond to regions II_+ and II_θ of the diagram of states, Fig. 5.5, where unscreened intramolecular Coulomb repulsion predominates on a large scale over excluded volume interactions. The behavior of the weakly charged polyelectrolyte globule in a salt-added solution is a rather delicate problem and is not clearly understood at present.

At low salt concentrations, when the Debye screening length κ^{-1} exceeds by far the size R_e of a polyion in a salt-free solution, the chain conformation remains the same as in a salt-free solution.

In the intermediate range of salt concentrations, $\xi < \kappa^{-1} < R_e$, the chain conformation on scales smaller than κ^{-1} is still determined by unscreened Coulomb repulsion and coincides with that of the polyion in a salt-free solution.

It is obvious that on the large scale, the polyelectrolyte chain must obey excluded volume statistics and acquire the swollen coil conformation as it is determined by finite-range (screened Coulomb) repulsive interaction between large parts of the chain, Fig. 5.6.

The picture of chain conformation on the intermediate scale (larger than κ^{-1} but smaller than the total chain size R_s) is still not completely clear.

As was suggested in earlier work by Khokhlov and Khachaturian [23], even screened Coulomb interactions between electrostatic blobs are able to provide the stiffening of the chain of blobs on the scale of the order κ^{-2}/ξ. This idea of induced rigidity of the chain of electrostatic blobs originates from the Odijk-Skolnik-Fixman (OSF) concept of electrostatic persistent length formulated initially for strongly charged, $f \cong 1$, polyelectrolytes [11, 12]. As has been shown recently by Barrat and Joanny [20], this concept is applicable to strongly charged and intrinsically stiff chains, whereas in the chain of electrostatic blobs formed by weakly charged flexible polyelectrolytes, the local fluctuations of the

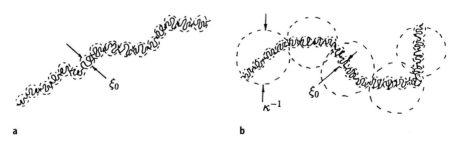

a   b

Fig. 5.6a, b. Polyelectrolyte molecule: **a** stretched in the salt-free solution; **b** swollen in the salt-added solution

chain conformation significantly reduces this induced rigidity effect. The variational approach used in [20] proves that the chain of electrostatic blobs becomes flexible on the scale of the order κ^{-1}. The latter dependence is strongly supported by the results of numerous experimental studies of solutions of intrinsically flexible polyelectrolytes.

In our further consideration, we shall not specify the index of the κ^{-1}-dependence of the effective segment l_e of a chain of blobs using the expression

$$l_e \cong \xi(\kappa\xi)^{-q} \tag{5.18}$$

where $q = 2$ in Khokhlov's model and $q = 1$ in the Barrat-Joanny model. One can argue that the actual value of q is between these two limiting cases, i.e., $1 < q < 2$. However, recent computer simulations [28] show that in some cases even the value of $q \leq 1$ can be obtained. Thus we keep q as an empirical parameter, supposing only that $q \geq 1$ that is essential for our purposes.

As we assume that the effective segments of the chain of blobs interact as hard cylinders of length l_e and thickness κ^{-1}, we obtain the following expression for the size of the polyion in a salt-added solution:

$$R_s \cong R_e^{3/5}\kappa^{-2/5}(\kappa\xi)^{(1-q)/5}, \tag{5.19}$$

where R_e is still given by Eqs. (5.11), (5.13), or (5.17). This conformation may be characterized as a "swollen polyelectrolyte coil". We note that locally (on the scale of electrostatic blobs) the chain obeys excluded volume or Gaussian statistics. As follows from Eq. (5.19), the size of a swollen polyelectrolyte coil decreases with an increase in the ionic strength of the solution as $\kappa^{-2/5}$ if no stiffening of the chain of blobs occurs, $q = 1$, and as $\kappa^{-3/5}$ if the Odijk-Skolnick-Fixmann-approach is applicable to the chain of electrostatic blobs, $q = 2$.

Finally, if the salt concentration is large enough to provide screening of Coulomb interactions on the scale of an electrostatic blob, i.e., at $\kappa\xi \gg 1$, the concept of electrostatic blobs loses its applicability, and the polyelectrolyte chain acquires the usual swollen coil conformation. Under good solvent conditions this swelling is determined by ordinary excluded volume interactions between all the monomers of the chain whereas in a θ-solvent the swelling is determined by screened Coulomb repulsion between charged monomers. The effective second viral coefficient ("the electrostatic excluded volume") in the latter case is equal to $v_{eff} \cong f^2 l_B \kappa^{-2}$. The reduced coil size is given by $\alpha_s \cong z_{eff}^{1/5} \cong v_{eff}^{1/5} N^{1/10}$. Further increase in salt concentration (in κ) results in a decrease in v_{eff} and at $z_{eff} \ll max\{|z|, 1\}$ the chain passes into one of the quasi-neutral regimes: a swollen coil at $z \gg 1$, a globule at $z \ll -1$, or a Gaussian coil at $|z| \ll 1$.

5.5
Stretching Transition in the Polyelectrolyte Chain

5.5.1
Stretching in a Salt-Free Solution

In this section we present a scaling theory of stretching of a weakly charged polyelectrolyte in a salt-free solution. We shall follow the lines of [3, 4] and our aim is to determine the limit of stability of weakly perturbed conformation in the flow field. In other words, we are looking for the spinodal line of the dynamic first order phase transition from weakly perturbed to completely stretched conformation. Corresponding "critical" value of the flow field gradient s_{crit} is usually obtained experimentally (see Fig. 5.1b).

As the polyion is immersed in an elongational flow, two forces, the Coulomb intramolecular repulsion and the hydrodynamic friction force, endeavour to stretch the chain, whereas the chain elasticity prevents this stretching.

We assume the following form for the stationary elongational flow:

$$v_x = sx , \quad v_y = - sy/2 , \quad v_z = - sz/2 \tag{5.20}$$

and suppose that the polyelectrolyte chain stays in the flow with constant gradient s for a period of time much longer than the characteristic time of the stretching transition.

In order to determine the conformation of a polyelectrolyte in the flow we extend a scaling-type approach suggested by de Gennes [1, 18] for the study of coil-stretch transition in neutral polymers. This approach is based on the analysis of the shape of an effective potential of the chain in flow

$$U_{eff}(R) = F_{Kramers}(R) + F_{elastic}(R) + F_{Coulomb}(R) . \tag{5.21}$$

This includes, in addition to the elastic and the Coulomb free energy terms discussed above, the Kramers potential

$$F_{Kramers}(\alpha)/kT \cong - \eta R^3 s \cong - \tau_g s\alpha^3 , \tag{5.22}$$

whose derivative determines the friction force applied to a polymer in an elongational flow described by Eq. (5.20), and stretching the chain in the x-direction [29, 30]. We have introduced the fundamental relaxation time of an uncharged chain under θ-solvent conditions $\tau_g \cong N^{3/2}a^3\eta/kT$ [1], where η is the viscosity of the solvent. The use of Eq. (5.22) for the description of the effect of the flow on the chain means:

(i) substitution of the three-dimensional flow field, Eq. (5.20), by a one-dimensional one. The correctness of such a substitution is discussed elsewhere [18];

(ii) friction coefficient of a chain is assumed to be $\zeta \cong \zeta_g \alpha$ where ζ_g is the friction coefficient of a Gaussian chain $\zeta_g \cong \eta N^{1/2}a$. Such an assumption means that we neglect the logarithmic term in the expression for ζ at large extension [31].

As follows from Eq. (5.21), the shape of the effective potential curve $U_{eff}(\alpha)$ and, consequently, the conformation of the polyelectrolyte chain in the flow are determined by three dimensionless parameters: z, w, and $\tau_g s$. The former two describe intramolecular interactions whereas the latter one describes interaction with the external flow field.

Let us consider first the behavior of quasi-neutral macromolecules in the flow at z and w corresponding to the regions I of the diagram of states. In this case the contribution of the Coulomb term, Eq. (5.7), to the free energy is negligible and if the flow velocity gradient is small then the chain retains the conformation of a swollen or of a Gaussian coil (regions I_+ and I_θ, respectively) or the conformation of a collapsed globule (region I_-).

The stretching of the chain by the flow makes intramolecular electrostatic repulsion even weaker so that the chain behavior in the flow coincides with that of the uncharged one.

The total effective potential curves $U_{eff}(\alpha)$ exhibit at sufficiently large s two minima. The first of them corresponds to weakly perturbed state, $\alpha_{\min} \cong z^{1/5}$, 1 or $|z|^{-1/3}$ under good, θ- or poor solvent conditions, respectively, whereas the second minimum, $\alpha'_{\min} \lesssim N^{1/2}$, corresponds to completely stretched conformation. The position of a maximum separating these minima is given by

$$\alpha_{max} \cong \begin{Bmatrix} (s\tau_g)^{-2} z^{-1} \\ (s\tau_g)^{-1} \\ (s\tau_g)^{-1/2} |z|^{1/2} \end{Bmatrix} \tag{5.23}$$

whereas the height of this maximum is

$$(U_{eff}/kT)_{max} \cong \begin{Bmatrix} (s\tau_g)^{-5} z^{-3} \\ (s\tau_g)^{-2} \\ (s\tau_g)^{-1/2} |z|^{3/2} \end{Bmatrix} \tag{5.24}$$

It is easy to see that the crossover between the lines in Eqs. (5.23) and (5.24) occurs at $s\tau_g \cong |z|^{-1}$.

At larger elongations, $\alpha \gg \alpha_{max}$, the effective potential $U_{eff}(\alpha)$ decreases rapidly due to the growth of absolute value of the Kramers friction term. Only at extremely large elongations when $\alpha \cong N^{1/2}$ does the non-linear elasticity come into play and the free energy begin to grow again. The deep minimum appearing on the $U_{eff}(\alpha)$ curve at $\alpha \cong N^{1/2}$ corresponds to the state of completely stretched chain, $R \cong Na$.

As the flow field gradient s increases, the height of the potential barrier U_{max} separating weakly perturbed and completely stretched states decreases and its position α_{max} moves towards smaller values of α. The critical value of the flow field gradient corresponding to the spinodal transition from the swollen or Gaussian coil or collapsed globular conformation to the completely stretched state can be estimated from the condition of disappearance of the potential barrier. That happens when α_{max} defined by Eq. (5.23) reaches an unperturbed value $\alpha \cong z^{1/5}$, 1, or $|z|^{-1/3}$, respectively. As a result, we obtain

$$s_{crit}^{(n)} \cong \tau_g^{-1} \cdot \begin{Bmatrix} z^{-3/5} \\ 1, \\ |z|^{5/3} \end{Bmatrix} \tag{5.25}$$

(superscript "n" indicates the regime of quasi-neutral behavior). It is remarkable that under poor solvent conditions this result can be obtained by equating the Stokes friction force applied to a spherical globule to a critical tension, $\cong kT|v|$, required for the unraveling of a neutral globule into a string of thermal blobs [24, 25].

The dependence of the critical value of the flow velocity gradient on the solvent strength corresponding to the cross-section 1 of the diagram of states (Fig. 5.5, $w < 1$, quasi-neutral behavior) is presented in Fig. 5.7, curve 1. As follows from Eq. (5.25) and from Fig. 5.7, an increase in the solvent strength (in z) leads at $|z| > 1$ to monotonic decrease in the value of s_{crit}.

It is instructive to obtain the values of s_{crit}, Eq. (5.25), using another set of arguments. As is known, the critical value of the flow field gradient can be obtained from the equation

$$s_{crit}\tau(\alpha, z) \cong 1 ,$$

(5.26)

where

$$\tau(\alpha, z) = \frac{\zeta(\alpha)}{K(\alpha, z)}$$

(5.27)

in the fundamental relaxation time of swollen ($\alpha > 1$), or Gaussian chain. (In the latter case $\tau(\alpha = 1) \equiv \tau_g$.)

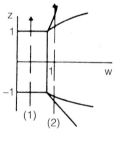

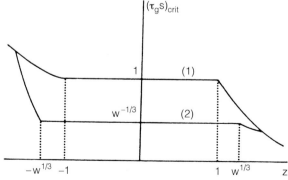

Fig. 5.7. Schematic dependence of the critical value of the flow field gradient s_{crit} on the solvent strength; curves 1 and 2 correspond to cross-sections 1 (quasi-neutral regime) and 2 (polyelectrolyte regime) of the diagram of states, respectively

The chain friction coefficient in our approximation grows linearly with α,

$$\zeta(\alpha) \cong \zeta_g \alpha ,\tag{5.28}$$

whereas the elastic modulus of the chain is given by

$$K(\alpha) \cong (Na^2)^{-1} \frac{\partial^2 F_{elastic}(\alpha, z)}{\partial \alpha^2}\tag{5.29}$$

where $K_g = (Na^2)^{-1}$ is the elastic modulus of a Gaussian chain. Taking into account Eqs. (5.9) and (5.26), we recover the results of Eqs. (5.25).

Let us consider now the case when the charge of the chain is sufficiently large to ensure partial chain stretching due to the intramolecular Coulomb repulsion ($w > 1$, regimes II of the diagram of states).

If the polyelectrolyte molecule is immersed in the elongational flow then two stretching forces, the intramolecular Coulomb repulsion and the friction force, are applied to the chain. The potential curves $U_{eff}(\alpha)$ schematically presented in Fig. 5.8 exhibit at sufficiently large s two minima: the first one at $\alpha \gtrsim \alpha_e$ given by Eq. (5.3) corresponds a to partially stretched chain extended by intramolecular Coulomb repulsion chain which is only weakly perturbed by the flow field whereas the second one at $\alpha \cong N^{1/2}$ corresponds to completely stretched state. The potential barrier separating these minima is localized at $\alpha \lesssim \alpha_{max}$.

The condition of disappearance of the potential barrier, $U_{max} \cong F_e$, or equivalently, $\alpha_{max} \cong \alpha_e$, corresponds to a spinodal transition from a partially stretched conformation (a chain of electrostatic blobs) to a completely stretched state. This transition occurs at $s \cong s_{crit}^{(sf)}$ given by

$$s_{crit}^{(sf)} \cong \tau_g^{-1} \cdot \left\{ \begin{array}{l} w^{-1/7} z^{-4/7}, \\ w^{-1/3}, \\ w^{-1} |z|^2 \end{array} \right\} .\tag{5.30}$$

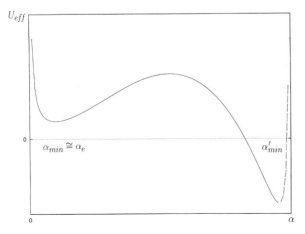

U_{eff}

$\alpha_{min} \cong \alpha_e$

α'_{min}

α

Fig. 5.8. Schematic representation of the effective potential $U_{eff}(\alpha)$ curve for the polyelectrolyte molecule in an elongational flow; the minimum at $\alpha_{min} \cong \alpha_e$ corresponds to the polyelectrolyte conformation weakly perturbed by the flow whereas the minimum at α'_{min} corresponds to the completely stretched state; the part of the curve where finite extensibility effects come into play is shown by dashed line

whereas if $z \gg w^5$ or $z \ll -w$ then Eqs. (5.25) applies; superscript (sf) in Eq. (5.30) refers to a salt-free solution.

The dependence of the critical flow field gradient for a polyelectrolyte chain on the solvent strength is presented in Fig. 5.7, curve 2, corresponding to the cross-section (2) of the diagram of states.

We note that the value of s_{crit} corresponding to the stretching transition in a polyelectrolyte globule, Eq. (5.30), third line, can also be obtained by equating the friction force $\cong \eta s \alpha^2 N$ applied to a necklace globule in the flow to the critical tension $\cong kT|v|$ required for the unraveling of a globule.

Equation (5.30) can be written also in the scaling form

$$s_{crit}^{(sf)} \cong \tau_g^{-1} w^{-1/3} (\xi_t/\xi_\theta)^x, \tag{5.31}$$

where $x = 4/7, 0,$ and -2 under the conditions of good, θ-, and poor solvents, respectively. As follows from Eqs. (5.30) and (5.31), the critical value of the flow field gradient decreases with an increase in the solvent strength, z, and in the charge parameter w. If $w \gg 1$ then the critical stretching flow is much weaker for a polyelectrolyte chain than for a neutral one in a wide range of the solvent strength around the θ-point. The effect is most pronounced for a θ-solvent (weak non-electrostatic interactions), whereas the difference between neutral and polyelectrolyte chains becomes less significant under extremely good or extremely poor solvent conditions. The width of the θ-range increases with an increase in w as $\sim w^{1/3}$.

Figure 5.9 presents the dependence of $s_{crit}^{(sf)}$ on the parameter w, characterizing the strength of electrostatic interactions in the chain. Curves 3, 4, and 5 and corresponding cross-sections of the diagram of states refer to good ($z \gg 1$), θ- ($|z| \ll 1$), and poor ($z \ll -1$) solvent conditions, respectively. As follows from Fig. 5.9, at large w corresponding to polyelectrolyte regimes $II_{+,\theta,-}$ of the diagram of states, Fig. 5.1, the value of $s_{crit}^{(sf)}$ decreases monotonically with an increase in the chain charge parameter w. In the range of small w corresponding to quasi-neutral regimes $I_{+,\theta,-}$ the critical velocity gradient remains (in the scaling approximation) virtually independent of w; this interval of the w value becomes larger as $|z|$ increases.

Of course, Eq. (5.30) can be obtained using the relation $(s\tau(\alpha_e))_{crit}^{(sf)} \cong 1$ between critical velocity gradient and the fundamental relaxation time of a polyelectrolyte chain partially stretched by intramolecular Coulomb repulsion. This relaxation time can be calculated using Eqs. (5.27)–(5.29) and Eq. (5.9) at $\alpha = \alpha_e$ given by Eq. (5.39).

5.5.2
Stretching in a Salt-Added Solution

Let us turn now to the analysis of stretching transition in a swollen polyelectrolyte coil in a salt-added solution. We assume in this section that the solvent is good or θ-solvent for uncharged monomers.

The elasticity of a swollen polyelectrolyte coil in a salt-added solution depends on the scale of deformation: at intermediate elongation, $R_s \ll R \ll R_e$, the

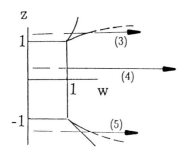

Fig. 5.9. Schematic dependence of the critical value of the flow field gradient on the polyelectrolyte charge via parameter w under the conditions of good (3), θ- (4) and poor (5) solvents; the corresponding cross-sections of the diagram of states are marked by the same numbers

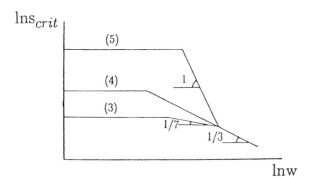

deformation of the chain occurs on a scale larger than the size of an electrostatic blob ξ so that the local blob structure remains unperturbed. An increase in the free energy due to the deformation of a swollen polyelectrolyte coil on this scale can be presented as

$$\frac{F_{elastic}(\alpha)}{kT} \cong \left(\frac{\alpha}{\alpha_s}\right)^{5/2} \tag{5.32}$$

where $\alpha_s = R_s/N^{1/2}a$ is a reduced size of a swollen polyelectrolyte coil (see Eq. 5.19).

At higher elongations, $R_e \ll R \ll Na$, the chain structure on the scale smaller than ξ is re-arranged and the elastic free energy $F_{elastic}(\alpha)$ is given by Eq. (5.9).

The critical value of the flow field gradient corresponding to the stretching of a swollen polyelectrolyte coil by the flow can be obtained from the condition $s_{crit}^{(s)}\tau_s \cong 1$, where $\tau_s \cong \zeta_s/K_s$ is the fundamental relaxation time of a swollen polyelectrolyte coil; $\zeta_s \cong \zeta_g\alpha_s$ is its friction coefficient and K_s is the elastic modulus which can be obtained from Eqs. (5.29) and (5.32).

Using Eqs. (5.28), (5.29) and (5.32) at $\alpha = \alpha_s$ one gets $s_{crit}^{(s)} \cong \tau_g^{-1}\alpha_s^{-3}$. Taking into account Eqs. (5.3) and (5.32), the following equation is obtained

$$s_{crit}^{(s)} \cong \tau_g^{-1}(N^{1/2}\kappa a)^{6/5}(\kappa\xi)^{3(q-1)/5} \left\{ \begin{array}{l} w^{-18/35}z^{-9/35} \\ w^{-3/5} \end{array} \right\} \tag{5.33}$$

(under the conditions of good and θ-solvents respectively; the superscript (s) refers to a salt-added solution).

Equation (5.33) gives the critical value of the flow velocity gradient corresponding to the spinodal transition from a swollen polyelectrolyte coil to completely stretched string of swollen or Gaussian electrostatic blobs. The latter conformation coincides in general with the conformation of a polyion in a salt-free solution.

As follows from Eq. (5.33) the critical value of the flow field gradient corresponding to the stretching of a swollen polyelectrolyte coil decreases with an increase in the solvent strength (in z) and in the polyion charge (in w). It grows with an increase in the ionic strength of a solution as $\kappa^{3(q+1)/5}$ regardless of the solvent strength.

The analysis of Eqs. (5.30) and (5.33) shows that at moderate screening (at moderate ionic strength of a solution) the range $s_{crit}^{(s)} < s < s_{crit}^{(sf)}$ of the flow field gradient exists. In this range the conformation of stretched string of electrostatic blobs remains at least metastable. Thus, at moderate screening two successive transitions occur as the flow field gradient, s, increases: the first one, at $s \cong s_{crit}^{(s)}$ defined by Eq. (5.33), is the transition from a swollen polyelectrolyte coil conformation to a stretched chain of electrostatic blobs, and the second one, at $s \cong s_{crit}^{(sf)}$ defined by Eq. (5.30), is the transition from a completely stretched string of electrostatic blobs to a completely stretched chain of monomers. The range of the ionic strength corresponding to the manifestation of two successive stretching transitions is determined by the condition $s_{crit}^{(s)} < s_{crit}^{(sf)}$ or, taking into account Eqs. (5.30) and (5.33),

$$1 \ll \kappa R_e \ll (\kappa \xi)^{(1-q)/2} \begin{cases} w^{25/42} z^{-5/42} \\ w^{5/9} \end{cases} \tag{5.34}$$

As can be seen from Eq. (5.34) the width of this range depends on the number of chain segments, N, on the fraction f of charged monomers and on the solvent strength v. If we set $q = 1$, that corresponds to upper estimation for $s_{crit}^{(s)}$, then we obtain for the width of two-stage transition region

$$(l_B/a)^{-13/42} N^{1/6} f^{-13/21} v^{1/7} \ll (\kappa a)^{-1} \ll (l_B/a)^{2/7} N f^{4/7} v^{1/7}$$

$$(l_B/a)^{-2/9} N^{1/6} f^{-4/9} \ll (\kappa a)^{-1} \ll (l_B/a)^{1/3} N f^{2/3} \tag{5.35}$$

under good and θ-solvent conditions, respectively. For more general case of arbitrary q ($q > 1$) the left sides of the inequalities at Eq. (5.35) take the forms

$$(l_B/a)^{-\frac{9q+4}{21(q+1)}} f^{-\frac{9q+17}{21(q+1)}} v^{\frac{6q+5}{21(q+1)}} N^{\frac{1}{3(q+1)}} \ll (\kappa a)^{-1}$$

$$(l_B/af^2)^{\frac{-3q+1}{9(q+1)}} N^{\frac{1}{3q+1}} \ll (\kappa a)^{-1} \tag{5.36}$$

As follows from the inequalities in Eq. (5.35), the two-step transition can be observed only if the macromolecules are sufficiently long (at $f \sim 0.1$, $N > 10^3$). At larger salt concentration when the inequality in Eq. (5.34) is violated the value of s required for stretching of a swollen polyelectrolyte coil becomes large enough. The state of a completely stretched string of electrostatic blobs becomes unstable at $s > s_{crit}^{(s)}$ and the polyion acquires completely extended conformation as a result of one-stage stretching transition. It is easy to prove that the finite range of salt concentration corresponding to two-stage stretching transition always exists for w, z corresponding to regions II of the diagram of states at Fig. 5.5.

Further increase in the ionic strength of a solution results at $\kappa^{-1} \ll \xi$ in the screening of electrostatic interactions on a scale smaller than the electrostatic blob size ξ.

Under the conditions of good solvent (swollen electrostatic blobs, region II$_+$ of the diagram in Fig. 5.5) the polyelectrolyte chain acquires at $\kappa^{-1} \ll \xi_+$ the conformation of a quasi-neutral swollen coil in which non-electrostatic excluded volume interactions between uncharged monomers predominate over Coulomb interactions and determine the chain conformation. The value of the critical flow field gradient, s_{crit}, required for the stretching of the swollen coil is given by Eq. (5.25), top line.

Finally, under θ-solvent conditions (Gaussian electrostatic blobs, region II$_\theta$ of the diagram of states, Fig. 5.5) the polyelectrolyte chain also acquires at $\kappa^{-1} < \xi_\theta$ the conformation of a swollen coil but the swelling is determined by screened Coulomb repulsion between charged monomers. The effective "electrostatic" second virial coefficient is equal to $v_{eff} \cong f^2 l_B \kappa^{-2}$ and the critical flow field gradient in this case is given by the top line of Eq. (5.25), with z substituted by $z_{eff} \cong v_{eff} N^{1/2}$.

5.6
Conclusions and Discussion

We have analyzed the influence of the intramolecular Coulomb and excluded volume interactions on the behavior of a polyelectrolyte molecule in an extensional flow field following a simple scaling approach. We have shown, that the interplay between three main parameters, w, z, and s, characterizing the magnitude of the Coulomb, excluded volume, and hydrodynamic interactions, respectively, determines the stability of different conformations of a polyelectrolyte molecule in the elongational flow.

Let us give a brief summary of our knowledge about the behavior of charged polymers in elongational flows.

If the longitudinal component of the flow field gradient is small, the hydrodynamic friction force weakly affects the conformation of a charged polymer which is already strongly perturbed by intramolecular Coulomb repulsion. In this case the local conformational structure of the macromolecule is determined by short-range excluded volume interactions, whereas on the larger scale it is partially stretched due to intramolecular Coulomb repulsion. If the electrostatic screening length is much smaller than the overall dimension of the polyion, then the chain acquires the conformation of a swollen polyelectrolyte coil. In the opposite limit of weak screening, the chain as a whole is stretched proportionally to the total number of monomers N. On the local scale the polyelectrolyte chain can be represented as a chain of swollen or Gaussian electrostatic blobs under good or θ-solvent conditions, respectively, or as an alternating succession of collapsed electrostatic blobs and strings of thermal blobs (the "necklace globule") under poor solvent conditions.

We stress, that the range of quasi-static (unperturbed by the flow) behavior for polyelectrolytes is restricted by much smaller shear rates than the corresponding range for neutral polymers. The reason for stronger sensitivity of polyelectrolytes to the extensional flow field is, of course, the equilibrium extension of the chain due to the intramolecular Coulomb repulsion.

An increase in the flow field gradient s induces an abrupt conformational stretching transition in the polyelectrolyte chain. In the weak screening limit, this is a transition from the stretched chain of blobs to completely stretched chain of monomers. In the moderate screening regime the transition from swollen polyelectrolyte coil to the stretched chain of blobs precedes the transition to completely stretched state. A certain region of kinetic stability of the stretched chain of blobs in the flow exists. The physical reason for this stability is a strong dependence of the elasticity of a polyelectrolyte chain in a moderate screening regime on the scale of deformation.

Thus we predicted a qualitative difference in the coil-stretching transition induced by the elongational shear flow in weakly charged polyelectrolytes, in a salt-added solution, and in neutral chains: a two-stage transition is expected in the former case in contrast to the one-stage transition in the latter case. In both cases the transition has the character of a dynamic first-order phase transition.

The critical values of the flow field gradient appear to be strongly dependent on the degree of chain ionization and on the salt concentration (Eqs. 5.22 and 5.26). These power dependencies can be checked in the experiments on polyelectrolyte stretching in the elongational flow if the degree of chain ionization and the ionic strength of a solution are varied.

We note that our analysis refers to "quenched" polyelectrolytes where the number and the distribution of charges along the chain are independent both of the chain conformation and of the external conditions. However, another important class of polyelectrolytes, "annealed" polyelectrolytes, exists. In the latter case the fraction of charged monomers or the degree of ionization adjust to the variation of external conditions. Weak polyacids or polybases are examples of these types of annealed polyelectrolytes. Their degree of ionization is strongly affected by local pH of the solution. As the conformational transition may be accompanied by the variation of local pH in the volume occupied by the polyelectrolyte chain, the fraction of charged monomers and, as a result, the charge parameter w, may become increasing functions of α. In this case the stretching transition is expected to become even more abrupt.

Although we have restricted our theoretical analysis to the consideration of weakly-charged polyelectrolytes, the qualitative picture of the behavior of charged polymers in the flow remains valid for strongly charged but intrinsically flexible chains as well. The essential point for the manifestation of an abrupt two-stage stretching transition is the remaining local flexibility of the chain. However, qualitative description of local flexibility and related entropic elasticity in strongly charged polyelectrolytes as well as an interplay between Coulomb and hydrodynamic interactions on the monomer scale requires more detailed microscopic analysis and, as a result, the use of a model with "higher resolution". Perhaps future computer simulations of polyelectrolytes in flow will give better insight to this problem.

Let us note that our analysis makes it possible to determine the range of stability of different chain conformations: spinodal rather than binodal corresponding to the dynamic phase transitions are obtained. This is meaningful from the experimental point of view, because the time spent by a macromolecule in the region of the stretching flow field is always limited by experimental conditions. All of the preceding results were based on the quasi-equilibrium approach which presumes that macromolecules stay in the flow long enough to accomplish the full transition into the stretched state. Consideration of finite time effects for macromolecules in stationary elongational flow as well as the description of polyelectrolyte behavior in transient flows needs to include dynamic aspects. This is a special problem which is beyond the scope of this chapter (see, for example [10, 32]).

The range of validity of the theory is restricted to dilute solution conditions (a single chain approximation has been used above). Such conditions are realized for polyelectrolytes at much lower concentrations than for neutral polymers. More complicated effects arising in a semi-dilute solution are related to the screening of both Coulomb and hydrodynamic interactions due to the presence of other polyelectrolyte chains and their counterions. These effects are beyond our present considerations.

Acknowledgments. We are grateful to E.B. Zhulina for numerous inspiring discussions. We greatly appreciate the hospitality of Prof. K. Binder, Johannes Gutenberg Universitat, Mainz, during the preparation of this manuscript. O.V.B. was supported by the Alexander von Humboldt Foundation. A.A.D. acknowledges partial support from the Deutsche Forschungsgemeinschaft (DFG) under Sonderforschungsbereich (SFB) 262/02. This work was performed partially (A.A.D.) by the financial support of the Russian Fund of Fundamental Research, Grant 93-03-5797, ISF Grant NT9000 and NATO Collaborative Research Grant HTECH.GRG 940365.

List of Symbols and Abbrevations

NaPSS	Sodium poly(styrene sulphonate)
BF	birefringence
QELS	quasi-elastic light scattering
IV	intrinsic viscosity
$\alpha = R/N^{1/2}a$	reduced dimension of a polyelectrolyte chain
ζ	friction coefficient of a polyelectrolyte chain
ζ_g	friction coefficient of a Gaussian coil

η	viscosity of a solvent
κ	inverse Debye screening length
ξ	electrostatic blob size, in particular
$-\ \xi_\theta$	under θ-solvent conditions
$-\ \xi_+$	under good solvent conditions
ξ_t	thermal correlation length
τ	fundamental relaxation time
$\tau_g \cong N^{3/2}a^3\eta/T$	fundamental relaxation time of a Gaussian coil
a	monomer length
e	electron charge
f	fraction of charged monomers
g	number of monomers in an electrostatic blob
$K(\alpha, z)$	elastic modulus of a polyelectrolyte chain
K_g	elastic modulus of a Gaussian coil
$l_B = e^2/\varepsilon kT$	Bjerrum length
N	number of monomers in the chain
R_e	equilibrium dimension of a polyelectrolyte chain in a salt-free solution in the absence of flow
R_s	the same in the salt-added solution
s	flow field gradient
$s_{crit}^{(n)}$	critical value of the flow field gradient corresponding to the stretching transition in a neutral chain
$s_{crit}^{(sf)}$	the same for a polyelectrolyte chain in a salt-free solution
$s_{crit}^{(s)}$	the same for a polyelectrolyte chain in a salt-added solution
va^3	second virial coefficient
$v_{eff}a^3$	effective electrostatic second virial coefficient
$w \cong l_B(fN)^2/N^{1/2}a$	parameter of electrostatic intramolecular interaction
$z \cong vN^{1/2}$	parameter of excluded volume intramolecular interaction

References

1. de Gennes PG (1979) Scaling concepts in polymer physics. Cornell University Press, Ithaca
2. Pincus P (1977) Macromolecules 10:210
3. Darinskii AA, Borisov OV (1995) Europhys Lett 29:365
4. Borisov OV, Darinskii AA, Zhulina EB (1995) Macromolecules 28:7180
5. Virk PS (1975) Nature 253:109
6. Ouibrahim A, Fruman DH (1980) J Non-Newtonian Fluid Mech 7:315
7. Merrill EW, Horn AF (1984) Polym Comm 25:144
8. Durst F, Haas R, Interthal W (1982) Rheol Acta 21:572
9. Miles MJ, Tanaka K, Keller A (1983) Polymer 24:1082
10. Dunlap PN, Leal LG (1984) Rheologica Acta 23:283
11. Odijk T (1977) J Polym Sci 15:477
12. Skolnick J, Fixman M (1977) Macromolecules 10:944
13. Stevens MJ, Kremer K (1993) Phys Rev Lett 71:2228
14. Stevens MJ, Kremer K (1993) Macromolecules 26:4717
15. Stevens MJ, Kremer K (1995) J Chem Phys 103:1669
16. Warner HR (1972) Ind Eng Chem Fund 11:375

17. Dunlap PN, Wang C-H, Leal LG (1987) J Pol Sci B; Pol Phys 25:2211
18. de Gennes PG (1974) J Chem Phys 60:5030
19. Hinch EJ (1977) Phys Fluids 20:S22
20. Barrat J-L, Joanny J-F (1993) Europhys Lett 24:333
21. Barrat J-L, Joanny J-F (1996) Advances in Chemical Physics 94, edited by I Prigogine and SA Rice, John Wiley
22. de Gennes PG, Pincus P, Velasco RM, Broachard F (1976) J Phys France 37:1461
23. Khokhlov AR, Khachaturian KA (1982) Polymer 23:1742
24. Halperin A, Zhulina EB (1991) Europhys Lett 15:417
25. Halperin A, Zhulina EB (1991) Macromolecules 24:5393
26. Dobrynin AV, Rubinstein M, Obukhov SP (1996) Macromolecules, 29:2974
27. Khokhlov AR (1980) J Phys A 13:979
28. Micka U, Kremer K, Phys Rev E, submitted
29. Rabin Y, Henyey FS, Pathria RK (1985) Phys Rev Lett 55:201
30. Kramers HA (1946) J Chem Phys 14:415
31. Brestkin YuV, Gotlib YuYa, Klushin LI (1989) Polymer Science USSR 31:1143
32. Darinskii AA, Safjannikova MG (1996) Vysokomolec Soed 38:236

Calculation of Flows with Large Elongational Components: CONNFFESSIT Calculation of the Flow of a FENE Fluid in a Planar 10:1 Contraction

Manuel Laso, Marco Picasso and Hans Christian Öttinger

6.1
Introduction

During the last two decades, a large number of flow calculations in planar and axisymmetric contractions (also known as slit-die or die entries) have been performed. It is no exaggeration to say that this type of geometry (Fig. 6.1) is probably the one most thoroughly investigated in viscoelastic flow calculations. Interest in die entry flow calculations is partly motivated by their relevance in processing procedures such as extrusion and injection molding. Apart from this practical aspect, the presence of a reentrant corner in the die entry geometry has been associated with non-convergence and other numerical difficulties collectively known as the HWNP (high Weissenberg number problem) which have made contraction flow a popular benchmark problem. Keunings [20] (Sect. 9.7) presents a comprehensive overview of numerical work on contraction geometry up to 1989, as well as a discussion of possible causes of the HWNP. More recent results are reported in [17] and [28].

In the context of this book, flows in planar or cylindrical contractions have the additional interest of possessing a strong elongational character in the region of the reentrant corner. Furthermore, the intensity of the elongational component of the flow varies smoothly from zero for Poiseuille flow at the inlet to a maximum in the region of the die entry itself and then decreases back to zero for Poiseuille flow downstream.

The importance of rheological characterization of polymer solutions and melts in elongational flows has been repeatedly stressed. It has been suggested that poor quality of comparisons between simulations and experiments in complex flows can often be at least partially attributed to missing or incomplete rheological characterization of the fluid under extensional flows. Unfortunately, measurements of material functions in purely elongational flows are often unreliable or even impossible. It has therefore been suggested that, apart from simple shear flows, complex flows should be used to characterize viscoelastic liquids. Recent publications [3, 38] represent two such attempts in which point-wise measured stress and velocity data in a complex flow are used to determine the adequacy of constitutive equations. The geometries (confined cylinder and planar contraction) can therefore be considered as rheometers of sorts in which the quality of both a constitutive equation and its parameters (determined under simple shear flow) conditions is tested.

If we confine our attention to contraction flows, the continual improvements in algorithms and the increasingly larger range of flow rates accessible to numerical work have made it possible in the last years to perform detailed comparisons between experimental and calculated velocity fields in contraction of both circular [11] and rectangular [15, 26] cross sections. These comparisons include extensive and careful analyses of both velocity fields and trajectories of macroscopic tracer particles, usually based on laser doppler velocimetry. Very satisfactory agreement between experiments and numerical results can be achieved, provided an accurate rheological characterization of the fluid is available.

All these computational pieces of work share the feature of being based on a continuum-mechanical description of the polymer solution or of the melt. They therefore yield a full description of the field variables at the macroscopic level typical of discretization methods (e.g., finite elements, volumes, differences, etc.). They are however unable to provide any information on a molecular level ("molecular" in the sense of the relatively simple mechanical models such as dumbbells, networks, etc. which underlie some macroscopic constitutive equations (CEs) [4, 29]; for those working in the field of molecular modeling [1] these models would not be considered as truly molecular but as coarse-grained at best. The terms "molecule" and "molecular" will be used in the coarse-grained sense throughout this chapter).

Therefore, in spite of the substantial progress achieved in flow calculations, most current numerical methods are inherently unable to offer insights into molecular behavior: in traditional isothermal and incompressible calculations, the fluid is entirely and exclusively characterized by the relationship between stress and strain history expressed as a differential or integral constitutive equation. Even for those CEs which allow a microscopic interpretation in terms of simple molecular models, the averaging procedure required to arrive at a macroscopic expression for the stress erases all microscopic (molecular) information. The primary variables involved in classical flow calculations are only velocity, pressure and stress; there is no way to retrieve any details of the lost molecular picture[1].

Furthermore, since the emphasis of the present work is on elongational flows, besides the choice of a suitable geometry, it is important to select a fluid with as realistic and non-trivial elongational behavior as possible. The patent inability of CEs like Oldroyd-B at describing realistic elongational behavior prompted us to use a fluid which admits a molecular interpretation of finite extensibility. While retaining the conceptual simplicity of Hookean dumbbells, a molecular model with finite extensibility will display markedly different and more physically realistic behavior in the extensional part of the channel flow. The use of a FENE fluid (see Sect. 6.4 below), for which no closed-form constitutive equation exists, makes it necessary to resort to non-standard numerical methods to solve the viscoelastic flow problem (see next section). Approximations and linearizations like the FENE-P and FENE-CR models are clearly less adequate: although they set limits on the value of the maximum *average* molecular extension, there is nothing in them preventing individual molecules from extending to an arbitrarily large extension.

---

1 Unless of course, one were to perform a posteriori a separate stochastic simulation of the molecular model using the velocity field obtained from the classical calculation.

6.2
CONNFFESSIT

Partly motivated by the idea of extracting molecular information from visco-elastic flow calculations, an alternative computational method has recently been introduced that combines stochastic and traditional discretization methods [14, 21–23, 30, 31]. In the CONNFFESSIT (Calculation Of Non-Newtonian Flows: Finite Elements & Stochastic Simulation Techniques), the primary variables in isothermal, incompressible calculations are velocity, pressure, and molecular configurations (degrees of freedom), stress having only a subordinate function.

As in most viscoelastic flow calculations, the objective of a CONNFFESSIT calculation is to solve the equations of mass and momentum conservation for an incompressible, isothermal non-Newtonian fluid:

$$(\nabla \cdot v) = 0$$

$$\varrho \frac{\partial v}{\partial t} + [\nabla \cdot \varrho vv] + [\nabla \cdot \underset{\sim}{\pi}] = 0$$

where $\underset{\sim}{\pi}$ is the total momentum-flux or total stress tensor which can be split in the following way:

$$\underset{\sim}{\pi} = p\underset{\sim}{\delta} + \underset{\sim}{\tau}$$

This set of partial differential equations must be closed by a third equation (the CE) relating $\underset{\sim}{\tau}$ to the history of the flow.

From the mathematical point of view, CEs of differential and integral type have been traditionally used (see [4], part III); additionally, some CEs can be written in both integral and differential forms. The numerical methods required to solve differential and integral CEs differ considerably: while it is relatively straightforward to develop coupled techniques for differential CEs, it is not a simple task to devise coupled algorithms for integral CEs, partly because the polymer contribution to the stress must be obtained by integration along particle paths which are not known a priori and partly because the Lagrangian formulation of integral models involves a complex functional of the velocity field.

From the methodological point of view, CEs have been developed along two not entirely separate routes: an essentially empirical one, in which a postulated CE contains an arbitrary number of terms and is basically meant to satisfy material objectivity and to do a regression and interpolation job of rheological data, and a second route based on solving the kinetic theory of a suitable microscopic molecular model (elastic or rigid dumbbells, bead-spring-rod chains, networks, reptation models, etc.). The second route offers the advantages of a consistent physical representation of the polymeric solution or melt at the molecular level, which is not necessarily the case when an empirical CE is used. This second method sometimes leads to CEs which have already been obtained in an empirical manner.

The main disadvantage of this second route is its complication: the way from the molecular model to the CE gets forbiddingly complex as the degree of realism of the molecular representation increases. Moreover, even for some exceedingly simple molecular models it is impossible to derive a closed-form CE. The FENE model (see Sect. 4.6) is a typical example of a very simple molecular model for which no closed-form CE can be derived. Even in cases where a CE can be obtained, its numerical evaluation can be so cumbersome as to render it impractical for numerical flow calculations in a continuum-mechanical framework (Doi and Edwards [10, 11], Bird et al. [4]).

In a CONNFFESSIT calculation, the fundamental difficulty associated with kinetic models for which no closed-form CE can be derived is by-passed, since all that is required is a consistent microscopic description of polymer dynamics, even if its kinetic theory is analytically unsolvable. The contribution of the polymer to the stress is obtained from a stochastic simulation of the molecular model. The rigorous correspondence between microscopic molecular models and macroscopic CEs has been treated extensively [4, 29].

The basic idea of CONNFFESSIT is to use a discretization method such as finite elements to solve the mass and momentum conservation equations while the polymer contribution to the stress is obtained by averaging the individual contributions of a great many molecules and not from a CE. These molecules are entrained by the fluid much as it happens in real flow situations. In addition to their macroscopic degrees of freedom (spatial coordinates), they possess internal degrees of freedom, the time evolution of which is governed by stochastic differential equations. The internal degrees of freedom fully describe the molecules; simple molecular models, like the FENE dumbbells to be considered in this work, are described by few degrees of freedom: it suffices to specify the three components of the vector joining the two masses that make up the dumbbell.

A typical CONNFFESSIT flow calculation is time-dependent, although it need not be so. It starts from a given velocity field (in the domain and at the boundaries) and a given set of molecules, usually distributed uniformly over the integration domain (in the following, the set of dumbbells that are located inside a given element at a given instant will be called the *local ensemble* of that element; the set of all dumbbells in the integration domain; i.e., the union of all local ensembles, will be called the *global ensemble*). In this case of uniform polymer concentration, each element contains a number of molecules proportional to its area (length or volume in 1-D or 3-D).

In viscoelastic flow calculations, in addition to the initial conditions, it is necessary to assume a flow history in some way ([20], Sect. 9.2.6). The configurations of the molecules that initially fill the domain must reflect what has happened to them during the infinite time interval before $t = 0$ s. Initial conditions correspond typically (but not necessarily) to zero velocity and to a flow history of zero strain throughout the domain. The initial configurations of the dumbbells are then drawn from a suitable equilibrium configurational distribution function (see for example Table 11.5–1 in [4]). Additionally, if open boundaries exist and a non-zero strain history is specified, it is first necessary to determine analytically or numerically what the molecular configurations at the inlet of the integration domain are. This departs radically from the Newtonian situation, where only the instantaneous values of the velocity are required to determine the stress. Whereas viscoelastic fluids possess a memory which fades at a rate inversely proportional to a typical relaxation time, a Newtonian fluid has an infinitely fast relaxation, i.e., possesses no memory.

CONNFFESSIT has already been used to solve one- and two-dimensional steady-state and time-dependent viscoelastic flow problems. Within the error bars due to its stochastic nature, CONNFFESSIT yields results in complete agreement with traditional techniques. In addition, CONNFFESSIT has been used to perform flow calculations with molecular models for which there is no equivalent closed-form constitutive equation (FENE [41]) or which are too complex [4, 10, 11] for traditional numerical methods.

Although molecular information was always available in CONNFFESSIT calculations, previous publications focused on the macroscopic description of the flow. The main objective of the present contribution is to demonstrate the feasibility of extracting from CONNFFESSIT flow calculations not only the macroscopic field variables but detailed molecular information (such as molecular extension) as well.

6.3
Geometry of the Planar Contraction Flow Problem

All but one previous CONNFFESSIT calculation were performed for flows of mainly shearing character. The only CONNFFESSIT calculation of flow in the axisymmetrical 4:1 contraction [13, 14] was aimed at demonstrating the feasibility of two-dimensional calculations. No molecular information was reported.

We concentrate now on the 10:1 slit-die or planar contraction flow (Fig. 6.1). The flow in this geometry is essentially plane Poiseuille flow far away upstream and downstream from the contraction, but has a significant extensional component close to the contraction itself. It is therefore well suited to an investigation of molecular behavior in extensional flow.

The actual size of the slit-die was chosen to fall within the range usually encountered in industrial practice. The lengths of the upstream and downstream

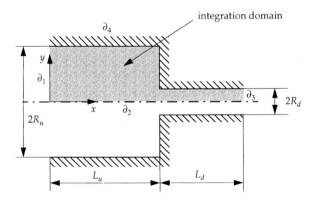

Fig. 6.1. Integration domain for contraction flow

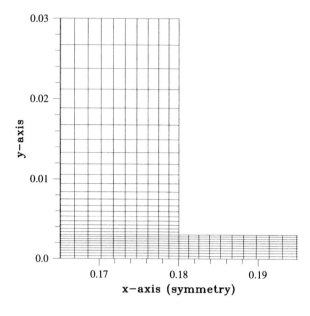

Fig. 6.2. Close-up of the finite element mesh in the region of the reentrant corner

channel were however both set to a generous $60R_d$ in order to minimize entry effects and to allow the flow to develop fully [5, 8, 9, 24, 25, 32]. The origin of Cartesian coordinates is placed centered at the inlet of the channel and the channel has a total length of 0.36 m (SI units are used throughout) in the x-direction. Given the symmetry of the domain, calculations are performed on only half of it. This integration domain is discretized in a mesh made up of 1920 quadrilaterals and 2047 corner nodes (i.e., vertices of the quadrilaterals). Coarser meshes were investigated as well, mainly to verify convergence with mesh refinement, but all results to be presented refer to the mesh just described. The portion of this mesh closest to the reentrant corner is shown in Fig. 6.2. The smallest elements in this figure (and in the whole domain) have typical sizes of 1.2 mm × 0.36 mm, which is coarse compared with some of the meshes used in traditional calculations (for example, the finest elements used in [15] are 60 and 10 times smaller in the x and y directions respectively). Our relatively large element size is dictated primarily by available computational resources as will be seen in Sect. 6.8.

6.4
The Polymeric Fluid

Since the emphasis of the present book is on molecular behavior in extensional flows, we have chosen a molecular model with finite extensibility. The *Finitely Extensible Nonlinear Elastic* (FENE) model [4, 41] is one of the simplest that fulfills this condition. It was first proposed for dumbbells in an attempt to improve on the Hookean dumbbell by including finite molecular extensibility.

Assuming a Hookean dumbbell molecular model would be tantamount to using an Oldroyd-B or a Maxwell-B constitutive equation (CE), depending on whether there is a Newtonian contribution or not.

The FENE fluid consists of a suspension of neutrally buoyant non-interacting dumbbells in a Newtonian fluid. The force law of the spring is nonlinear and displays asymptotic behavior as the elongation grows: at small elongation, there is little deviation from the Hookean behavior; as the elongation approaches the maximum, the restoring force grows without bounds. Since real polymer molecules deviate from Gaussian behavior as their elongation approaches the maximum (for example all-*trans* in a linear alkane), the finite extensibility of the FENE model makes it somewhat more realistic than the Hookean dumbbell.

The force law for the FENE model is given by

$$F^C = \frac{HQ}{1 - \left(\dfrac{Q}{Q_0}\right)^2} \tag{6.1}$$

This model has the additional interest (in the context of CONNFFESSIT) that no macroscopic constitutive equation can be derived for it: whereas the kinetic theory of some microscopic models for polymer dynamics can be solved analytically to yield a macroscopic CE, for the FENE fluid such an expression can be found only for steady state and only up to terms cubic in the velocity gradient [4]. Lacking a CE, traditional continuum-mechanical methods are unable to calculate the behavior of a polymeric fluid described by a FENE model under general non-homogeneous flow.

This is in stark contrast with the Hookean dumbbell: a dilute (in the sense of noninteracting) suspension of Hookean dumbbells can be, as far as its rheological behavior is concerned, exactly represented by the well-known Oldroyd-B CE. Conversely, the use of this CE in viscoelastic flow calculations tacitly implies a molecular description by means of Hookean dumbbells or molecules (although this interpretation is not the only one). For the time being, CONNFFESSIT is probably the only viable way to perform flow calculations for models for which no closed-form CE can be derived[2].

---

2 At least in principle, it would be possible to solve the diffusion equation for the configurational distribution function ([4], Sect. 13.2) in each element and use it to obtain the corresponding stresses. Since this equation is a partial differential equation in Q-space, this would involve the use, e.g., of Finite Elements, but this time at the element level. Such a combination of Finite Elements at the macroscopic level (to solve the equations of continuity and motion) with Finite Elements at the element level to solve the N_{el} diffusion equations for the configurational distribution function is not practical. Such a computational tour de force has actually already been successfully carried out [2, 12] for molecular models with a few degrees of freedom. The application of this strategy to models with a larger number of internal degrees of freedom, like bead-spring chains, would probably be impractical (consider for example mesh generation and shape functions in a space of a hundred dimensions).

Analytically tractable approximations to this FENE dumbbell model have been proposed by Peterlin (FENE-P model) [33, 34] and Chilcott and Rallison (FENE-CR model) [7]. These simplified models show a fair degree of agreement with the true FENE. They are weakest at reproducing the dynamic behavior following onset of shear [21] or extensional flow and their ability to reproduce true FENE behavior degrades as the (dimensionless) shear rate is increased. Quantitative comparisons of the FENE-P and the FENE-CR models with the true FENE can be found in [19, 40]. The linearized FENE-P model leads to an analytical CE (Eq. 13.5 – 56 of [4]) of considerable complexity for numerical viscoelastic flow calculations. Although it was used by Mochimaru to solve the start-up of planar Couette flow [27], it has only recently started to be used in complex geometries [18, 37].

The FENE dumbbell models can be characterized by three constants: the first, $\lambda = \zeta/(4H)$, is a characteristic relaxation time, similar to the one defined for Hookean dumbbells. The second is the dimensionless finite extensibility parameter

$$b = \frac{HQ_0^2}{kT}$$

which can be interpreted as the (dimensionless) maximum elongation squared. Although from the mathematical point of view the value of b could be chosen at will, physically meaningful values for vinyl polymers are in the range 30 – 300 [19]. The third constant α specifies what fraction of the total viscosity of the polymer solution at zero shear rate is due to the Newtonian solvent:

$$\alpha = \frac{\eta_s}{\eta_s + \eta_p^0}$$

6.5
The Time-Marching Procedure

A typical CONNFFESSIT calculation is time dependent: given an initial ensemble and initial conditions for the velocity (and therefore for the velocity gradient), the macroscopic and microscopic degrees of freedom of all molecules are propagated in time by a small interval using suitable numerical integration schemes. After this step, the local ensembles are determined, i.e., we find out in which element each dumbbell is located, and the contributions of the dumbbells to the stress is computed at the element level using these local ensembles. The molecules therefore act as stress computers. The stresses are then used to advance in time the macroscopic momentum conservation equation by finite element methods in which the stress due to the polymer is treated as a constant, right-hand-side term.

The macroscopic degrees of freedom of the FENE model are the components of the position vector of the center of mass of the dumbbell. The molecules follow streamlines and the coordinates of the center of mass of a given molecule are given by

$$\frac{dx}{dt} = v_x(t, x, y) \tag{6.2}$$

$$\frac{dy}{dt} = v_y(t, x, y) \tag{6.3}$$

The dynamics of the internal degrees of freedom of the dumbbell are described by stochastic differential equations which can be derived from the diffusion equation in configuration space. A description of the correspondence between Fokker-Planck equations and stochastic differential equations can be found in [29]. For FENE dumbbells (in the absence of external forces), the diffusion equation can be written as [4]

$$\frac{\partial \psi}{\partial t} = \frac{2}{\zeta} \left(\frac{\partial}{\partial Q} \cdot F^C \psi \right) - \left(\frac{\partial}{\partial Q} \cdot [\kappa \cdot Q] \, \psi \right) + \frac{2kT}{\zeta} \left(\frac{\partial}{\partial Q} \cdot \frac{\partial}{\partial Q} \psi \right)$$

where F^C is given by Eq. (6.1).

This equation is formally a Fokker-Planck equation [39], which has its counterpart in a stochastic differential equation (SDE), which for the force law of Eq. (6.1) has the form

$$dQ = \left([\kappa \cdot Q] - \frac{2H}{\zeta} \frac{Q}{1 - \left(\frac{Q}{Q_0}\right)^2} \right) dt + \sqrt{\frac{4kT}{\zeta}} \, dw \tag{6.4}$$

where w is a three-dimensional Wiener process. Equation (6.4) can be made dimensionless by dividing by $\sqrt{\dfrac{kT}{H}}$ and using the definition of λ:

$$dQ' = \left([\kappa \cdot Q'] - \frac{1}{2\lambda} \frac{Q'}{1 - \frac{Q'^2}{b}} \right) dt + \sqrt{\frac{1}{\lambda}} \, dw \tag{6.5}$$

In this SDE the first term in the parenthesis corresponds to the deterministic effect of the underlying macroscopic flow of the solvent, which tends to rotate and extend the dumbbell; the second represents the restoring force of the FENE spring. The last term in Eq. (6.5) corresponds to the Brownian agitation to which the dumbbells are subjected due to bombardment by solvent molecules. Equation (6.5) contains all the dynamic information required to perform the stochastic calculation for a given starting configurational distribution.

At equilibrium ($\kappa = 0$), the distribution of the connector vector Q' for FENE dumbbells is

$$\psi(Q') = \frac{1}{2\pi b^{\frac{3}{2}} B\left(\frac{3}{2}, \frac{b+2}{2}\right)} \left[1 - \frac{Q'^2}{b}\right]^{b/2} \tag{6.6}$$

for $(Q')^2 < b$, and 0 otherwise. Q' vectors are easily drawn from the equilibrium distribution: since it is spherically symmetric Q' is isotropic and is drawn from a uniform distribution over the surface of a sphere. Its length can be drawn from the distribution of its modulus (proportional to Eq. (6.6) times $4\pi r^2$) by simple rejection [36].

After all internal and external degrees of freedom have been advanced, the local ensembles are determined and the contribution of the polymer molecules to the stress tensor for each ensemble is evaluated using Kramers' form for the polymer contribution to the stress in terms of the normalized connector vector:

$$\underset{\sim}{\tau_p} = - nkT \left(\left\langle \frac{Q'Q'}{1 - \frac{Q'^2}{b}} \right\rangle - \underset{\sim}{\delta}\right) \tag{6.7}$$

In a CONNFFESSIT calculation, Eqs. (6.5) and (6.7) are valid for the local ensembles (of size N_{loc}^j) in each element j. Assuming the fluid to be initially at rest, the ensembles are initialized by distributing the total number of dumbbells in the global ensemble (N_g) over all elements with a uniform density of dumbbells, the configurations (i.e., the Q's) of which are drawn from the corresponding equilibrium distribution, Eq. (6.6).

If the fluid has not been at rest during $t < 0$, it is necessary to specify a past for the flow history. This history is used to generate the configurations of the dumbbells occupying the domain at $t = 0$ and the configurations of dumbbells entering the domain through a boundary.

The initial conditions define the velocity field (possibly zero everywhere) and therefore the velocity gradient κ at $t = 0$, which are used to advance the position of the centers of mass (Eqs. 6.2 and 6.3) and of the internal degrees of freedom of the dumbbells (Eq. 6.5) by a small time interval (the first step $\Delta t_1 = t_1 - t_0$ in a partition of time into intervals $[t_i, t_{i+1}]$ where $t_i < t_{i+1}, i = 0, 1, 2, ...$). After this step, the stress in each element (or in a suitable sub-element) is computed using Eq. (6.7) and used as a constant term or body force in the integration of the equation of momentum:

$$\varrho \frac{\partial v}{\partial t} + \varrho(v \cdot \nabla)v - 2\eta_s \nabla \cdot (\nabla v + (\nabla v)^T) - \bar{z}\nabla(\nabla v) = - \nabla \cdot \underset{\sim}{\tau^p} \tag{6.8}$$

where the contribution to the stress of the Newtonian suspending fluid has been written explicitly. Additionally, the incompressibility condition has been replaced with the penalty equation

$$p = - \bar{z}(\nabla \cdot v)$$

where $\bar{z}$ is a large penalty parameter (typically 10^8). The fluid is thus treated as being slightly compressible. This is a popular method for solving incompressible fluid problems since it eliminates not only the incompressibility condition but also pressure as an unknown.

6.6
Algorithms

In the present work, the integrations required in a two-dimensional CONNFFESSIT calculation were performed in the following way.

6.6.1
Integration of the Trajectories of the Dumbbells (External Degrees of Freedom)

We have integrated the trajectories of the centers of mass of the dumbbells using an explicit first-order Euler method:

$$x_{i+1}^j = x_i^j + v_x(x_i^j, y_i^j, t_i)\, \Delta t_i \tag{6.9}$$

$$y_{i+1}^j = y_i^j + v_y(x_i^j, y_i^j, t_i)\, \Delta t_i \tag{6.10}$$

the z-component being ignored in planar flow. The local value of the velocity is obtained by interpolating within the element in which the dumbbell is located by means of the corresponding shape functions below (Sect. 6.6.3). In CONNFFESSIT, particle trajectory integration closely mimics the flow of real polymer molecules.

6.6.2
Integration of the Internal Degrees of Freedom of the Dumbbells

Integration of Eq. (6.5) (FENE) was performed using a semi-implicit second order algorithm [29] in which the spring-force law is treated explicitly in the predictor step and implicitly in the corrector step.

This predictor-corrector scheme requires the velocity gradient at t_{i+1} to be known at t_i, which is the case in the simulation of rheological homogeneous flows, but not in viscoelastic flow calculations in complex geometries. A proper way of applying the previous predictor-corrector scheme in a CONNFFESSIT calculation is to approximate $\underset{\sim}{\kappa}(t_{i+1})$ in Eq. (6.12) for example by a backwards difference formula:

$$\underset{\sim}{\bar{\kappa}}(t_{i+1}) = \underset{\sim}{\kappa}(t_i) + \underset{\sim}{\dot{\kappa}}(t_i)\Delta t_i \approx \underset{\sim}{\kappa}(t_i) + \left(\frac{\underset{\sim}{\kappa}(t_i) - \underset{\sim}{\kappa}(t_{i-1})}{\Delta t_{i-1}} \right) \Delta t_i$$

and then to apply Eqs. (6.11) and (6.12) in succession for each dumbbell without intermediate updates of the velocity field.

6.6.3
Integration of the Momentum Conservation Equation

The integration of Eq. (6.8) was performed by a standard semi-implicit time-marching finite element technique. In the present work, bilinear, quadrilateral finite elements were used so that the velocity was continuous across two quadrilateral elements. Quadrilaterals were split in two triangles within which the stress was constant, the value of the stress being assigned to the central Gaussian integration point. Although this finite element does not satisfy the *inf-sup* stability condition and is known to produce spurious checker-board effects in the pressure [35], it is widely used because of its simplicity and because the velocity field (in which we are mainly interested) can be obtained accurately and is not subject to artifacts. The resulting system of equations was solved by a direct method. In order to prevent the possible appearance of spurious oscillations of the classical Galerkin method at moderate Reynolds numbers, the Streamline Upwind Petrov-Galerkin (SUPG) scheme was implemented [6, 16].

Time marching was performed by means of a scheme implicit in the velocity and explicit in the polymer contribution to the stress, i.e., the polymer contribution to the stress was treated as a right-hand-side constant at each step. This stress was computed element-wise as a local ensemble average. For the FENE model:

$$\underline{\tau}_p^j(t_i) = nkT \left[\frac{1}{N_{loc}^j} \left(\sum_k \frac{\boldsymbol{Q}'^k(t_i)\boldsymbol{Q}'^k(t_i)}{1 - \frac{(\boldsymbol{Q}'^k(t_i))^2}{b}} \right) - \underset{\sim}{\delta} \right] \tag{6.11}$$

where the counter k runs over the indices of all dumbbells belonging to the j-th element (j-th local ensemble).

6.6.4
Local Ensembles

The integration of the internal and external degrees of freedom and of the momentum conservation equation requires knowledge of the local ensembles. Due to the flow of dumbbells across element boundaries and across the boundaries of the integration domain, it is necessary to determine at every time step in which element each of the dumbbells is located. Additionally, the lists of molecules belonging to the local ensembles must be reordered in such a way that efficient program execution is guaranteed. These ancillary tasks take up in practice a large share of computation time. This overhead is however more than compensated for by the speed-up afforded in the remaining calculations at the molecule level (dumbbell velocities, configurations) and at the element level (stress). Dumbbell tracking and relocation was performed by means of velocity-biased element neighbor lists, the details of which have been described elsewhere [23].

6.7
Initial and Boundary Conditions

As already mentioned, the specification of initial conditions in CONNFFESSIT requires the specification of the primary variables, i.e., of the velocity at every nodal point and of the configurations of the dumbbells or of the configurational distribution function in each element.

Similarly, the specification of boundary conditions requires the usual specification of components of velocities at impenetrable boundaries (like the walls of the die), symmetry axes, and inlet and outlet boundaries. In addition, just as it is necessary to specify a flow history for $t < 0$ for the evaluation of integral CEs, in CONNFFESSIT we have to specify the configurations of the dumbbells that enter the domain.

The configurations of the dumbbells located in the integration domain at $t = 0$ and of those entering it through the inlet boundary must be consistent with the respective flow histories.

The boundary ∂ of the domain is defined as

$$\partial = \partial_1 \cup \partial_2 \cup \partial_3 \cup \partial_4$$

where the individual sub-boundaries ∂_1 through ∂_4 are specified in Fig. 6.1.

In the present case, we have as initial conditions for the velocity:

$$\boldsymbol{v}(x, y) = \boldsymbol{0} \quad \text{at} \quad t = 0 \text{ on } \Omega \tag{6.12}$$

and as boundary conditions for the velocity:

$$v_y(x, y) = 0 \quad \text{at} \quad t > 0 \text{ on } \partial_1$$

$$v_y(x, y) = 0 \quad \text{at} \quad t > 0 \text{ on } \partial_2$$

$$v_y(x, y) = 0 \quad \text{at} \quad t > 0 \text{ on } \partial_3$$

$$\boldsymbol{v}(x, y) = \boldsymbol{0} \quad \text{at} \quad t > 0 \text{ on } \partial_4 \tag{6.13}$$

where the x-component of $\boldsymbol{v}$ on ∂_2 and ∂_3 is unspecified, whereas the x-component of $\boldsymbol{v}$ on ∂_1 is specified as will be described below.

The initial conditions for the dumbbell configurations correspond to equilibrium, i.e., $\underset{\sim}{\kappa} = 0$ for all $t < 0$. The configurations are therefore drawn from Eq. (6.6) and are consistent with the initial conditions for the velocity at Eq. (6.12).

The boundary conditions for the dumbbell configurations correspond to steady-state or fully developed Poiseuile flow, i.e., the dumbbells entering the domain through ∂_1 are assumed to come from a channel of the same width and infinite length in which a steady-state Poiseuille flow exists. Since it is not possible to find a closed-form solution for the Poiseuille flow of the FENE fluid, the velocity profile and the dumbbell configurations must be determined in a previous CONNFFESSIT calculation (which we will call Problem I, while the contraction flow will be Problem II). This is similar to the common practice in viscoelastic flow calculations with integral CEs: the inlet (and possibly outlet) velocity profile is determined first in a separate one-dimensional calculation. This calculation is performed at the same volumetric flow rate that will be used in the contraction flow.

In the present work, the inlet velocity profile and dumbbell configurations for Problem II were obtained from a two-dimensional CONNFFESSIT calculation on a rectangular domain of the same half-width R_u as the upstream channel in the contraction flow (see (Fig. 6.3) and of a length $10R_u$. This could have been done by means of a one-dimensional calculation; the main reason for solving an essentially one-dimensional problem using a two-dimensional method was to investigate the effect of the stochastic noise in the stresses on the resulting noise in the velocity field in a two-dimensional calculation with open boundaries (there is a second reason which will be mentioned below). Except for the different shape of the domain, Problem I is very similar to Problem II. It seems that specifying its boundary conditions would in turn require solving another previous "Problem 0". However, this seemingly endless recurrence terminates in this case with Problem I, because here we are interested in the steady-state velocity profile only. Problem I was then solved similarly, except that:

- initially, the rectangular domain was filled with dumbbells drawn from the equilibrium distribution function Eq. (6.6) and a parabolic (Newtonian) velocity profile was imposed at the inlet;
- the outlet velocity profile was used at every time step to specify the inlet velocity profile (i.e., a time-dependent boundary condition, where the inlet velocity profile is set equal to the outlet velocity profile);
- dumbbells leaving the domain through the outlet were reintroduced at the inlet at the same value of y and without altering their configuration.

This strategy is tantamount to calculating a Poiseuille flow for the FENE fluid in a folded or periodic domain. Time-marching was carried out for long enough time to ensure that velocities and configurations were uniform throughout the

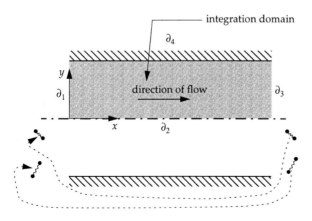

Fig. 6.3. Integration domain for planar Poiseuille flow

channel and constant in time (within stochastic noise). Different starting velocity profiles and dumbbell configurations were tried. In all cases the same steady-state configurations and velocity fields were obtained, thus showing that the integration time was sufficient to erase the memory of the initial dumbbell configurations and velocity fields, i.e., it was, for practical purposes, infinite.

The velocity profile obtained in this way was then imposed as time-independent velocity boundary condition for Problem II at the inlet ∂_1. The configurations and the y-component of the position of the dumbbells exiting the domain of Problem I were written to a file. This information was then used as boundary condition for the dumbbells[3]. This procedure is a straightforward and intuitive way of implementing boundary conditions consistent with the assumed flow history.

On the other hand, it is obvious that an integration of Problem II over long times requires a large number of configurations to be fed at the inlet. The most direct solution is first to perform a comparably long integration of Problem I. In particular, assuming an equal computational concentration of dumbbells in both integration domains, we have to ensure that the volume of fluid leaving the outlet of Problem I at steady-state be at least equal to the volume required to feed the inlet of Problem II during the entire length of the run. This may require the storage of a very large number of configurations, which may not be practical.

An alternative would be to perform a parametrization of the configurational distribution functions in each of the elements having an edge on the outlet boundary of Problem I, and then to sample these fitted functions to generate as many Q's as required in Problem II. This is however rather cumbersome and, above all, inaccurate.

In the present work we have resorted to a third method, which consists of storing a relatively small collection of dumbbell configurations from Problem I and re-using them as many times as necessary (i.e., rewinding the file where they are stored and rereading it from the beginning) Problem II. This idea is related to the use of *shifted random numbers*, a known numerical stochastic technique [29] (Sect. 4.1.4).

In test runs, we compared this method with the first alternative mentioned above, namely the storage of a very large number of distinct configurations, and found no systematic deviations between both alternatives in the sense that the velocity and stress fields were in agreement within error bars. Furthermore, although the set of configurations was used several dozen times in the course of a run, there was no detectable periodicity in the velocities nor in the stresses of a frequency consistent with the rewind frequency of the configuration file.

This is not surprising, since even though the set of configurations that enter the domain through ∂_1 is finite, i.e., after the configuration file is exhausted, the same Q's start entering the domain at the same positions; their subsequent trajectories are dictated by truly independent random numbers and are thus inde-

---

3 Since in Poiseuille flow the velocity gradient is not constant across the width, it is necessary to store not only the dumbbell configurations, but the exact position at which they leave the domain, so that they can be introduced in the domain of Problem II at the same value of y.

pendent. The resulting stress and velocity fields and molecule paths through the domain are therefore independent as well.

The outlet boundary condition of zero transversal velocity is imposed by a weak formulation. Dumbbells exiting the domain are removed from the lists of local ensembles, their indirect addressing pointers are reset, and their degrees of freedom are initialized afresh by reading new Q, y from the configuration file and relocating them in the corresponding local ensemble at the inlet.

6.8
Continuum-Mechanical and Molecular Results

A CONNFFESSIT calculation as described above was performed for a FENE fluid characterized by the following parameters:

$\eta_s = 0.01$ Pa.s $\varrho = 1000$ kg/m^3

$\lambda = 2$ s $b = 50$ $\alpha = 0.2$

in the die entry geometry described in Sect. 6.3. A global ensemble size of $N_g = 5 \times 10^5$ dumbbells and a constant time step of $\Delta t = 10^{-2} = 0.005\,\lambda$ for the integrations in Sects. 6.6.1–6.6.4 were used. Integration was continued up to $10\,\lambda$ (20 s).

The velocity profile imposed as time-independent boundary condition on ∂_1 is represented in Fig. 6.4 and was obtained as steady-state solution to a plane Poiseuille flow problem for the same fluid on a mesh of the same resolution in the y-direction (the velocity profile corresponding to a volumetric flow rate was $V = 2 \times 10^{-5}$ m^3/s.m). Integration in this case was continued up to $20\,\lambda$. The profile presented in Fig. 6.4 was obtained as the average over the last 2000 iterations ($10\,\lambda$). The noise in the velocity profile was further reduced by averaging over all (20) grid lines perpendicular to the channel axis of symmetry.

The finite slope of this profile at the symmetry axis (horizontal axis) is a consequence of the boundary condition $v_y(x, y) = 0$ at $t > 0$ on ∂_2 being imposed only weakly, i.e., only an infinitely fine mesh will produce a profile normal to the axis. This is the second reason why Problem I was solved in a two-dimensional fashion: within the weak formulation of the two-dimensional finite element method and for the mesh used, the solution to the steady-state plane Poiseuille flow problem is given in Fig. 6.4. Since the same finite element method and a mesh of the same refinement in the y-direction are used for the contraction flow, it is more consistent (albeit only asymptotically accurate) to use a boundary condition of the same type of weakness.

At this flow rate, the stress ratio

$$S_R = - \frac{\tau_{xx} - \tau_{yy}}{\tau_{xy}}$$

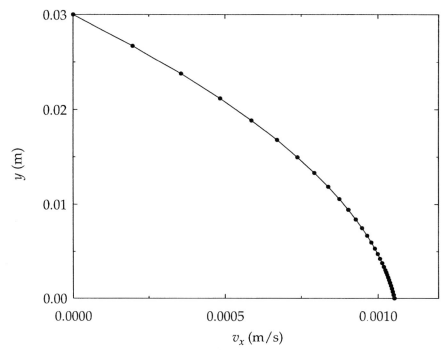

Fig. 6.4. Velocity profile at channel inlet

based on outlet values at the channel wall had a value of 4.6. The Deborah number

$$De = \frac{\lambda \langle v \rangle}{R_d}$$

was 4.1 for an average axial velocity of 0.00621 m/s.

The approach to the steady state can be judged from Fig. 6.6, which depicts the temporal evolution of the x-component of the velocity at the four sampling points P1 through P4 defined in Table 6.1 and Fig. 6.5. After about $t = 5$ s there is no detectable change in the velocity at any of the points. The steady state seems to take slightly longer to be reached at P2 than at the other locations, although by $t = 5$ s any systematic trend has fallen below the stochastic noise level at all locations. The results at steady state to be reported in subsequent sections are based on velocity and stress fields averaged over the last 10 s of the run.

Figure 6.7 presents the streamline pattern at steady-state (the starting points of the open streamlines were equally spaced at the inlet) and Fig. 6.8 a vector plot of the velocity field in the neighborhood of the reentrant corner. From both plots the existence of a small recirculation region is clearly seen. Two additional computations at $V = 1 \times 10^{-5}$ and 0.5×10^{-5} m^3/s.m showed a very

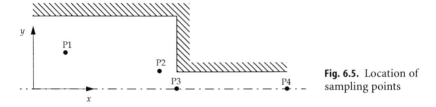

Fig. 6.5. Location of sampling points

Table 6.1. Coordinates of sampling points

Point	x	y
P1	0.10049	0.014956
P2	0.17023	0.00300
P3	0.18000	0.00000
P4	0.36000	0.00000

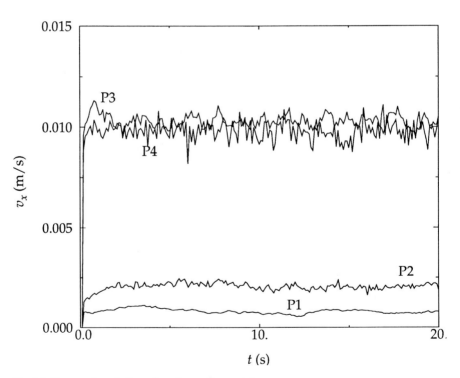

Fig. 6.6. Temporal evolution of v_x at sampling points

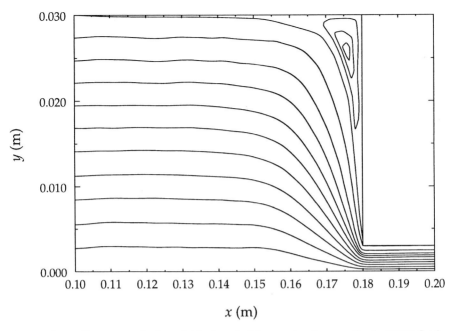

Fig. 6.7. Streamline pattern in the neighborhood of the reentrant corner for the FENE fluid

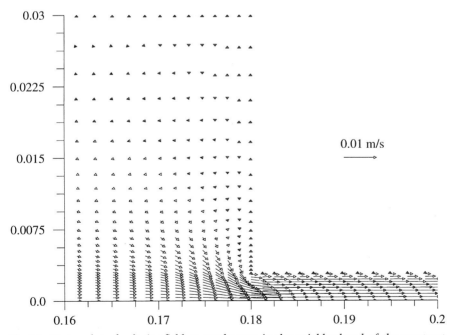

Fig. 6.8. Vector plot of velocity field at steady-state in the neighborhood of the reentrant corner. The magnitude is indicated by the reference arrow

weak dependence of vortex size on flow rate. Both the small vortex size and the insensitivity of its size on flow rate are consistent with previous results for the planar contraction.

The type of information reflected in Figs. 6.9 and 6.10 is rather unusual in traditional viscoelastic flow calculations; the contour lines in these figures correspond to constant levels of fluctuation intensity in the velocity. In Fig. 6.9a the absolute value of the fluctuation σ_v is depicted (one standard deviation in the norm of the velocity). Even though the fluctuation in v and v itself are not colinear, Fig. 6.9a is a compact way of representing noise levels. Fluctuations are strongest closest to the reentrant corner, although the largest fluctuations do not occur at the corner itself, as can be seen in the magnified view of Fig. 6.9b. The relative value of the fluctuation $\sigma_v/|v|$ is shown in Fig. 6.10. Relative noise levels are of course highest in the recirculation zone, where the average velocity is very low but fluctuations do not vanish.

Needless to say, this level of noise in the velocity field is totally unheard of in traditional viscoelastic flow calculations, but is typical of small-scale CONNF-FESSIT calculations. It is thus obvious that whenever traditional methods are applicable, they will be superior to CONNFFESSIT in accuracy. Typical relative fluctuations outside the recirculation region in Fig. 6.10 are less than 20% with local maxima of up to 40% close to the corner. A reduction of average fluctuations down to 2% would imply a global ensemble roughly 100 times larger, something perfectly feasible on today's supercomputers (the present calculation was performed on a modest RISC workstation). Above all, at the accuracy level of a few percent, the question of statistical noise in the numerical solution ceases to be relevant when compared with the uncertainties in the experimental characterization of the polymeric fluid. On the other hand, if sophisticated molecular models are to be used, or if the continuum-mechanical approach does not hold (as may be the case very close to a reentrant corner), CONNFFESSIT may currently represent the only alternative. For example, the accuracy achieved in the present CONNFFESSIT calculation is perfectly adequate to distinguish between the behaviors of the true FENE fluid and the linearized FENE-P approximation.

The outlet velocity profile is represented in Fig. 6.11. The same comments as for the inlet boundary condition apply to the outlet boundary condition at the symmetry axis. Since the mesh is in relative terms coarser at the outlet, the weakness of the approximation of the boundary condition is even more apparent. The dotted line in this figure represents the inlet velocity profile scaled by a factor of ten. In spite of the typical rate of strain being ten times higher, the velocity profiles scale almost perfectly with the ratio of reciprocal channel widths (or of average velocities). The dimensionless average shear rate $\lambda\dot{\gamma}$ at the outlet is 7.2 (maximum value 14.3), which for the FENE fluid (with $b = 50$) implies an average decrease in the polymer contribution to the viscosity of approx. 50% with respect to zero shear rate [19]. For $\alpha = 0.2$, this corresponds to an average decrease of 40% in the apparent viscosity with respect to the average at the inlet (which, as far as shear-thinning is concerned, is virtually at zero shear rate). The effect of this relatively large decrease in viscosity on the velocity profile is very moderate. As a matter of fact, a separate calculation of planar

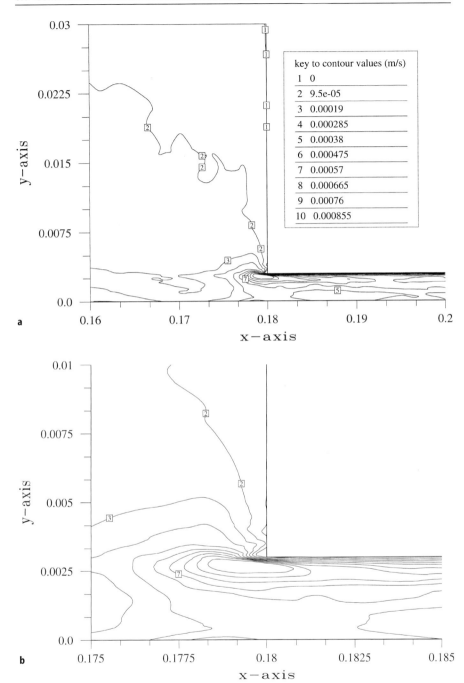

key to contour values (m/s)

1	0
2	9.5e-05
3	0.00019
4	0.000285
5	0.00038
6	0.000475
7	0.00057
8	0.000665
9	0.00076
10	0.000855

Fig. 6.9a, b. **a** Contour plot of noise level (one standard deviation) in $\sigma_v/|v|$. **b** Magnification of region closest to reentrant corner

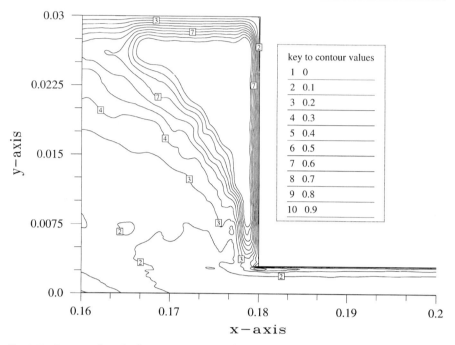

Fig. 6.10. Contour plot of relative noise level $\sigma_v/|\boldsymbol{v}|$ in the velocity

Poiseuille flow in the outlet channel at one-tenth of the flow rate with a finer mesh shows that a good fraction of the difference between the outlet profile and the scaled inlet profile is attributable to the coarseness of the mesh.

We now turn our attention to the molecular information delivered by CONNFFESSIT. As already mentioned in Sect. 6.2, the primary variables in a CONNFFESSIT calculation are the velocities, pressures (in this case eliminated by the penalty formulation of the incompressibility condition), and configurations.

Figure 6.12 conveys an idea of the computational density of dumbbells in the region of the reentrant corner. Each dot corresponds to a dumbbell or "molecule". The size of the global ensemble N_g is chosen so that the number of dumbbells in the smallest elements $\min(N_{loc}^j)$ is still sufficient to yield meaningful values. For reasons of accuracy, this means in practice $\min(N_{loc}^j) \approx 20$ to 50, although it may be substantially different for other models and geometries. Apart from accuracy considerations, lower values are not recommended since they are prone to drop to zero due to fluctuations.

Since dumbbells follow streamlines deterministically and the fluid is incompressible, an initially uniform dumbbell concentration should remain so throughout the calculation. This is seen to be very approximately the case in the snapshot presented in Fig. 6.12 which was taken at $t = 20$ s, after 2000 integration steps. It is however possible to discern regions in this figure where the concentration is either clearly higher (A) or lower (B) than the average. This effect is observable only after a pretty large number of integration steps and is clearly

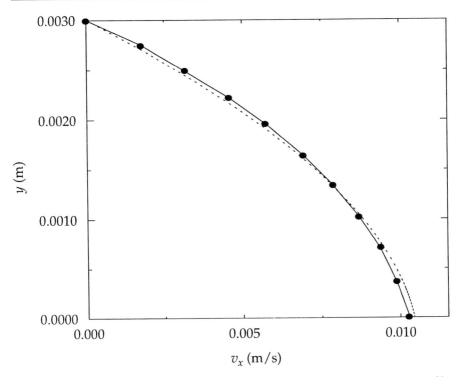

Fig. 6.11. Velocity profile at channel outlet. The dotted line is the scaled inlet velocity profile

not due to a statistical fluctuation. Region A is enriched in dumbbells that, due to discretization errors in the integration of their paths, are so close to the boundary ∂_4 (on which $v(x, y) = 0$ holds) that their velocity is effectively zero and they are therefore unable to turn around the corner. Region B is depleted in precisely such particles. As can be seen in Fig. 6.7, the streamlines that traverse the uppermost part of the narrow channel (approx. $0.0027 < y < 0.003$) must make a very sharp turn round the reentrant corner. Individual particles, subjected to random fluctuations in their velocity (Fig. 6.7 reflects a velocity field *averaged* over the second half of the run) have a non-negligible chance of drifting toward the wall and remaining stuck there. An obvious cure to such a malady is a more accurate integration scheme than the simple Euler scheme of Sect. 6.6.1.

Another consequence of the spatially uniform distribution of dumbbells is that a very refined mesh automatically displays a large N_g in order for the criterion $\min(N_{loc}^j) \approx 20$ to 50 to be fulfilled. For given computational resources, the smallest elements in the mesh determine what the concentration of dumbbells and therefore the value of N_{el} and N_g must be. This is the reason for our mesh being relatively coarse.

It would be possible to assign an initial non-uniform dumbbell concentration in order to reduce fluctuations in the smallest elements. However, the

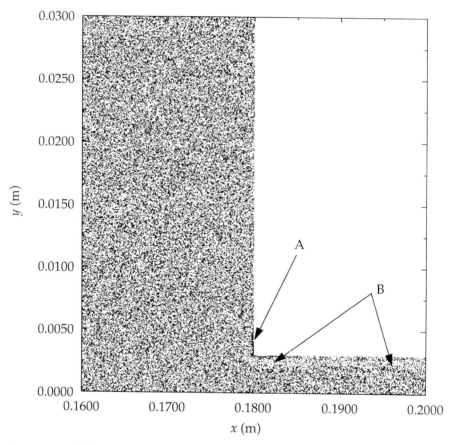

Fig. 6.12. Actual location of dumbbells at $t = 20$ s. Each dot represents a single dumbbell

macroscopic flow of dumbbells would soon spoil this naive scheme. This is another respect in which CONNFFESSIT differs from continuum-mechanical approaches, in which a high local resolution seems to be achieved. It must however be considered that for a mesh of given fineness, CONNFFESSIT evaluates the stress in an element as an average over the molecules in the local ensemble. Since elements tend to be finest where gradients are largest, the flow histories of the dumbbells in a local ensemble can be very different. Even the simplest averaging scheme, like Eq. (6.11), thus takes into account a wide spectrum of flow histories. In a traditional continuum-mechanical calculation, integration over flow histories is done for a few Gaussian integration points and is therefore much less representative of the flow history of all the material of the element. Figure 6.9 of [14] illustrates how dramatically different the flow histories and the resulting stresses can be for different points within a single element. In this sense, a relatively large element in a CONNFFESSIT calculation contains information (albeit quite crudely averaged) that requires several finer elements in a continuum-mechanical scheme to be collected. The fineness of a mesh in a CONNFFESSIT calculation cannot therefore be judged by the same standard as in a classical calculation.

Finally, it is perhaps worth saying that the concentration of "computational" molecules just discussed (in this 2-D case, number of dumbbells per unit area) has nothing to do with the value of n in Eq. (6.7), which is the actual concentration of polymer molecules in the solution and which is derived from the given value of the viscosity of the polymer at zero shear rate.

Turning our attention now to the internal degrees of freedom, we plot in Fig. 6.13, as a simple measure of molecular extension, the average value $\langle Q'^2 \rangle$ in each element in the immediate neighborhood of the reentrant corner. As intuitively expected, dumbbells that approach the corner very closely are subjected to strong extension, as indicated by the narrowly spaced contour lines in that area. The most elongated dumbbells (darkest grey) are located in the uppermost part of the narrow channel downstream of the corner. The smoothness of the contours is remarkable if we consider that the values of $\langle Q'^2 \rangle$ are not averaged over time, but are instantaneous values taken at $t = 10$ s (due to the choice of scales, the vertical dimension and therefore the waviness of the contour lines in Fig. 6.13 is magnified by a factor of three).

Figure 6.14 depicts the values of Q'^2_x and Q'^2_y for 1000 dumbbells flowing out of the domain as a function of the position (y-coordinate) at which they exit (over the whole run approx. 32,000 dumbbells left the domain). Dumbbells close to

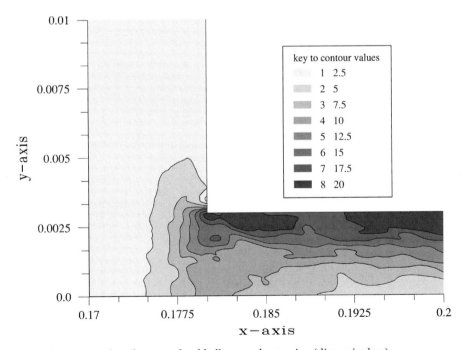

Fig. 6.13. Contour plot of average dumbbell squared extension (dimensionless)

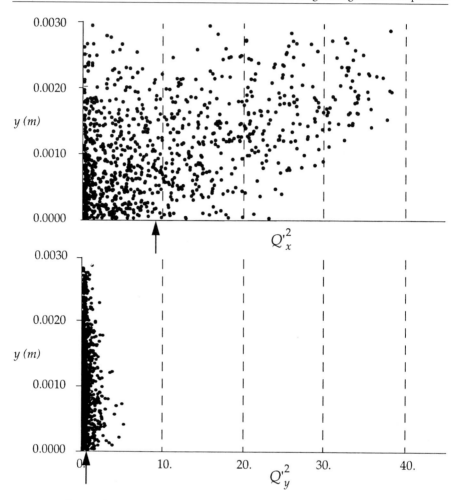

Fig. 6.14. Q'^2_x and Q'^2_y for dumbbells leaving the domain. Arrows indicate average values over the whole cross section

the center of the channel ($y = 0$) experience a milder deformation than those close to the wall ($y = 0.003$). There are, however, more of the former (because of the velocity profile) and they shift the overall average value $\langle Q'^2 \rangle$ towards lower values (the averages of Q'^2_x and Q'^2_y over the whole cross-section are represented by the arrows in that figure). Thus, if the polymer solution were to be collected at the outlet, instantaneously homogenized, and frozen, the average squared molecular extension in the x- and y-directions would be given by the arrows. If we had a polymer melt (and a calculation using an appropriate molecular model) and this melt were cooled immediately after exiting the die, or a solution were coagulated to produce a fiber, the properties of the material at the center of the fiber would differ from those of peripheral material.

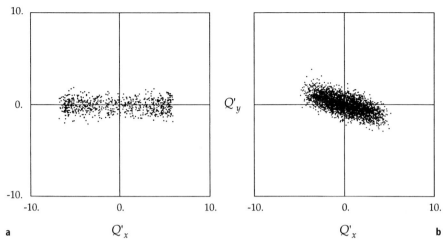

Fig. 6.15a, b. Q'_x and Q'_y for dumbbells leaving the domain: **a** dumbbells closest to the channel wall ($0.0027 < y < 0.003$; **b** dumbbells nearest to the axis of symmetry ($0 < y < 0.0003$)

Figure 6.15a shows the configurations of a few hundred dumbbells that have left the domain at the upper tenth of the channel width, while Fig. 6.15b corresponds to the region closest to the axis of symmetry; the number of dots in Fig. 6.15a, b is proportional to the flow rates at the corresponding location and therefore to their respective weights in the calculation of average extension. Each dot in these plots is the tip of a vector with origin at (0,0) and components (Q'_x, Q'_y) representing an individual dumbbell configuration. The very high degree of orientation of the dumbbells closest to the wall is apparent from Fig. 6.15a. Even molecular configurations closest to the center of the channel show a strong departure from isotropy as can be observed in Fig. 6.15b. The prediction of resulting anisotropies in macroscopic properties (mechanical, optical, etc.) from known molecular configurations is the next logical step. Such a fine-graining task is a fascinating possibility to which access can be gained through a CONNFFESSIT calculation.

Figure 6.16 is a plot of axial velocity and elongation rate along the centerline of the channel. Both the characteristic velocity overshoot and the asymmetric peak in elongation at the die entry are clearly observable. Both are in good qualitative agreement with Fig. 6.5 of [15]. In Fig. 6.16b, time is represented in the abscissa. This is the elongation history a polymer molecule would experience as it were entrained by the solvent along the centerline of the channel.

Similarly, Figs. 6.17 and 6.18 present shear and elongation rates along three streamlines, the origins of which are at $x = 0$ and the value of y indicated on the plot. The shear and elongation rates along streamlines were determined by transforming the rate-of-strain tensor to a rotating local coordinate system attached to a particle flowing along the streamline. As the particle was tracked along the streamline, the local coordinate system was rotated so that the positive 1-direction was always in the direction of flow, the 2-direction was the

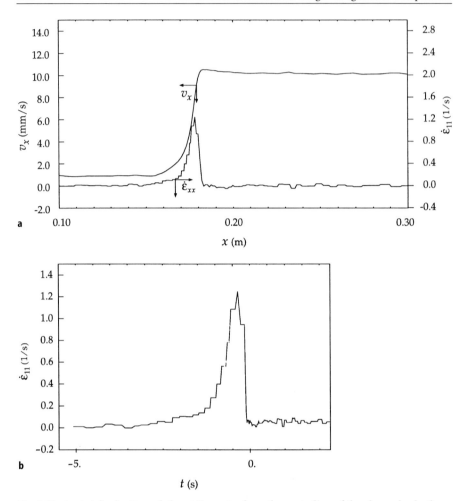

Fig. 6.16a, b. Axial velocity and elongation rate along the centerline of the channel: **a** both are plotted as a function of position along the centerline; **b** elongation rate is plotted as a function of time and represents the temporal evolution of the elongation as observed by a particle that moves along with the fluid

other direction in which the velocity changed, and the 3-direction was chosen to form a right-handed coordinate system. In our figures, the rate of strain and the elongation rate are plotted as a function of time, where $t = 0$ has arbitrarily been taken to be the time at which the particle reaches the abscissa of the entry itself ($x = 0.18$ m). The steps in the curves stem from the type of approximation used for the velocity, which implies constant velocity gradients in each half-quadrilateral.

If we now follow the configuration of dumbbells along the streamline that starts at ($x = 0, y = 0.025$) the plot shown in Fig. 6.19 is obtained. The curve was

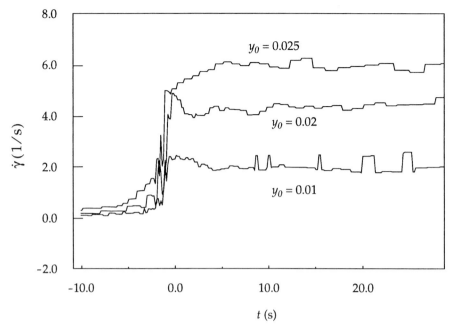

Fig. 6.17. Shear rate along several streamlines as a function of time (measured with respect to the instant when the streamline reaches the abscissa of the die entry $x = 0.18$)

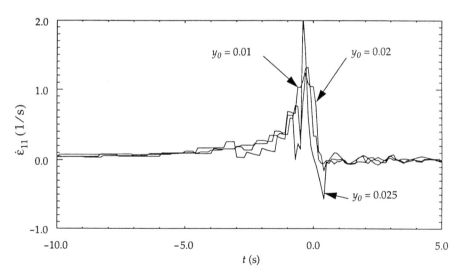

Fig. 6.18. Elongation rate along several streamlines as a function of time

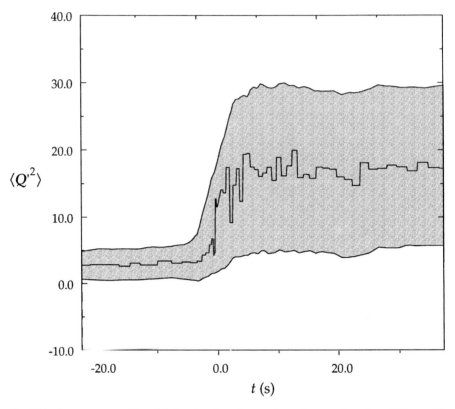

Fig. 6.19. Average squared dumbbell extension along a streamline as a function of time with respect to $t = 0$ as defined in Fig. 6.17. The value of $\langle Q'^2 \rangle$ is given by the line in the shaded zone. The width of the shaded zone is twice the standard deviation in $\langle Q'^2 \rangle$

obtained by measuring $\langle Q'^2 \rangle$ in each element the streamline traverses at steady state. The steady state of Q'^2 was obtained by averaging over only three global ensembles (stored at $t = 10, 15$, and 20 s). Hence the considerable noise level. Of course it would have been possible to evaluate $\langle Q'^2 \rangle$ a posteriori, once the velocity field had been determined, by sending a large number of dumbbells along the streamline, integrating their internal degrees of freedom along the streamline, and evaluating $\langle Q'^2 \rangle$ at each time step, just as it was done in [14]. We preferred the "noisy" alternative to show that this information is available in a CONNFFESSIT calculation and does not require any such postprocessing. If more precise values of $\langle Q'^2 \rangle$ were required, noise level could be reduced by simply storing the global ensemble more frequently.

The average squared dumbbell extension $\langle Q'^2 \rangle$ along the streamline experiences a drastic increase as it approaches the reentrant corner. The dumbbells subsequently enter the narrow channel region and stay close to the wall where they are subjected to strong shearing. Besides the increase in $\langle Q'^2 \rangle$, the distribu-

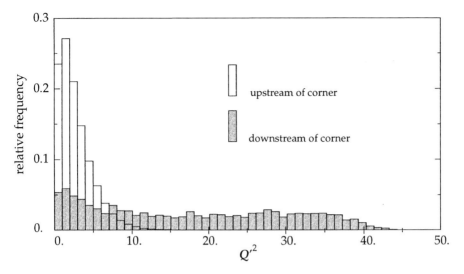

Fig. 6.20. Histograms of $\langle Q'^2 \rangle$ probability distribution functions along the streamline of Fig. 6.19, upstream and downstream of the corner, "upstream" being defined as the portion of the streamline in $x < 0.10$ m and downstream as the portion in $x > 0.25$ m. The histograms were computed using all dumbbells lying in a stripe of width 0.002 (upstream histogram) or of width 0.0002 (downstream histogram) about the streamline

tion of Q'^2 widens, as indicated by the width of the shaded area, which is twice the standard deviation in $\langle Q'^2 \rangle$. This plot provides only partial information about the evolution of the dumbbell configurations, since the distribution of Q'^2 is not symmetrical. The actual probability distribution function of Q'^2 is shown in Fig. 6.20 as two histograms: one for the upstream and another for the downstream portions of the streamline, as defined in the figure caption. As the fluid turns round the corner and enters the narrow channel, the distribution of Q'^2 develops a very extended tail at high Q'. While upstream of the corner virtually no dumbbells have an average square extension greater than about 15, Q'^2 is greater than that value for roughly 50% of all dumbbells downstream of the corner. A non-negligible fraction even goes beyond 40, which is 80% of the maximum value ($b = 50$). The nonlinear law of Eq. (6.1) implies that the force in the FENE spring is five times the force in an equivalent Hookean spring ($b = \infty$) for dumbbells with $Q'^2 = 40$ and ten times for those with $Q'^2 = 45$. An interesting possibility, which can be implemented in a CONNFFESSIT calculation with minimum effort, is the introduction of a critical tension for molecular break-up. In its simplest form this can be done by defining a threshold extension (and therefore threshold force in the spring) beyond which the dumbbell splits into two. While the parent chain disappears, the two daughter chains are incorporated into the global ensemble and are dealt with in the same fashion as all the other molecules.

The exact formulation of break-up criteria and of the properties which the daughter dumbbells inherit from their parent dumbbell is an almost entirely uncharted area of research. But irrespective of the details of such formulations, CONNFFESSIT opens the possibility of studying the influence of flow and processing on the configurations of, and on the forces acting on, these coarse-grained "molecules". Polymerization or chemical degradation of polymer solutions and melts during flow and processing could be embedded in such a calculation in a similar fashion.

As a further measure of dumbbell extension, the eigenvalues of

$$
\begin{bmatrix} \langle Q_x Q_x \rangle & \langle Q_x Q_y \rangle \\ \langle Q_y Q_x \rangle & \langle Q_y Q_y \rangle \end{bmatrix}
$$

along the same streamline are plotted as a function of time in Fig. 6.21a. Figure 6.21b is a parametric plot of the eigenvalues (notice the differences in axis scales). For large negative times both are quite similar, the slight anisotropy being a consequence of the small shear to which dumbbells are subjected. However, as the dumbbells approach the narrow channel, an abrupt increase in the ratio of eigenvalues takes place. The transition from approximate isotropy to high anisotropy has already started at $t = -2$ s and is completed by $t = 0.9$ s. Most of the change however happens at negative times (relative to the instant when the trajectory reaches $x = 0.18$ m), i.e., before the dumbbells enter the high shear region of the narrow channel. Since the effective channel width before the entry proper ($x = 0.18$ m) is considerably larger than the width of the narrow channel, the local shear is necessarily less than in the narrow channel. The increase in eigenvalue ratio must therefore be attributed to the extensional component of the flow just before reaching $x = 0.18$ m. The curve of $\dot{\varepsilon}_{11}$ for y_0 in Fig. 6.18 shows that there is indeed a strong elongational component in the flow before the entry proper. While the rate of elongation rapidly decays back to zero, the shear component remains high and is responsible for the permanent anisotropy.

In the present calculation, the time scale of dumbbell relaxation is very similar to the time required for a dumbbell to enter the narrow channel and to the time during which it undergoes elongational flow. It would be worthwhile to investigate the effect of a much larger dumbbell relaxation time, i.e., at a correspondingly larger De.

6.9
Summary

We have studied the time-dependent flow of a FENE fluid in a planar 10:1 contraction using CONNFFESSIT. This flow has a strong elongational character in the region of the reentrant corner, and it has actually been used for determining elongational viscosities. Besides the continuum-mechanical information which traditional methods deliver, CONNFFESSIT offers a quantitative insight into molecular behavior even when a closed-form constitutive equation is not available. As illustrative examples, we have described basic molecular properties like extension and orientation along streamlines.

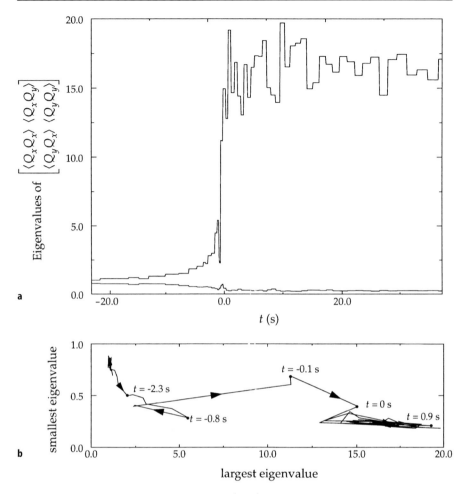

Fig. 6.21 a, b. Evolution of the eigenvalues of $\langle Q'Q'\rangle$ (only x and y components considered) along the same streamline as in the previous figure: **a** individual eigenvalues as a function of time; **b** joint evolution with time as parameter along the curve

CONNFFESSIT offers a general and flexible framework for incorporating processes taking place at the coarse-grained molecular level into macroscopic viscoelastic flow calculations and for investigating the effect of processing on molecular characteristics.

Acknowledgments. H.C.Ö. and M.L. gratefully acknowledge funding from the BRITE/EuRam project BRPR-CT96-0145. M.L. wishes to acknowledge additional funding by the PATI Program of the Spanish Ministry of Energy and Industry.

List of Symbols and Abbreviations

B()	beta function	
b	constant in FENE and FENE-P models	(–)
De	Deborah number	(–)
F^C	connector force	(N)
H	spring constant	(N/m)
k	Boltzmann's constant	(J/K)
N_{el}	number of mesh cells or elements	(–)
N_g	number of dumbbells in the global ensemble	(–)
N_{loc}^j	number of dumbbells in element j (size of j-th local ensemble)	(–)
n	number density of molecules	(m^{-3})
p	pressure	(Pa)
Q	connector vector	(m)
Q_0	maximum value of connector vector	(m)
Q'	dimensionless or normalized connector vector	(–)
$\overline{Q}'$	predicted dimensionless connector vector in predictor-corrector scheme	(–)
R_d	half width of channel downstream of contraction	(m)
R_u	half width of channel upstream of contraction	(m)
S_R	stress ratio	(–)
T	temperature	(K)
t	time	(s)
V	volumetric flow rate per unit depth of channel	(m^3/s.m)
v	velocity	(m/s)
$\langle v \rangle$	average velocity at channel outlet	(m/s)
W	three-dimensional vector of normal random deviates	(–)
w	three-dimensional Wiener process	(–)
x^j	x-component of position vector of j-th dumbbell	(m)
y^j	y-component of position vector of j-th dumbbell	(m)
$\overline{z}$	penalty parameter	(m)

Greek letters

α	ratio of solvent viscosity to total viscosity at zero shear rate	(–)
δ	unit tensor	
$\dot{\varepsilon}$	elongation rate	(s^{-1})
η_s	viscosity of Newtonian fluid	(kg/m/s)
κ	transposed velocity gradient $(\nabla v)^T$	(s^{-1})
$\tilde{\kappa}$	approximated transposed velocity gradient	(s^{-1})
λ	time constant of FENE dumbbells	(s)
π	total stress tensor	(Pa)
ϱ	density	(kg/m^3)
σ	covariance matrix	
σ	standard deviation	
τ_p	polymer contribution to the extra-stress	(Pa)
ζ	friction coefficient	(N.s/m)
ψ	configurational distribution function	

Ω integration domain
∂_i i-th sub-boundary of the integration domain

Subscripts
i time step index or coefficient index
p polymer
v velocity

Superscripts
j dumbbell index or element index
0 at zero shear rate

References

1. Allen MP, Tildesley DJ (1987) Computer simulation of liquids. Clarendon Press, Oxford
2. Armstrong RC (1998) Numerical simulation of viscoelastic fibre spinning. To appear in Proceedings of the Royal Society-Unilever Indo/UK Forum on Dynamics of Complex Fluids
3. Baaijens HP, Peters GWM, Baaijens FPTJ (1995) Rheol 39:1243
4. Bird RB, Curtiss CF, Armstrong RC, Hassager O (1987) Dynamics of polymeric liquids, vol II. Wiley
5. Boger DV, Crochet MJ, Keiller RA (1992) J Non-Newtonian Fluid Mech 44:267
6. Brookes AN, Hughes TJR (1982) Comp Methods Appl Mech Eng 32:199
7. Chilcott MD, Rallison JM (1988) J Non-Newtonian Fluid Mech 29:381
8. Debae F, Legat V, Crochet MJ (1994) J Rheol 38:421
9. Debbaut B, Marchal JM, Crochet MJ (1988) J Non-Newtonian Fluid Mech 29:119
10. Doi M, Edwards SF (1978) J Chem Soc, Faraday Trans II 74:1789
11. Doi M, Edwards SF (1979) J Chem Soc, Faraday Trans II 75:38
12. Fan XJ (1989) Acta Mechan Sin 5:216
13. Feigl K, Laso M, Öttinger HC (1994) In: Gallegos C, Guerreo A, Muñoz J, Berjano M (Eds) Proc IVth European Rheology Conference, Sevilla
14. Feigl K, Laso M, Öttinger HC (1995) Macromolecules 28(9):3261
15. Feigl K, Öttinger HC (1996) J Rheol 40:21
16. Franca LP, Hughes TJR (1993) Comp Methods Appl Mech Eng 105:285
17. Gallegos C, Guerrero A, Muñoz J, Berjano M (1994) (eds) Proc IVth European Rheology Conference, Sevilla
18. Halin P, Keunings R, Laso M, Öttinger HC, Picasso M (1996) Paper presented at the XIIth International Rheology Conference, Quebec
19. Herrchen M, Öttinger HC (1997) I Non-Newtonian Fluid Mech 68:17
20. Keunings R (1989) in Fundamentals of computer modeling for polymer processing, Tucker CL (ed) Carl Hanser Verlag 402
21. Laso M, Öttinger HC (1993) J Non-Newtonian Fluid Mech 47:1
22. Laso M, Öttinger HC (1993) Physikalische Blätter 49(2):121
23. Laso M, Picasso M, Öttinger HC (1997) AIChEJ 43:877
24. Luo X-L, Mitsoulis E (1990) Int J Num Methods Fluids 11:1015
25. Marchal JM, Crochet MJ (1987) J Non-Newtonian Fluid Mech 26:77
26. Mitsoulis E (1993) J Rheol 37:1029
27. Mochimaru YJ (1983) J Non-Newtonian Fluid Mech 12:135
28. Moldenaers P, Keunings R (eds) (1992) Theoretical and applied rheology, Proc XIth Int Congress on Rheology, Brussels, Elsevier
29. Öttinger HC (1996) Stochastic processes in polymeric fluids, Springer

30. Öttinger HC, Laso M (1992) In: Keunings R, Moldenaers P (eds), Proc XIth Int Congress on Rheology, Elsevier, Amsterdam
31. Öttinger HC, Laso M (1994) Bridging the gap between molecular models and viscoleastic flow calculations in Lectures on Thermodynamics and Statistical VIII, 139–153, World Scientific
32. Park HJ, Mitsoulis E (1992) J Non-Newtonian Fluid Mech 42:301
33. Peterlin A (1961) Makr Chem 44–46:338
34. Peterlin A (1962) Kolloid-Zeitschrift 182:110
35. Pironneau O (1989) Finite element methods for fluids, John Wiley, New York
36. Press WH, Teukolsky SA, Vetterling WT, Flannery BP (1992) Numerical Recipes, Cambridge University Press
37. Purnode B (1996) Vortices and Change of Type in Contraction Flows of Viscoelastic Fluids, PhD Thesis, Université Catholique de Louvain
38. Quinzani LM, Armstrong RC, Brown RA (1995) J Rheol 39:1201
39. Risken H (1989) The Fokker-Planck Equation, Springer Series in Synergetics
40. van den Brule BHAA (1993) J Non-Newtonian Fluid Mech 47:357
41. Warner HR (1972) Ind Eng Chem Fundamentals 11:379

CHAPTER 7

Polymer Solutions in Strong Stagnation Point Extensional Flows

J. A. Odell and S. P. Carrington

7.1
Introduction

It has long been empirically established that solutions of parts per million of high molecular weight polymers can exhibit remarkable non-Newtonian behaviour. The most striking examples being the observation of turbulent drag reduction [1] and dramatic flow thickening beyond a critical flow-rate in porous media flows [2]. Such phenomena, as discussed elsewhere in this book, have tremendous commercial importance and have been attributed to perhaps the most fundamental physical property of high polymers, that of extensibility. In stretching flows especially, a transition from an ambient coil state to a stretched out conformation might be expected to modify flow behaviour dramatically.

Extensional flows are of considerable importance and generality, occurring in flows through orifices, filters, porous media, constrictions in pipes, and in any flow possessing turbulence or vorticity. In fact, most real flow systems contain an extensional component.

For a polymer molecule to achieve high chain extension in a stretching flow-field, there must be sufficient strain-rate maintained over a long enough time for a given fluid element to attain the required strain. In any flow-field which incorporates a stagnation point, some fluid elements remain trapped within the flow-field over a long period of time. Such fluid elements are subject to the very high strains necessary for flexible coils to extend.

This chapter will examine the response of macromolecules to such flow-fields. First a brief summary of the most salient theories and models of the stretching macromolecule are presented. This is followed by a discussion of the techniques for the realisation and assessment of stretching. We then review the major experimental works for dilute solutions in the light of theoretical expectations. This is extended to include a discussion of the experimental observations for semi-dilute solutions.

The useful properties of polymer additives are severely restricted by the onset of thermomechanical degradation, and this is discussed for both dilute and semi-dilute solutions. Finally we discuss stagnation point extensional viscometry and attempts to relate the extreme extension thickening to molecular observations.

7.2
Theory and Modelling of Stretching Macromolecules

Early theoretical works by James and Guth [3] and Flory [4] investigated the forces required to produce extension of individual chains. In 1961, Peterlin asserted that modelling of non-Newtonian properties required the introduction of a finite extensibility [5]. In 1974, de Gennes predicted that the coil-stretch phenomenon in dilute solutions in stretching flow-fields could be critical, with a sudden transition from a coil to a close to fully extended state as the strain-rate is increased, illustrated schematically in Fig. 7.1 [6]. This he envisaged to be due to an increase in frictional interaction of the moving solvent with the coil as it begins to extend and become more free-draining – resulting in a run-away process. Hysteresis is predicted in the coil-stretch-coil cycle. Similar ideas were advanced in the same year by Hinch [7]. An enormous increase in extensional viscosity would be anticipated for the stretched out molecules.

Brestkin used a non-linear dumb-bell model with varying friction coefficient on extension [8]. This method incorporated a system of spherical coordinates to avoid the Peterlin approximation used by de Gennes. The outcome was again an S-shaped transition (i.e. showing hysteresis) critical in velocity gradient.

Magda et al. used a computational scheme for the distribution function of a beadspring model with conformation dependent hydrodynamic interaction [9]. Their model also predicts a coil-stretch transition in steady extensional flow, with hysteresis in the value of extensional viscosity. This is in agreement with the dumb-bell models of de Gennes [6] and Hinch [7].

Wiest et al. modelled polymer molecules as bead-spring chains with finitely extensible non-linear elastic springs, using the Peterlin approximation [10]. For a constant high strain rate, they observed a gradual transition from the coiled equilibrium chain to the stretched state, via locally deformed states. S-shaped curves were not observed, and these were attributed to a non-physical anomaly which arises as a consequence of the use of the Peterlin approximation for a dumb-bell [11]. The inclusion of hydrodynamic interaction did not qualitatively change the molecular response.

Liu used a stochastic approach as a basis for a Brownian dynamics simulation for a freely jointed bead-rod chain, with constant hydrodynamic inter-

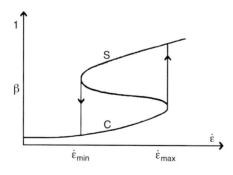

Fig. 7.1. Hysteresis of the extension (β) as a function of strain-rate ($\dot{\varepsilon}$) (taken from [6])

action [12]. Liu found no evidence of an S-shaped (hysteresis) curve for the mean square end-to-end distance ($\langle r^2 \rangle$) as a function of elongation rate, but by neglecting the changing friction coefficient on extension, this would not be totally unexpected. This work also attempted to calculate the extensional viscosity as the chain is stretched out. The models predicted that the viscosity would increase roughly linearly with N, the number of flexible links.

Darinskii et al. simulated the motion of a dumb-bell with a conformation dependent friction coefficient, using the Brownian dynamics technique [13]. In contrast to Wiest et al. [10] they observed the S-shaped (hysteresis) curve for the extension ratio of the dumb-bell as a function of the velocity gradient (in agreement with de Gennes), when simulation times were comparable with the transition times.

Rallison and Hinch modelled the extensional viscosity for a FENE dumb-bell with and without a conformation dependant friction [14]. Without conformational friction dependence they again predicted extensional viscosity to scale with N, and including the conformational dependence they predicted that the viscosity would scale with N^3. Only in the latter case was hysteresis predicted in the stretching behaviour.

Hinch has reported computer simulations of chain uncoiling via "kinks dynamics" [15]. Such a process could provide high birefringence levels at relatively low molecular strains. However, these flow simulations were carried out at extremely high stretching rates, and involve timescales for which local segmental orientation is likely.

There is clearly still no general consensus by theory or modelling as to the criticality and hysteresis of the coil-stretch transition. Neither has a single picture of the mode or the final extent of stretching emerged. Furthermore there is no universal agreement about the anticipated increase in the extensional viscosity.

7.3
Experimental Realization

7.3.1
Elongational Flow Devices

To extend a long flexible chain in solution by flow the fluid must be persistently extending, i.e. the extensional component of the velocity gradient should dominate over the rotational component [16]. Preceding theory has also anticipated that, in order to fully extend a random coiled molecule, the molecule should be deformed at a stretching rate high enough to exceed its rate of relaxation, i.e. the strain-rate should be higher than the critical value:

$$\dot{\varepsilon}_c \tau > 1 \tag{7.1}$$

where τ is the longest relaxation time of the molecule [6, 7]. The coil to stretch transition has been predicted as sudden because of the hysteresis of molecular

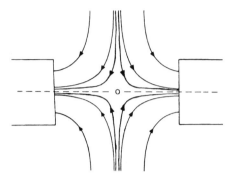

Fig. 7.2. A schematic view of the opposed jets and the flow-field created between them

relaxation time with chain extension. Furthermore, the coil should remain in the flow-field at $\dot\varepsilon > \dot\varepsilon_c$ for a long enough time so that it can accumulate the required molecular strain.

Flows which are purely extensional can be realised by uniaxial extension and pure shear. Frank et al. [18] suggested that localised areas of longitudinal velocity gradients could be produced by two opposed jets. The system consists of two cylindrical jets, which are immersed facing each other in a polymer solution. The fluid is then sucked simultaneously through both jets. A schematic of the flow field is given in Fig. 7.2. A stagnation point (denoted "o" in Fig. 7.2) exists at the centre of the system, where the velocity is zero. Along the axial region, a fluid element will be accelerated from the central zone towards the jet entrance. Inside the jet, the velocity becomes constant in the axial direction (i.e. Poiseuille flow).

The strain rate in the jets system is controlled by the flow rate achieved through a continuous pumping device. The value of the strain rate is determined from the volume flow rate and the macroscopic geometry of the system. To a first approximation, the flow field created by sucking fluid into two opposed jets can be considered to be uniaxial extension with a nominal strain rate, $\dot\varepsilon_{nom}$, at volume flow rate, Q, defined as [19–21]

$$\dot\varepsilon_{nom} = \frac{Q}{Ad} \tag{7.2}$$

where A is the area of each jet and d is the jet separation. A pure uniaxial extensional flow would have a constant strain rate throughout.

The flow flow field created is simple uniaxial extension to a first approximation, and is described by the strain rate tensor

$$\dot\varepsilon_{ij} = \begin{vmatrix} S & 0 & 0 \\ 0 & -\dfrac{1}{2}S & 0 \\ 0 & 0 & -\dfrac{1}{2}S \end{vmatrix} \tag{7.3}$$

Velocities at the jet entrance can be very high, whilst the fluid is forced to accelerate from the stagnation point at the centre of the symmetry axis. As discussed in the introduction, a second condition must be realised if a polymer molecule is to achieve full chain extension in an extensional flow field. The flow field must provide a sufficient strain rate and maintain it for a long enough time for a given fluid element to attain the required strain. A strain of $100\times$ or more may be necessary to stretch a long, highly flexible polymer. However, if the molecule slips with respect to the fluid during the uncoiling process (i.e. the molecular deformation may be *non-affine*), then the fluid strain would need to be even greater. The presence of a stagnation point provides long residence times for the stream lines that pass close to it. Therefore a region of quasi-steady-state flow (QSSF), where steady-state molecular conformations can be achieved, surrounds the stagnation point [21].

By making the jets slit shaped with the slit directions parallel, the flow field corresponds to pure shear and is described by the strain rate tensor

$$\dot{\varepsilon}_{ij} = \begin{vmatrix} S & 0 & 0 \\ 0 & -S & 0 \\ 0 & 0 & 0 \end{vmatrix} \tag{7.4}$$

However, this can be adequately approximated by the cross-slot device of Scrivener et al. [22]. Figure 7.3 shows the cross-slot system of Miles and Keller [23], constructed from four glass blocks. The arrows indicate the fluid direction and the pure shear flow is realised along the outgoing direction within the central cross-over region. The exact centre is again, a locality of zero velocity (stagnation point, or more property stagnation line).

The "four-roll mill" of Taylor [24] is a system which also produces a pure shearing flow and is shown in Fig. 7.4. The strain rate is obtained by a knowledge of the rotation speed of the rollers and the scale of the system.

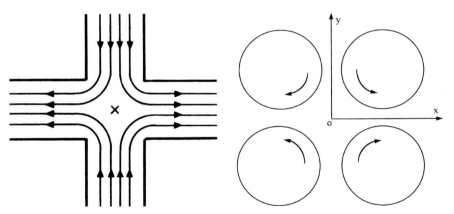

Fig. 7.3. A schematic representation of the cross-slots device used to create a planar extensional flow

Fig. 7.4. A schematic view along the rollers of the G.I. Taylor 4-roll mill

Each apparatus has unique advantages and disadvantages. The jets offer very high strain-rates at low Reynolds number with relative freedom from wall effects, but the cylindrical symmetry makes the detailed assessment of birefringence problematic. The cross-slots offer easier birefringence measurements, but the effects of walls are likely to be pronounced. In the four-roll mill the developed flow field has a more uniform strain rate throughout, than for both the jets and the cross-slots (which contain stationary surfaces and sharp corners at the edge of the flow). Pure shear does not occur where the circular roller surfaces cannot match the required hyperbolic streamlines. Turbulence sets in at high velocity gradients and therefore the mill tends to be used for low strain rate work. The optical path length of the mill is many times that of the jets, which may lead to advantages with weakly birefringent materials.

7.3.2
Assessment of Orientation and Stretching

Experimentally, the coil-stretch transition has been followed by measuring the birefringent intensity against strain rate in a well defined extensional flow field. Early work was carried out by Keller and co-workers [19–21]. Notable studies have also been made by Fuller, Leal and co-workers [25–27] and by Brestkin and co-workers [28].

A typical experiment involves scanning through a range of strain rates and simultaneously observing the region of extensional flow through crossed polarisers at $\pm 45°$ to the axis of extension. Any subsequent extension and alignment of optically anisotropic macromolecules along this axis will produce a birefringent signal. The measurement of the retardation within the birefringent region will provide quantitative information on the degree of molecular orientation. The measurement can be made directly with a compensator, which provides details of the absolute value of the orientation. However, since changes in retardation are being considered, it is sufficient to record the intensity of transmitted light between crossed polars as a function of strain rate for a given experiment. In dilute solutions retardations are much less than λ and the recorded intensity is proportional to the square of the retardation. Hence a plot of I (or $I^{1/2}$) vs $\dot{\varepsilon}$ indicates orientational variations with strain rate.

7.3.2.1
Birefringence Observation and Measurement

For a more detailed analysis, a detection system providing information about line widths and local intensity variations across the birefringent region is required. This may be achieved by scanning a sharply focused beam across the birefringent line and using a photodiode or photomultiplier tube (PMT) to record intensity as a function of position. The retardation profiles are then fitted to a Gaussian distribution, which can be analytically Abel transformed to yield the radial distribution of birefringence. Fuller and Leal [26] and Cathey and

Fuller [25] have used this method in the four-roll mill and the opposed jets respectively, with the Fast Polarisation Modulation (FPM) technique for improved sensitivity in the measurement of retardation. They report profiles that are approximately Gaussian in shape. Focused laser scanning coupled with FPM and the use of the Abel transform has also been used by Nguyen et al. [29].

However, this method has drawbacks. Mechanical scanning means that an instantaneous observation of the whole flow field is not possible. The deconvolution procedure contains uncertainties, since the laser beam profile must be estimated, and this results in reduced resolution. Since Fuller and Leal [26] used a width of between 30–50% of the deconvoluted line width, such errors may be considerable, although it would be of less significance for Cathey and Fuller [25], who used a beam width ≈ 6% of the line width. This sharp focusing of the beam means that it is non-parallel as it passes through the birefringent region, adding further complexity to the resultant signal analysis. Cathey and Fuller did not discuss this point. Even when the laser is relatively narrow compared to the line width, it may significantly modify the resulting profiles when the line is non-Gaussian due to flow perturbation. The relevance of this has been discussed by Carrington and Odell [30].

An alternative to the scanning method is the direct imaging technique, which uses a stationary detector containing many independent segments measuring the incident light intensity. If an image of the birefringent line is focused onto the plane of such a detector, line profiles may be digitally recorded.

A typical optical train for birefringence observation using the direct imaging technique, together with a flow control system is shown in Fig. 7.5. The laser beam is focused down to pass between the jets, such that the resulting beam diameter is still large enough to span the region of extension. The use of a cylindrical lens to further focus the horizontal direction gives an even greater illuminating intensity, and this method has been found to be particularly useful for projecting the signal onto the linear CCD array. Residual birefringence (e.g. from the cell windows) can be removed with a Sénarmont compensator.

7.3.2.2
Theoretical Calculations of Maximum Birefringence

In order to assess the degree of stretching or orientation from optical anisotropy, one must estimate the maximum birefringence (corresponding to perfectly oriented chains). This is an area fraught with difficulty.

The intrinsic birefringence of a fully stretched molecule, Δn_o, may be calculated by using the following equation [31, 32]:

$$\Delta n_o = \frac{2\pi}{9} \frac{(n_o^2 + 2)^2}{n_o} \frac{N_A c}{N_{m/s} m} (\alpha_1 - \alpha_2) \tag{7.5}$$

where n_o is the mean refractive index of the polymer, N_A is Avogadro's number, c is the concentration, $N_{m/s}$ is the number of monomers per segment, m is the molecular weight of a single monomer unit and $(\alpha_1 - \alpha_2)$ is the optical ani-

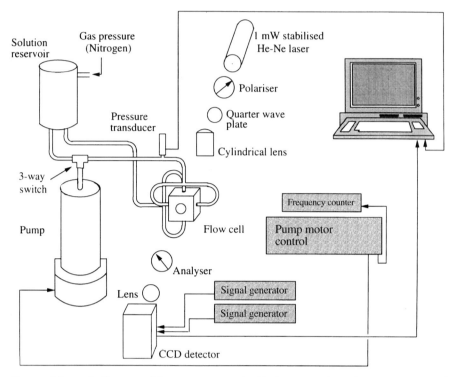

Fig. 7.5. Schematic diagram of the apparatus used with the opposed jets for measurement of extensional birefringence and flow resistance

sotropy of a segment. For atactic poly(styrene) (aPS), the most widely studied polymer, n_o is quoted as 1.6 [33] and $(\alpha_1 - \alpha_2)$ is quoted as $- 145 \times 10^{-25}$ [32]. Δn_o depends on the concentration of polymer in solution, but not on the molecular weight of the polymer. For aPS, $m = 104$ and $N_{m/s} = 7.9$ in an ideal solvent [34].

Peterlin has pointed out that this expression neglects any contribution from the form anisotropy [35]. Form birefringence arises if the refractive index of the macromolecule differs from that of the solvent, creating a directional dependence of the velocity of light through the medium as a whole. Flow birefringence measurements are therefore composed of contributions from intrinsic and form birefringence. The form anisotropy acts to increase the overall birefringence (i.e. make it more positive). The form anisotropy can have a significant effect in decalin solutions. Including the form birefringence the theoretical maximum birefringence for aPS in decalin is predicted to be $0.078\,c$ [36, 37].

However, the form effect may not be the only complication. Frisman and Dadivanian have measured the segmental anisotropy, $(\alpha_1 - \alpha_2)$, of a series of macromolecules in different solvents by streaming birefringence and point out that solvent orientation can make a significant contribution to overall birefringence [38].

Clearly the prediction of birefringence values of fully stretched molecules in solution is a complex subject. Furthermore it is important to note that birefringence shows us the mean segmental orientation, rather than the degree of stretching. To estimate the latter we must provide some model of the stretching process. Other techniques are discussed below which hold the promise of more direct assessment of the shape of the stretching macromolecule.

7.3.2.3
The Retardation – Birefringence Transform for Cylindrical Symmetry

As was seen earlier the jets have experimental advantages but, because of the cylindrical symmetry, it is necessary to deconvolute the central maximum birefringence (Δn_o) from the observed projected retardation behaviour.

An Abel transform has commonly been used to calculate the central birefringence [29]. The use of the Abel transform poses problems; previous workers have fitted a Gaussian distribution which enables an analytical solution to the transform [25], but reduces to the long-standing practice of dividing the retardation by the line width [30, 36]. Furthermore, as will be shown later, the assumption of a Gaussian fit can be seriously flawed.

For these reasons, a numerical technique has been used by Carrington et al., which can handle any arbitrary distribution of birefringence around the stagnation point as long as it has cylindrical symmetry (as it must) [36]. By an iterative procedure, the true birefringence profile is generated. Tatham has assessed the performance of the iteration procedure by using a noise-free Gaussian test function, and found the error in $(\Delta n)_{max}$ to be 0.69% and that in $(\Delta n)_{width}$ to be 0.64% [39].

7.4
Chain Stretching in Dilute Solutions

7.4.1
The Coil → Stretch Transition

Experimental results for dilute solutions of flexible linear polymers generally show no evidence of molecular extension until a critical strain rate, $\dot{\varepsilon}_c$, is exceeded. The subsequent birefringent signal is observed as a highly localised, narrow line along the central out flow axis (Fig. 7.6). The narrowness of the line arises because only fluid elements passing near the stagnation point provide sufficient residence times to allow the accumulation of the necessary strain for chain extension. This localisation within the flow-field interior has led to the assumption that non-uniform flow at the walls can be neglected.

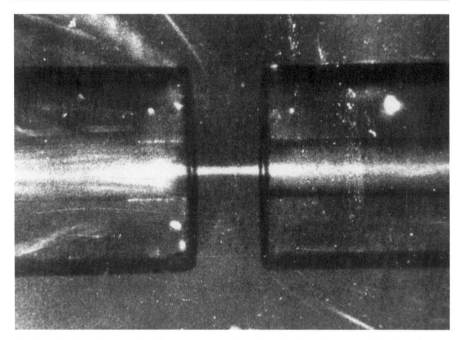

Fig. 7.6. The birefringent line between opposed jets for a 0.1% solution of aPS

7.4.2
The Functional Dependence of Molecular Relaxation Time (τ)

Assuming perfect criticality for the coil-stretch transition, a perfectly mono-disperse, flexible polymer in dilute solution ideally should show a "step function" in Δn vs $\dot{\varepsilon}$ behaviour, though it should be noted that de Gennes envisaged both pre- and post-transition stretching (see later). At $\dot{\varepsilon}_c$ there is expected to be a dramatic increase in birefringence as the molecules extend rapidly to practically full extension, giving a final plateau. Pope and Keller [19] first generated experimental curves which showed this behaviour, using an aPS fraction of $\bar{M}_w = 2 \times 10^6$. These data allowed a value of τ to be determined, from the approximation $\tau = 1/\dot{\varepsilon}_c$. A direct proportionality for the dependence of τ on the solvent viscosity has been found to exist by Odell et al. [40]. This is in accordance with both the Rouse free-draining model and the Zimm non-free-draining model.

Farrell and Keller [41] and Fuller and Leal [42] used various different narrow molecular weight fractions of aPS in dilute solution, to obtain a series of τ values. The experiments revealed that τ was approximately proportional to $M^{1.5}$, which is consistent with the Zimm non-free-draining model. These experiments with aPS have been repeated in a wide variety of solvents, details of which are given in Table 7.1.

Table 7.1. Experimental determination of the scaling of $\dot{\varepsilon}_c$ on molecular weight for aPS fractions

Solvent	Solvent Quality	Exponent	Authors' Reference
Decalin	θ-solvent	1.5	Odell et al. [40]
Decalin	θ-solvent	1.5	Narh [113]
Polychlorinated biphenyl	Good	1.5	Fuller and Leal [42]
Toluene	Good	1.5	Odell et al. [40]
Toluene	Good	1.5	Menasveta and Hoagland [114]
Benzyl benzoate	Good	1.5	Narh [113]
Tricresyl (tolyl) phosphate	Good	1.5	Cathey and Fuller [25]
Dioctyl phthalate	θ-solvent	1.5	Cathey and Fuller [25]
Bromoform	Good	1.5–1.8[a]	Brestkin et al. [28]
Decalin	θ-solvent	1.5	Nguyen et al. [29]
Toluene	Good	1.8	
1-Methyl naphthalene	Good	1.77	

[a] Note that Brestkin et al. [28] claimed to have determined a higher exponent, but it has been suggested that this has arisen due to the way in which the data was evaluated.

In most of these works, the same power law relationship is obtained:

$$\tau \propto \eta_s M^{1.5}/(kT) \qquad (7.6)$$

where η_s is the solvent viscosity, k the Boltzman constant and T the absolute temperature. This is surprising, since a good solvent expands the polymer coil and makes it more free-draining, giving the expectation of an increase in the M exponent (to 1.8 in the limit of no hydrodynamic screening). Odell and Keller [43] have found that the 1.5 power law relationship also applies to another flexible polymer, poly(ethylene oxide) (PEO).

Rabin et al. [44] have given some theoretical attention to the apparent universality of the 1.5 exponent, based upon the hypothesis that the transition probes the dynamics of the partially stretched chain, but the matter is not resolved.

7.4.3
Combinations of Rotational and Extensional Flows

Dunlap and Leal have systematically examined the effects of combinations of rotational and extensional flows [45]. This was achieved by the use of the four-roll mill, but with independent motor control of the speeds of rotation of the rollers. This enabled combinations of shear and pure extension characterised by the tensor

$$v = \gamma \begin{vmatrix} 0 & 1 & 0 \\ \lambda & 0 & 0 \\ 0 & 0 & 0 \end{vmatrix} \cdot r \qquad \text{for: } -1 < \lambda < +1 \qquad (7.7)$$

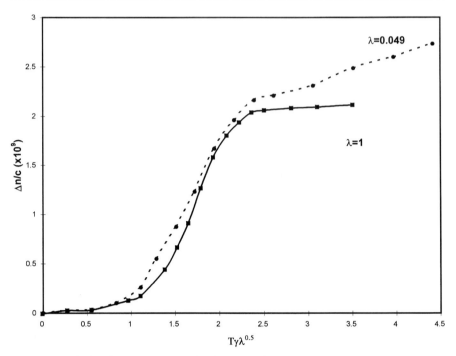

Fig. 7.7. Specific birefringence Δn/c vs the dimensionless eigenvalue of the velocity gradient tensor, $T\gamma\lambda^{0.5}$ for 0.15% $M_w = 2 \times 10^6$, $M_w/M_n = 1.3$ (a-ps) dissolved in 40% tricresylphosphate and 60% chlorinated paraffin (from [45, 65])

where v is the velocity at the point r, γ is the strain rate and λ is a parameter which characterises the flow type, ranging from simple shear (− 1) to pure shear (+ 1). The region of strong or persistently extending flows corresponds to $\lambda > 0$. Leal investigated this region and showed that the coil-stretch transition occurred for all values of λ greater than zero, but at a strain-rate that was a function of flow type. The onset of birefringence corresponded to $\tau\gamma\lambda^{0.5} \approx 1$. They further demonstrated that the transitions could be fitted to universal curve (Fig. 7.7) when plotted against the non-dimensionalised eigenvalue of the velocity gradient, $\tau\gamma\lambda^{0.5}$, where τ is the conformational relaxation time of the macromolecule.

7.4.4
The Evolution of Molecular Strain Around the Stagnation Point

Carrington et al. [36, 37] have undertaken a detailed fluid flow analysis of the jets geometry. The use of a CCD detector coupled with a numerical transform enabled the investigation of the spatial development of retardation and true birefringence around the stagnation point with increasing strain-rate. The observed birefringence is analysed in terms of molecular strain. This enables a comparison of the molecular strain with the fluid strain. This has been compared with expectations from the non-linear hydrodynamic friction dumb-bell model of Hinch [7] and Harlen et al. [46], particularly with regard to the affinity of deformation of the macromolecule as a function of molecular weight and solvent type [36, 37].

At all but the lowest concentrations, pronounced screening effects are observed, where the local increase in extensional viscosity produces significant flow modification with a reduced strain-rate around the stagnation point. This is illustrated by Fig. 7.8, which shows the change in retardation and birefringence profiles with strain-rate for a 0.02 % $M_p = 8 \times 10^6$ aPS/decalin solution. As the strain-rate increases, the retardation profiles increase in height and width. However, the CCD profiles show detail which has not been previously available from other detection methods. At higher strain-rates, the initial peak profile of the retardation traces starts to develop a flat top. The retardation profile at $\dot{\varepsilon}_{nom} = 5500\,\text{s}^{-1}$ has steep sides leading to a substantial plateau, indicating an almost "block-like" area of retardation. A further increase in strain-rate causes the flat tops of the retardation profiles to evolve into a central dip. The true birefringence recovered from the numerical transform shows even more striking dips around the stagnation point, often approaching zero. The screening is particularly strong for high molecular weights and good solvents.

As concentration and molecular weight are reduced, so the screening effect diminishes. Figure 7.9 shows the retardation and birefringence profiles for a 0.01 % $M_p = 4 \times 10^6$ aPS/decalin solution. The single peak nature of the profiles throughout the strain-rate range is very clear. It appears the flow modification is reduced to almost zero at this molecular weight. It is notable that the concentration required to see such truly dilute behaviour is of the order of $c*/100$, depending upon the method of estimation of $c*$.

Figure 7.10 shows the Δn vs $\dot{\varepsilon}$ plot for the same solution. The plateau in birefringence occurs at a strain-rate of $\dot{\varepsilon}_{nom} \approx 20,000\,\text{s}^{-1}$, corresponding to a value of $\Delta n_{max} = -(1.12 \pm 0.02) \times 10^{-5}$. This final plateau level of birefringence is rather greater than the theoretical values expected for fully oriented polymers, including form birefringence but subject to the uncertainties of solvent effects. It has been suggested that the contribution to the optical anisotropy due to solvent effects for decalin may be negative in sign, so that the theoretical maximum birefringence would more closely agree with the experimental maximum birefringence [36].

At such extremely low concentrations a well behaved, localised birefringence is observed. The onset of stretching shows a critical strain-rate and sigmoidal profile with increasing strain-rate. Figure 7.11 shows the birefringence as a function of strain-rate for the 0.01 % $M_p = 1.9 \times 10^6$ aPS (filled squares). The solid line in Fig. 7.11 represents a simulated transition, assuming perfect criticality and that strain-rate scales as $M^{-1.5}$, derived from the cumulative molecular weight distribution as assessed by GPC. The width of the transition is consistent with the known polydispersity of the polymers used.

Retardation

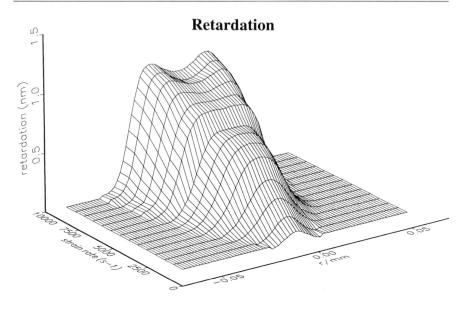

Birefringence

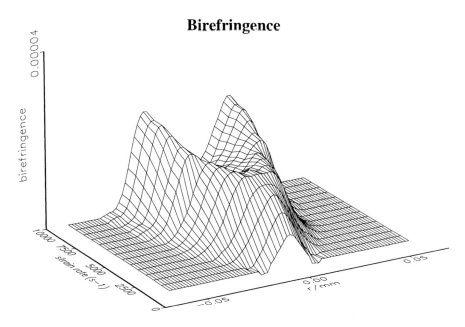

Fig. 7.8. The development of the retardation and birefringence profiles with strain rate, for a 0.02% $M_p = 8 \times 10^6$ aPS/decalin solution, in the opposed jets. $r = 0$ is the stagnation point

Retardation

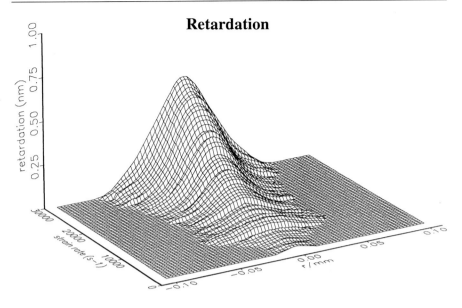

Birefringence

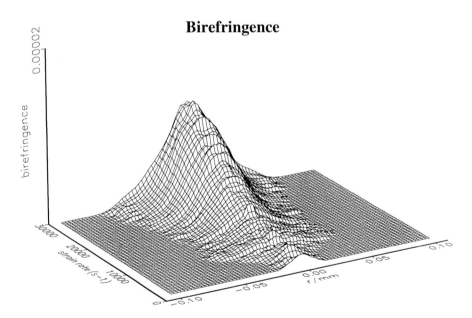

Fig. 7.9. The development of the retardation and birefringence profiles with strain rate, for a 0.01% $M_p = 4 \times 10^6$ aPS/decalin solution

Birefringence   **Retardation (nm)**

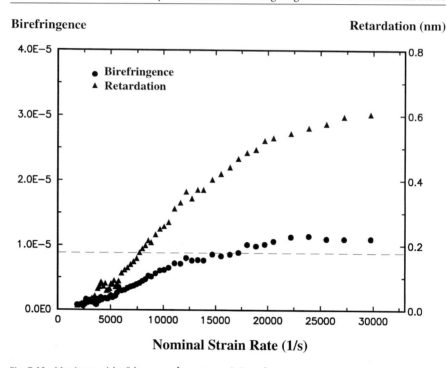

Nominal Strain Rate (1/s)

Fig. 7.10. Maximum birefringence data Δn_{max} (●) and maximum retardation data (▲) as a function of strain rate for a 0.01 % $M_p = 4 \times 10^6$ aPS/decalin solution. The dotted line represents the theoretically calculated value of maximum birefringence

7.4.4.1
The Derivation of Molecular Strain from Birefringence

Birefringence is a measure of the segmental orientation, rather than the molecular strain. However, the stretching of a molecule in an extensional flow field is likely to produce birefringence that is a monotonic function of molecular strain. The two parameters can be related if information concerning the most probable orientation of segments during molecular uncoiling is available.

Treloar [31] has described a model, based on rubber elasticity theory, which derives the optical properties of a strained network (and is applied here to extensional flow). Consider a single flexible chain consisting of n segments, each of length, ℓ, between two cross-links. The chain is held with its ends separated by a vector distance, r. Kuhn and Grün [47] derived the statistical distribution of link angles, measured with respect to the direction of vector, r. The longitudinal and transverse polarisabilities of the chain are denoted by γ_1 and γ_2 (respectively), and a relationship between optical anisotropy of the chain and the end-to-end distance, r, is shown to be

$$(\gamma_1 - \gamma_2) = n(\alpha_1 - \alpha_2)\left\{1 - \frac{3r/n\ell}{\mathcal{L}^{-1}(r/n\ell)}\right\} \qquad (7.8)$$

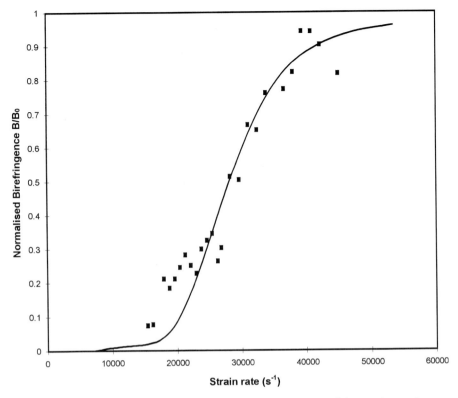

Fig. 7.11. Maximum birefringence data normalised by an estimate of the maximum theoretical birefringence value for fully stretched molecules plotted against nominal strain rate. 0.01% $M_p = 1.9 \times 10^6$ aPS/decalin solution, in the opposed jets apparatus. The solid line represents a simulated transition derived from the cumulative GPC trace (courtesy of Polymer Laboratories)

where α_1 and α_2 are the longitudinal and transverse polarisabilities of a single segment. $\mathscr{L}^{-1}$ is the inverse Langevin function.

The application of the Lorentz-Lorenz formula and a numerical approximation to $\mathscr{L}^{-1}$ results in

$$\frac{\Delta n}{\Delta n_o} = \frac{3}{5}\left(\frac{r}{n\ell}\right)^2 + \frac{1}{5}\left(\frac{r}{n\ell}\right)^4 + \frac{1}{5}\left(\frac{r}{n\ell}\right)^6 \tag{7.9}$$

Equation (7.9) is shown in graphical form by Fig. 7.12 and can be solved for the fractional extension, $r/n\ell = \beta$, using the Newton-Raphson method, providing Δn_o is known. The strain, t, is then given by $t = \beta t_{max}$ where t_{max} is the strain to full extension, i.e. the contour length divided by the end-to-end distance for the Gaussian chain.

This model implies that the mean molecular segmental orientation (denoted by the end-to-end separation, r), during each stage of uncoiling in the flow

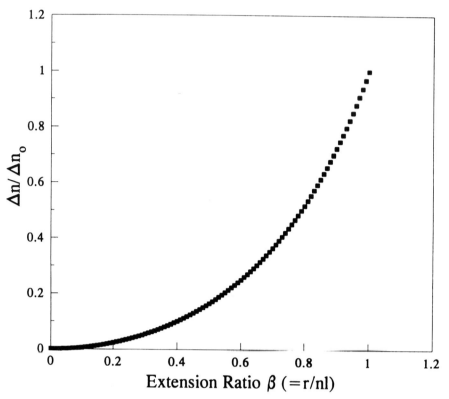

Fig. 7.12. Normalised birefringence vs fractional molecular extension (as modelled by Treloar [31])

field, is consistent with that of a molecule in stationary solution whose ends are fixed at a distance, r, apart. This is unlikely to be strictly correct because extra factors are produced by the moving solvent, which counter the maximum entropy description. Therefore, this model is likely produce an overestimate of t. Several other authors have formulated models of the uncoiling process. Peterlin [48] has calculated birefringence vs extension results similar to those of Treloar. The yo-yo model of Ryskin [49] or the kinks model of Hinch [15] are likely to produce birefringence vs extension curves very different to that of the Treloar model.

7.4.4.2
Flow-Field Modelling

In order to assess the strain rate history to which the polymer molecules have been subjected, a detailed knowledge of the flow field is required. This information allows the fluid strain accumulated at or above the critical strain-rate to be calculated, for comparison against molecular strain.

Mackley [50] found that the theoretical streamlines for an equivalent uni-axial extensional flow did not match closely the observed streamlines. The real flow contains a "nip" point (usually situated just outside the jets), at which the flow field changes sign from convergent to divergent. This limits the fluid strain that can be accumulated on passage through the jets. Mackley attempted to model the flow field in terms of two "point sinks" placed near to each other.

Schunk et al. [51, 52] solved the Navier Stokes and continuity equations by finite element analysis, and produced numerical simulations of the opposed jet flow field. They reported reasonable agreement with experimental measurements for Newtonian fluids at low Reynolds numbers (Mikkelsen et al. [53]).

Carrington et al. [36] have performed finite element analyses of the flow around the stagnation point for both potential and viscous flows, using finite element programs. Figure 7.13 shows the streamlines calculated for the region around the stagnation point (bottom left). They are appreciably different from those expected for a uniaxial extensional flow-field.

A contour plot of $\dot{\varepsilon}_{zz}$, the strain rate in the z-direction, in the region between the capillary walls has also been generated (Fig. 7.14). Previously, $\dot{\varepsilon}_{zz}$ has necessarily been assumed to be constant throughout the region between the jets, with an approximate value given by the nominal strain rate, $\dot{\varepsilon}_{nom}$. The contours are labelled with the value of the ratio between the real $\dot{\varepsilon}_{zz}$ and $\dot{\varepsilon}_{nom}$. The plot reveals that $\dot{\varepsilon}_{nom}$ is not reached at the stagnation point and that $\dot{\varepsilon}_{zz}$ can vary significantly across the region. Along the centre line, $r = 0$, the strain rate increases as the fluid element flows towards the capillary from the stagnation point, and then decreases when the fluid element approaches the capillary entrance. This indicates significant departures from uniaxial extension.

Figure 7.14 shows a decrease in $\dot{\varepsilon}_{zz}$ with radial distance, r, from the stagnation point. Therefore, an incoming molecule along the central streamline experiences an increasing strain-rate as the stagnation point is approached, until at a particular r value (r_o), the strain-rate becomes equal to $\dot{\varepsilon}_c$, the critical strain-rate. Molecular strain can begin to accumulate at this point, and therefore it is the starting point of the fluid strain analysis.

The pattern of variation of strain-rate around the stagnation point can vary greatly with individual jet geometry. The study of various different jet dimensions has shown that the controlling factor in the distribution of the strain-rates is the ratio of jet radius to separation. The higher this ratio, the closer the flow is to ideal uniaxial extension; however in practice, large ratios are difficult to achieve.

7.4.4.3
Comparison of Molecular and Fluid Strain

The strain analysis was applied to a 0.01 % $M_p = 4 \times 10^6$ aPS/decalin solution. Figure 7.15 shows the fluid and molecular strains at both high and low strain-rates. Despite the unavailability of low molecular strain information, Fig. 7.15 clearly reveals that the initial deformation is *non-affine* with the fluid strain. At a nominal strain-rate of $\dot{\varepsilon}_{nom} = 11,000$ s^{-1}, the molecular strain is appreciably

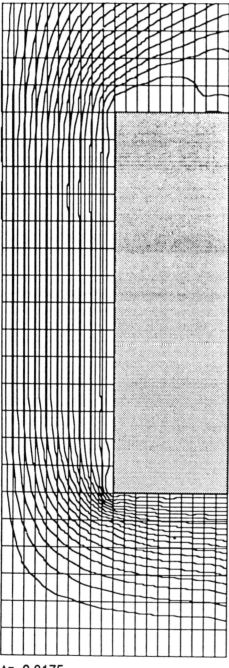

Fig. 7.13. Calculated streamlines around the jet entrance. One quarter of the field is shown, with the stagnation point at bottom left

$\Delta z = 0.0175$
$\Delta r = 0.0500$
$\Delta \psi = 0.05$

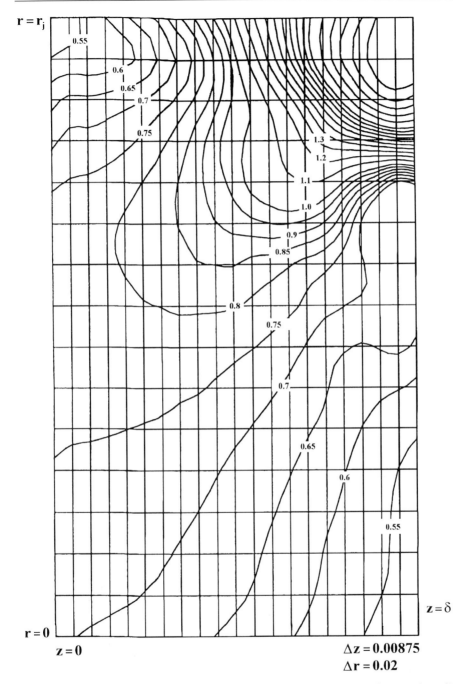

Fig. 7.14. Contour plot of strain rate calculated from the potential flow analysis, in the z-direction ($\dot{\varepsilon}_{zz}$) in the region between the jets ($0 < z < \delta, 0 < r < r_j$). The contours are labelled with the value of the ratio between the real $\dot{\varepsilon}_{zz}$ and the nominal strain rate, $\dot{\varepsilon}_{nom}$

Log ("strain")

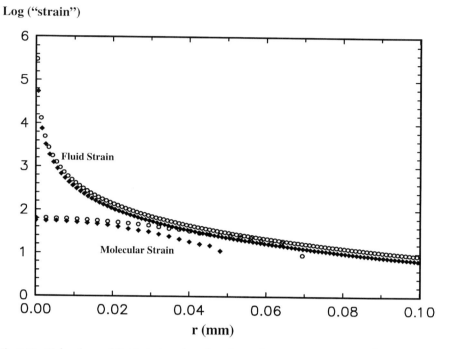

Fig. 7.15. Molecular and fluid strains plotted against r for a 0.01% $M_p = 4 \times 10^6$ aPS/decalin solution, at a nominal strain rate just above $\dot{\varepsilon}_c$ (◆ 11,000 s^{-1}) and at a nominal strain rate which shows the maximum experimental (plateau) birefringence (○ 25,500 s^{-1})

smaller than the fluid strain accumulated since $r = r_0$. This demonstrates that slip with respect to the fluid has taken place.

The region of significant molecular extension can be seen to broaden as the strain-rate increases. This effect largely arises from a higher available strain with increasing strain rate, but there is evidence to suggest a more affine mode of molecular deformation (i.e. there is less slip with respect to the fluid) at higher strain-rates.

The molecular and fluid strains have been plotted as a function of r, but it is also possible to create plots of strain as a function of residence time. Figure 7.16 is a plot of residence time vs log strain for several different strain-rates for a 0.01% $M_p = 4 \times 10^6$ aPS/decalin solution. Taking the data for $\dot{\varepsilon}_{nom} = 11,000$ s^{-1} as an example, it can be seen that the fluid and molecular strain curves are approximately parallel at the lowest molecular strains detectable, indicating an equal rate of deformation before saturation of the molecular strain occurs. The vertical separation of the two strain curves points to non-affine deformation at the beginning of molecular uncoiling. Low molecular strain data is unavailable, but at $t_{res} = 0$, the fluid and molecular strain curves must coincide by definition. The consequence of this is that the form of the molecular strain curve must be sigmoidal, which is consistent with a change in draining properties on extension [6].

Log ("strain")

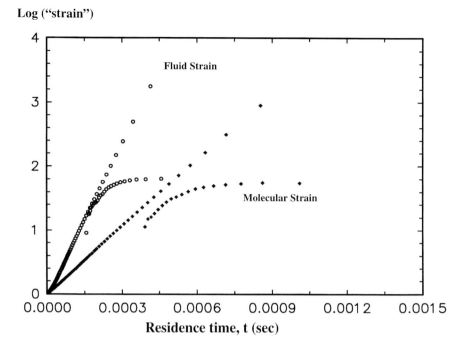

Fig. 7.16. Molecular and fluid strains plotted against residence time (t) for a 0.01 % $M_p = 4 \times 10^6$ aPS/decalin solution, at a nominal strain rate just above $\dot{\varepsilon}_c$ ($\blacklozenge$ 11,000 s^{-1}) and at a nominal strain rate which shows the maximum experimental (plateau) birefringence ($\bigcirc$ 25,500 s^{-1})

As the strain-rate increases, the vertical separation between the molecular and fluid strain curves reduces, and a mode of affine deformation is seen to initiate at lower residence times in the flow field. In fact at $\dot{\varepsilon}_{nom} = 25,500$ s^{-1}, the fluid and molecular strain curves appear to superimpose at low molecular strains, suggesting affine deformation (using the given birefringence-molecular strain relationship.

7.4.4.4
The Effect of Solvent Quality

The results obtained for an aPS/decalin solution can be compared with an aPS/TCP solution, using the same molecular weight polymer. At a temperature of 23 °C (experimental conditions), decalin is expected to be close to a θ-solvent for aPS, and TCP is expected to be a good solvent. The experimental results show that the region of high molecular extension for an aPS/TCP solution is significantly broader than the comparable region for an aPS/decalin solution.

The initial molecular conformation of aPS (prior to any perturbation by flow) will be more expanded in TCP than in decalin, which reduces the value of t_{max} (for a given molecular weight species), and the deformation should be more affine with the fluid, due to reduced hydrodynamic interaction in the more expanded coil. The outcome would be a broader region of high molecular extension in a good solvent, as compared to a θ-solvent, and a corresponding increase in the perturbation of the flow field [36].

7.4.4.5
The Effect of Chain Flexibility

Tatham et al. [39, 54] have studied the molecular behaviour of hydroxypropyl guar (HPG). HPG is an inherently less flexible molecule than either aPS or PEO, but maintains a coil-like structure (rather than being a completely rigid rod). The retardation and birefringence profiles for HPG/water solutions are found to be much broader than those for aPS, since less strain is required for HPG to attain full extension. The HPG also appears to stretch more affinely with the fluid at the start of the uncoiling process. This is consistent with predictions for a more expanded molecule, which has a less significant change in draining characteristics on extension, than would be the case for a highly flexible molecular coil with strong hydrodynamic screening.

7.4.4.6
Theoretical Simulation of Molecular Behaviour

Simulations of the response of the FENE dumb-bell model of Hinch [7] and Harlen et al. [46] to the modelled jets flow-field have been performed [36]. This model incorporates nonlinear friction on extension. The evolution of the dimensionless dumb-bell extension, R, with time, t, is given by

$$\frac{dR}{dt} = R\frac{\partial v}{\partial \zeta} - \frac{4(R-1)}{DR(1-(R^2/L^2))} \tag{7.10}$$

where D is the Deborah number (scaled so that $\dot{\varepsilon}_c$ corresponds to $D = 1$), and L is the extensibility of the dumb-bell. The first term on the right-hand side of Eq. (7.10) corresponds to affine deformation, and a "slip" term is subtracted from this. Generally, the FENE simulations reveal that as the flow strength (Deborah number) is increased, the molecular deformation becomes more affine with the fluid, and the width of the high molecular strain region increases, as shown in Fig. 7.17. This is in qualitative accordance with the experimental results.

7.4.5
The Equilibrium Stretched State

A recent paper by Nguyen et al. [29] reports experiments following the stretching of highly monodisperse atactic poly(styrene) solutions in the opposed jets apparatus. These authors report a discrepancy with earlier works and with the

Log ("strain")

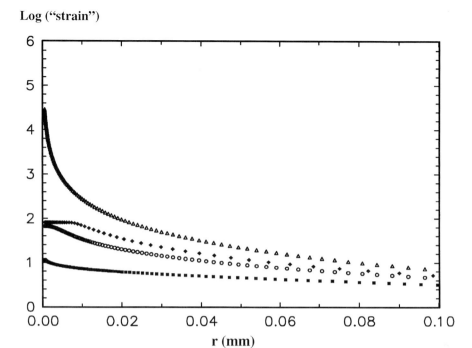

Fig. 7.17. Molecular and fluid strain profiles plotted against r, from the non-linear hydrodynamic friction FENE dumb-bell simulation for a polymer of molecular weight 8.5×10^6. ■ molecular strain at $D = 1$; ○ molecular strain at $D = 2$; ◆ molecular strain at $D = 4$; △ fluid strain

predictions of de Gennes, that the maximum birefringence is five times lower than the theoretically expected value for fully stretched chains. This has been interpreted as incomplete polymer extension. This reflects upon the state of the stretched molecule – is it essentially completely stretched, as originally envisaged by de Gennes, or does it only partially stretch?

Rallison and Hinch [14], Wiest et al. [10], Larson [55] and Hinch [15] have predicted kinks and hairpin folds during the uncoiling process, based on computer simulations. Ultimately, the completely extended chain conformation is reached, but birefringence values close to Δn_o could obviously occur at relatively low molecular strains.

The kinks dynamics simulations of chain unravelling in extensional flows [15, 55] can produce birefringence values which are close to Δn_o for comparatively low molecular strains. The form of the normalised birefringence against fractional molecular extension curve would be markedly different in the case of kinks dynamics, as compared to the Treloar function. However, the kinks dynamics simulations may not be appropriate for the flow regime that exists in the opposed jets experiments.

The flow strength can be described by the Deborah number, D, which is the dimensionless strain rate given by the product of $\dot{\varepsilon}\tau$. The Deborah number is scaled so that the critical strain rate, $\dot{\varepsilon}_c$, corresponds to $D = 1$. The results of the simulations presented by Hinch [15] are typically in the range of $W_c = 100 - 10^5$, where W_c is the chain Weissenberg number (approximately equivalent to Deborah number). In the opposed jets experiments, the Deborah numbers corresponding to the M_p of the polymer samples typically rise to $D = 3-4$ (although the Deborah numbers for the higher molecular weight components in the distribution will be larger). Thus, the flow regime in the opposed jets is not comparable with the flow fields of the kinks dynamics simulations. Note that the kinks dynamics simulations show that, as the flow strength is reduced, the proportion of chain segments that are oriented in the flow direction reduces for a given time in the flow field. Very strong flows are likely to have a significant effect on chain segments, because the segmental relaxation times are being approached.

Even allowing that the maximum theoretical value for the birefringence is achieved within the uncertainty of the assessment of this value, as other experiments show, this does not allow us to assess the degree of stretching unambiguously. In the next sections a number of alternative techniques for the measurement of stretching are discussed.

7.4.5.1
Light Scattering

Smith et al. [56] have used light scattering to assess the deformation of polymer molecules in a contraction flow (transient extensional flow). Low residence times in such flows limit the accumulation of molecular strain. The maximum available fluid strain arising from the contraction (the ratio of the maximum and minimum areas) was approximately 13.6. The maximum extension ratio for polyisobutylene (in decalin) was found to be 4.

Menasveta and Hoagland [57] have used light scattering to assess the deformation of polymer chains in dilute solution, subject to steady uniaxial extension. The study was made using a 0.01% $M_w = 20 \times 10^6$ poly(styrene)/toluene solution, in an opposed jets device. These authors claimed that the extension ratio of the polymer (defined as the ratio of the radii of gyration in flowing and static solution) was approximately 2, for $\dot{\varepsilon}\tau > 10$. The retardation intensity was found to saturate (but no absolute values were provided), and this was subsequently attributed to segmental orientation, rather than high chain extension.

The technique of light scattering would appear to be an attractive independent method of assessing molecular conformation in extensional flows. However, the authors of the above study reported discrepancies in the y-intercept of the Zimm plots with and without flow, and their empirically derived correction ratios were substantially greater (and exceeded any uncertainty) in the corresponding theoretical analysis. Thus, the interpretation of the light scattering results may not be straightforward.

As has been discussed, even at very low concentrations and especially with good solvents, the retardation and true birefringence profiles across the region of high molecular extension reveal that the stagnation point is screened by the local increase in extensional viscosity due to stretched molecules from the on-set of stretching [36, 37]. The outcome of this is a reduced degree of molecular extension around the stagnation point. The solution used in the light scattering study was also a good solvent system (poly(styrene)/toluene), but the concentration and molecular weight of the polymer sample were relatively high. Therefore, the modification of the flow field would be expected to be severe. The positioning of the scattering volume at the (perceived) stagnation point, for the light scattering experiment, may have resulted in the examination of only partially stretched molecules. The retardation profile presented by Menasveta and Hoagland [57] is claimed to be an approximately Gaussian profile (apparently indicating the absence of major flow field modification), which was obtained by moving the flow assembly relative to the incident laser beam. However, this profile may not be justified due to the poor spatial resolution of the plot, and a profile showing flow modification could easily have been masked.

7.4.5.2
Raman Spectroscopy

Cooper et al. have tried to estimate the molecular stresses in the stretched molecule around a stagnation point directly by the use of Raman spectroscopy [58]. The application of stress will result in shifts in the Raman bands associated with the polymer chain. The stresses reported were modest compared to those expected for a fully stretched molecule, which has led these authors to suggest a "kinked" molecular conformation of one third of the overall extended length. However, the solutions used were not in the dilute regime. It is also very difficult to obtain the necessary measurement precision using the Raman technique. Mechanical fracture was noted in this study and was well accounted for by the TABS theory (see Sect. 7.7) – but this requires the molecules to be extended in apparent contradiction to the "kinked" conformation. In any case, it should be recalled that the Hinch "kinks" [15] model is inappropriate for such flow regimes.

7.4.5.3
Direct Observation of DNA

Recently Larson et al. performed experiments in which the conformation of ultra high molecular weight DNA molecules (approximating to scaled up models of flexible random coils) has been directly viewed when subjected to an applied flow field [59]. These are based upon the DNA being held at one end in "optical tweezers". The DNA is stained with a fluorescing dye and this, coupled with the extreme length (up to 150 µm), enables direct optical observation of molecular conformation. Chain extension was reported to approach 90% completion, and Larson et al. point out that the light scattering results [57] are not compatible with these findings.

7.4.5.4
Conclusions

The ultimate degree of stretching of macromolecules around stagnation points is still controversial. The birefringence observed by most workers points to a high state of segmental alignment. Furthermore at high Deborah number there is almost a quantitative agreement between the fluid strain along any given stream-line and the calculated molecular strain. The maximum molecular strain (near the stagnation point) is also in agreement with the strain required to stretch the molecules from their initial random coil to the near fully stretched state [60]. Ultrahigh molecular weight DNA molecules have been directly observed to be highly stretched [59]. Indeed, it might be argued that the very observation of extreme non-Newtonian extension thickening would require a high extension. When the strain-rates are increased the universal observation is of quite precise central fracture. This can only be rationalised on the basis of a model which requires the molecules to be nearly fully stretched (see Sect. 7.7).

On the other hand, less than saturation birefringence has been reported by some authors [29] and light scattering [57] and Raman spectroscopy [58] have not yielded the results expected for fully stretched molecules.

7.4.6
Criticality and Hysteresis

As discussed in Sect. 7.2, the criticality of the coil-stretch transition is controversial from theory or modelling. Attempts at experimental establishment have proved to be ambiguous or mutually contradictory.

One possible experiment is to examine the observed width of the coil-stretch transition. From the experimentally obtained relationship $\tau \propto M^{1.5}$ and the approximation, $\tau \propto 1/\dot{\varepsilon}_c$, it can be seen that increasing the strain rate causes shorter molecules to become extended. Therefore, in an experimental strain rate sweep with a polydisperse sample, an incremental change of $\dot{\varepsilon}$ from $\dot{\varepsilon}$ to $\dot{\varepsilon} + \delta\dot{\varepsilon}$ leads to a corresponding range of molecular weight material between M and $M + \delta M$ to become extended. This effect adds to the birefringent intensity and broadens the coil-stretch transition. Thus, if the coil-stretch transition is ideally critical, the birefringent intensity vs strain rate curve represents a cumulative molecular weight distribution.

Carrington et al. [30, 36, 60] have shown that experimental extensional flow curves are consistent with the known polydispersity of narrow molecular weight fractions, but the apparent broadening of the transition disguises its nature. This is illustrated in Fig. 7.11. The major contributing factor to the apparent braodening of the coil-stretch transition is the broadening of the birefringence vs strain rate curve, due to the residual polydispersity in the sample. The study of chain relaxation from an extended state should show evidence of hysteresis, as predicted by de Gennes, if a first order transition occurs.

Nguyen et al. [29] used a comparison of extensional flow with GPC characterisation for a bimodal distribution as evidence for the claim that progressive segmental orientation causes a broadening of the coil-stretch transition, and that the critical strain rate has no physical meaning. (The integral molecular weight distributions from both techniques are shown in Fig. 7.16 in [29]).

The implication of the analyses of Carrington et al. and Nguyen et al. is that a totally monodisperse material undergoing a critical coil-stretch transition should show a step function in its strain. This was not the expectation of de Gennes, and the processes of pre-transition stretching and, perhaps more important in this context, post-transition stretching, are clearly indicated in his 1974 paper [6].

Relatively little work has been performed to attempt to detect the hysteresis of relaxation time directly. Tatham and Keeble [61] used a rapidly closing valve to stop the flow in the opposed jets apparatus. This allowed the chain retraction process to be monitored by recording the decay of the birefringent intensity. Two distinct relaxation times were reported. The first was a fast decay in birefringence ($\tau \approx \tau_{cs}$) and the second was ≈ 50 times longer. De Gennes predicted a fast relaxation time ($\tau \approx \tau_z$) from the extended state, followed by a slower decay and then a further fast relaxation at the end of the retraction process due to the coil returning of the Zimm non-free-draining regime. Tatham and Keeble were unable to confirm the third process because of noisy data at low strains.

The PEO/water system used by Tatham and Keeble gave problems because the sample was highly polydisperse and the relaxation times involved were ≈ 1 ms, making valve closure critical. An attempt to overcome these problems was made by Adams and Garnier [62], who used an aPS/tricresyl (tritolyl) phosphate system. The sample was highly monodisperse ($M_w/M_n = 1.07$) and the viscous solvent increased the relaxation time to ≈ 7 ms. The results again showed evidence of hysteresis, although the longest relaxation time upon retraction was found to be only ≈ 15 times that of the fast decay.

Dyakonova et al. [63] have used an alternative oscillating flow approach to study the nature of chain retraction. If there is hysteresis in chain retraction, the polymer molecules will not completely revert to their original degree of uncoiling in the reverse half of the oscillatory flow cycle and hence the birefringent intensity should increase with each extensional flow cycle. Two different geometries have been used to produce oscillatory flows, the first being an array of cylinders in which the fluid undergoes periodic expansion and contraction (Fig. 7.18). A dilute solution of poly(styrene) in tricresyl (tritolyl) phosphate flowed through the cell at constant flow rate and the birefringent intensity along the central axis was measured. The results show that the maximum molecular strain, in the contraction area between pairs of successive cylinders,

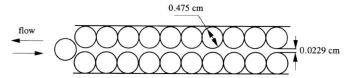

Fig. 7.18. A schematic representation of the array of cylinders. The extra cylinder on the symmetry axis provides an input stagnation point, depending upon the direction of flow

increases after 4–5 cycles to a plateau. This would be expected if hysteresis occurred in the region of the first contraction. It is noteworthy that when the flow incorporates a stagnation point at the inlet (achieved by simply reversing the flow) then levels of birefringence observed were much higher throughout the cell. This graphically illustrates the effectiveness of stagnation points at stretching molecules.

Similar results were obtained for periodic expansion and contraction of the flow achieved by pushing a polymer solution back and forth through a fixed contraction. The frequency and amplitude of the oscillatory flow were simultaneously increased and the birefringent intensity was measured. Hysteresis is indicated above certain critical flow conditions, since the intensity does not return to the background level after each cycle has ended [63].

The above evidence of hysteresis upon retraction promotes the de Gennes theory of a first order transition (Fig. 7.1) with a changing friction coefficient during molecular extension. It is clear, however, that no consensus has emerged on the issue of criticality, neither by theory nor modelling nor experiment. This represents perhaps the most intriguing open question in the subject.

7.5
Semi-Dilute Solution Behaviour

As the polymer solution concentration is increased from the dilute regime, it is to be expected that the increasing extensional viscosity will perturb the flow-field and that, ultimately, the chains would begin to overlap and hence behave as an entangled network. The change in extensional behaviour with increasing concentration was first hinted at by Pope and Keller [19]. A further systematic study was made by Odell et al. [40], using optical birefringence observations in the opposed jets apparatus. Müller et al. [64] also included flow resistance measurements and optical observations of the flow field from scattering at 90° to the optical axis, in order to detect entanglement effects in solution.

Figure 7.19 shows a typical sequence of developments for a 0.2% PEO ($M_w > 5 \times 10^6$) solution in water. The first effect observed on increasing $\dot{\varepsilon}$ for a semi-dilute solution is the onset of a narrow birefringent line (see Fig. 7.19a) along the jet (outflow) axis, which appears at a critical strain rate, $\dot{\varepsilon}_c$. This line is almost indistinguishable from the dilute solution behaviour (Fig. 7.6). An increase in concentration at this point causes a steady linear increase in τ and in the viscosity, η_s (η_s is now the solution viscosity, which increases to a first approximation, linearly with concentration). This viscosity dependence of τ follows the same trend for a dilute solution, when the solvent was varied for a constant concentration. Therefore, from this evidence, τ appears to be proportional to the viscosity of the medium in which the molecule is extending, irrespective

→

Fig. 7.19. Stages in the development of the birefringence patterns observed in semi-dilute solutions of flexible polymers as the strain-rate is increased in the opposed jets (0.2% PEO/-water $M_w > 5 \times 10^6$)

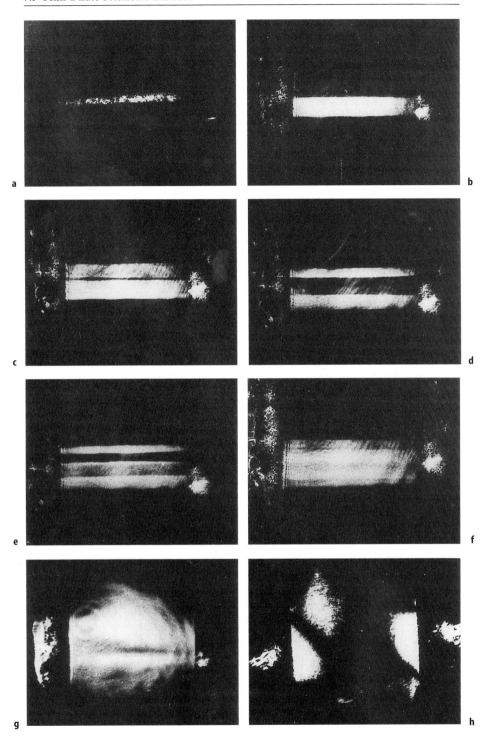

of how far this is constituted by molecules of the same or different species. This means that on a time scale of $\dot{\varepsilon}_c$, a given molecule will continue to extend as if in isolation, irrespective of coil overlap. The presence of other molecules of the same species acts as a source of viscous energy dissipation, and not as a source of physical entanglement.

Figure 7.19 is a sequence of individual stages, as the strain rate is increased beyond $\dot{\varepsilon}_c$. The highly localised birefringent line (Fig. 7.19a) is first seen to broaden (Fig. 7.19b). Note that even in the highest dilutions, such broadening is observed. However, the following sequence of effects require a critical lower concentration threshold (c^+).

As $\dot{\varepsilon}$ increases beyond $\dot{\varepsilon}_c$ a dark central line develops along the birefringent region (Fig. 7.19c), the dark line progressively broadens, both in absolute terms and relative to the overall width of the whole birefringent zone (Fig. 7.19d). The dark central line has been ascribed to a region of low birefringence (Keller and Odell [20]). The structure of the system (Fig. 7.19c, d) is a "pipe" with birefringent walls containing a non-birefringent interior. This is essentially a further development of the structures discussed in Sect. 7.4.4 and illustrated in Fig. 7.9.

A drop in Δn at the centre of the lines implies a reduction in chain extension, which in turn suggests that the molecules (in the centre of the birefringent zone) are no longer experiencing sufficient strain rate to remain fully stretched. Ng and Leal have recently highlighted the effects of concentration upon the development of birefringence around the stagnation point of the two roll mill [65]. Velocimetry shows directly the perturbation of the flow-field by the stretching molecules. They report reductions in birefringence greater than would be expected on the basis of flow-perturbation alone, and suggest the effect arises from entanglements.

Harlen et al. have studied the flow modification near a stagnation point using a FENE (finitely extensible non-linear elastic) dumb-bell model with nonlinear hydrodynamic friction [46]. They predict that pipe structures can occur at dilute polymer concentrations without direct mechanical interactions between neighbouring molecules. The process of chain extension is expected to generate massive extensional viscosities in localised regions, with an accompanying reduction in local flow velocity. Odell et al. have verified this effect experimentally in semi-dilute solution (0.2% $M_w = 4 \times 10^6$ aPS/decalin), but found no flow modification for a dilute (0.03%) solution of the same sample [21]. However, this was probably due to resolution limitations of the velocimetry technique used.

Beyond the widened pipe, a central bright line appears again (Fig. 7.19e). In some cases, it is possible to observe this central bright line developing a further dark central line within (i.e. a pipe within a pipe) (Fig. 7.19f). More usually, a further increase in $\dot{\varepsilon}$ causes the pipe system to become unstable (Fig. 7.19g).

The FENE dumb-bell with nonlinear hydrodynamic friction model of Harlen et al. [46] cannot provide an extension rate near the centre line which increases again, after dropping below the stretch-coil transition value and hence, more complex structures seen in experiments (the "pipe-within-a-pipe") are not predicted.

Beyond a yet higher critical strain-rate, $\dot{\varepsilon}_n$, a gross instability (flare) sets in (Fig. 7.19h) [20, 64]. In these circumstances, the stagnation point has been lost and the flow consists of twin unstable converging flows into each jet, shearing past each other at the centre of symmetry. This has been associated with the onset of entanglement effects, with the solution stretching as a network, though still well below any conventional value of c^*. This phase of the flow also corresponds to a great increase in flow-resistance (see Sect. 7.6). It has been hypothesised that such networks may be responsible for massive non-Newtonian extension thickening observed in circumstances where large strains may not be present, for instance in porous media flow [54].

A strong concentration dependence for this sequence of effects was found. This allows a "phase diagram" to be constructed, such as the example given in Fig. 7.20 for poly(ethylene oxide) of $M_w > 5 \times 10^6$. It can be seen that there is no limiting lower c for $\dot{\varepsilon}_c$, which extends to and levels off at infinite dilutions. The slow decrease of $\dot{\varepsilon}_c$ with c is due to the increase in solution viscosity.

At high concentrations, the $\dot{\varepsilon}_c$ and $\dot{\varepsilon}_n$ curves do not cross. Even in a highly overlapping environment, the polymer chains are extendable a individuals, provided that the experimental time scale is sufficient to allow disentanglements. The possible existence of an upper limiting concentration would be of great

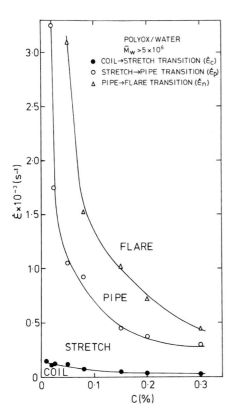

Fig. 7.20. A "phase diagram" of the development of birefringence patterns in the opposed jets as a function of strain-rate and concentration (PEO/water $M_w > 5 \times 10^6$)

interest for stretching chains in the condensed phase and for entanglement studies in the molten state.

7.6
Extensional Viscometry

Polymer melts and solutions show pronounced non-Newtonian flow behaviour. In rotational flow, polymer solutions are typically shear-thinning where the aparent viscosity reduced as the shear-rate is increased. In extensional or stretching flows, however, polymer solutions often show a marked increase in viscosity as the shear-rate is increased, often termed dilatancy.

The theoretical prediction of the magnitude of the extensional viscosity increase which can be attributed to molecular extension has been shown previously to be a subject of some controversy. Clearly, extensional viscosity is an important parameter, and so there has been much work to develop extensional viscometry techniques. Several different flow geometries have been used to investigate dilatant effects, notably the tubeless siphon, capillary entrance experiments, pulling threads and spinning techniques, and stagnation point flows [66].

However, it became apparent that the data obtained from the various experimental techniques was significantly different, with a major contributing factor being the use of different fluids. A specialist group of experimentalists decided that there was a need for the co-operative measurement of a single fluid to be undertaken [66].

The opposed jets system allows simultaneous optical birefringence and flow resistance measurements, enabling the correlation of molecular orientation with its effect upon the flow field [21, 64]. Odell et al. described the modification of a vacuum driven apparatus to include simultaneous flow resistance measurements, by monitoring the pressure drop across the jets (ΔP) [21, 64]. This data is used to determine effective extensional viscosities.

7.6.1
Newtonian Fluids

For a Newtonian fluid in a uniaxial extensional flow field, the pressure drop, ΔP_{ext}, due to viscous dissipation is given by

$$\Delta P_{ext} = \eta_e \dot{\varepsilon} \tag{7.11}$$

where $\dot{\varepsilon}$ is the nominal extensional strain rate and η_e is the extensional viscosity of the fluid. The ratio between the extensional viscosity, η_e, and the zero shear viscosity, η_s, of a Newtonian fluid (the Trouton ratio), is equal to 3 for a uniaxial flow geometry (Trouton [67], Bird et al. [68]):

$$\eta_e = 3\eta_s \tag{7.12}$$

Extensional viscosity values for any fluid can be obtained by using Eq. (7.11), and measuring ΔP_{ext} vs strain rate. The measured values of the pressure drop across the jets include contributions from other sources, in addition to ΔP_{ext}. Therefore, this technique can only be used to yield an *effective* extensional viscosity, η_e', and to allow a comparison of the behaviour of Newtonian solvents with non-Newtonian polymer solutions.

7.6.2
Dilute Solutions

Flow resistance measurements made for the aPS solutions should allow comparisons to be made with theoretical expectations for the increase in extensional viscosity due to stretched molecules, which is predicted to depend upon N to the power $1-3$ [12, 14, 69]. An assessment must be made of the effective area over which molecules experience significant extension, since a measured ΔP value will be dependent on this. Significant molecular extension is confined to a narrow region surrounding the stagnation point.

Müller et al. [64] have made measurements of the pressure drop across the jets (ΔP) as a function of $\dot{\varepsilon}$. The processing of these flow resistance measurements allows an effective extensional viscosity, η_e', to be obtained. An "effective" viscosity has to be measured since shear flows in the apparatus contribute to the overall flow resistance of the solution, and a precise assessment of this is difficult (although the shear contribution is probably insignificant when compared to typical increases in η_e). For an "absolute" extensional viscosity to be defined, the effects of non-uniform strain rates and residence times across the flow field would have to be taken into account.

The normalised extensional viscosity (H) compares the increase in extensional viscosity due to the polymer, to three times the corresponding increase in the simple shear viscosity, therefore:

$$H_{c\to 0} = \frac{(\eta_e - 3\eta)}{3(\eta - \eta_o)} \approx \frac{[\eta_e']}{3[\eta]} \tag{7.13}$$

where η_e is the extensional viscosity, η is the zero shear viscosity and η_o is the solvent viscosity. This value is thus close to the ratio of the intrinsic extensional viscosity to three times the intrinsic shear viscosity.

A correction in the measured value of $[\eta_e]$ is necessary to account for the relatively small fraction of molecules passing through the jets which are extended. Birefringence measurements may be used to assess the proportion of chains extending. Müller et al. [64] used a solution of 0.025% aPS ($M_w = 4.4 \times 10^6$, $M_w/M_n = 1.06$) in decalin, to yield a normalised extensional viscosity of 92, assuming a uniform intensity across the terminal line width. The conclusion of this work was that values of H of the order of N were unlikely to arise from the coil-stretch transition alone, and intermolecular entanglements would be necessary to provide large values of H. However, the detailing of line profiles has shown that a maximum in birefringence, and hence high molecular extension, occurs over only a restricted portion of the line. Thus, the actual contribution to the extensional viscosity by the polymer molecules may be significantly larger, and could bring measured values more in line with theoretical predictions.

Carrington et al. [30, 36, 37] introduced a linear CCD (Charged Coupled Device), to allow true birefringence profiles to be extracted from the retardation projections. As discussed in Sect. 7.4.4.1, this enables the determination of molecular extension as a function of position around the stagnation point. Such an assessment of the proportion of the stretching zone containing highly extended molecules allows a better estimation of their contribution to the observed extensional viscosity increase. The relative effective extensional viscosity obtained must thus be corrected to take into account the effective area of molecular extension.

For a 0.02% solution of 8×10^6 aPS/decalin, the correction factor for $[\eta_e']$ generated by this process is 225–332 [36, 60]. The division of $[\eta_e']_{corr}$ by three times the intrinsic viscosity ($[\eta] = 218$ ml g^{-1}) generates an estimate of the normalised extensional viscosity for aPS of approximately 2300–3400. For this aPS sample, $N \approx 8000$, and the estimate for H is of the order of N.

The use of a sophisticated birefringence analysis to provide a correction for the region of significant molecular extension has demonstrated that values of H which approach N may be obtained, but there is no firm evidence to suggest that higher powers of N may be generated.

Cathey and Fuller [25] have also used the opposed jets apparatus to study the mechanical responses of dilute polymer solutions. In their apparatus, one of the jets in the system was fixed and the other was mounted on a knife-edge fulcrum and attached to force transducer, with servo position control. The force required to maintain a constant separation of the jets gives a measure of the effective extensional viscosity. The advantage of this approach is that it only measures the force between the jets (and therefore pressure drops in the pipes do not have to be evaluated). This apparatus has become the basis of the commercial extensional rheometer, the Rheometrix RFX.

Dilute solutions of poly(styrene) in di-octyl phthalate were used and plots of η_e' vs $\dot{\varepsilon} M^{1.5}$ were constructed. For three different molecular weights, the rise in η_e' occurs at the same point on the horizontal axis, indicating that this increase is due to the coil-stretch transition (since $\tau \propto M^{1.5}$ and $\tau \propto 1/\dot{\varepsilon}_c$). The maximum normalised extensional viscosity was approximately 2 and scaled with $M^{0.8}$ (Hassager [69], Wiest et al. [10] and Liu [12] predict a linear dependence). Cathey and Fuller observed a drop in η_e' at high strain rates, which may affect this result, and this was attributed to a drop in residence times at high strain rates, restricting the maximum deformed length of the molecule [25].

Cathey and Fuller, with Schweizer and Mikkelsen (Schweizer et al. [70]) have used intensity profiles across the birefringent line and attempted to correct for non-uniform stretching in the region of molecular extension. They state that the normalised extensional viscosity would increase by a factor between 2 and 4.

7.6.3
Semi-Dilute Solutions

As the concentration increases, the development of interchain connectivity occurs and the extension behaviour is modified to show effects such as those previously described in Sect. 7.5. An effective extensional viscosity (η_e') vs strain-rate curve in this regime is given in Fig. 7.21, for the poly(ethylene oxide) shown in the phase diagram of Fig. 7.20. Complex dilatant (shear thickening) effects can be observed (curve (a)) when compared with pure solvent (curve (b)) and with the dilute regime. The broadening of the birefringent line and the initial formation of the pipe corresponds to η_e' increasing. Beyond this point, η_e' begins to drop until the flow develops an unstable nature and the stagnation point is lost. The effective elongational viscosity then increases enormously in the region of unstable flare.

A programme was instigated to correlate the results from the various elongational viscosity experiments. A test fluid was prepared at Monash University (hence the fluid's name, M1), which attempted to satisfy the constraints set by the widely different capabilities of the apparatus to be used. Nguyen and Sridhar [71] detail the preparation of the test fluid, which consisted of 0.244 % polyisobutylene in a mixed solvent consisting of 7 % kerosene in polybutene. Solution and melt viscosity measurements show the average molecular weight of the polyisobutylene to be approximately 3.8×10^6 (Laun and Hingmann [72]). Birefringence data, yielding the actual molecular weight distribution, shows that a significant fraction of the polymer has a molecular weight greater than 8×10^6 [73]. The findings of the return conference in Combloux (1989) are published in [74].

The extensional properties of M1 were studied in various laboratories and a summary of the findings is given in Table 7.2. The magnitudes of the apparent

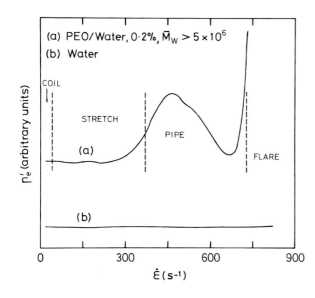

Fig. 7.21. Effective extensional viscosity plot: (a) for 0.2 % PEO/water $M_w > 5 \times 10^6$; (b) for water

Table 7.2. The extensional properties of M1 fluid, as reported by various laboratories. All data is taken from [74]

Technique	Strain Rate Range (s^{-1})	Apparent Extensional Viscosity (Pa s)
Fibre spinning	0.75–5	100–10,000
Fibre spinning	0.7–3	100–10^5
Fibre spinning	2–7	4000–20,000
Fibre spinning	1–4	20–5000
Open siphon	1–2	700–10,000
Filament stretching	20–60	10–100
Contraction flow	5–20	30000–10^8
Contraction flow	60–400	60–4000
Converging chanel	6–150	12–250
Opposed jets	0–7	100–600
Opposed jets	20–60	100–1000
Opposed jets	0.02–40	12–300
Climbing constants	–	10–20
Falling drop	14–17	10–1000

extensional viscosities measured vary over a wide range, and this exercise has highlighted the difficulty in determining absolute extensional viscosities.

The situation was now greatly simplified, as all the data were on the M1 sample. From our point of view, the salient issue was that all the various apparatuses imparted different strains to the molecules in solution, the difference being most striking between flows with and without stagnation points.

Stagnation points occur in most real flows, but do not feature in most extensional viscometry techniques. They are significant in that it is only with stagnation-point flows that there exists a region of virtually unlimited strain (near the stagnation-point). This is necessary to produce the coil-stretch transition in high molecular weight flexible molecules. Large strains are also necessary to induce effective entanglements in semi-dilute solutions, and we believe that such entanglements are fundamental to the explanation of strongly dilatant behaviour.

The effective extensional viscosity of the M1 fluid in the jets is shown as a function of strain-rate in Fig. 7.22. As the strain-rate was progressively increased, the first event was the coil-stretch transition. This is confirmed by the appearance of a narrow birefringent line passing through the stagnation point and an increase in η'_e (a). As the strain-rate was increased further, the molecules began to perturb the flow strongly, the birefringence patterns showed evidence of flow modification taking the form of pipes and the elongational viscosity further increased (b). The highest strain-rates produced birefringent flares which indicate a tangled network, with pronounced dilatancy (e). Ultimately, the network cannot comply with the infinite strains required around the stagnation point and elastic instabilities set in which destroy the stagnation point.

Another problem which arises when trying to correlate data from the various techniques is that in any geometry there is very little control over the pre-

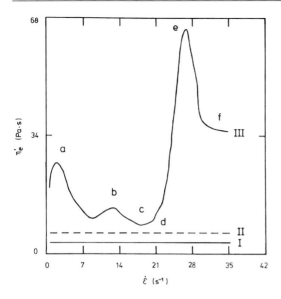

Fig. 7.22. Effective elongational viscosity of the M1 fluid as a function of strain-rate

cise strain-rates and strain histories throughout the sample. It is clear that the properties of the solution change significantly when the fluid is highly stretched. The total strain is thus an important parameter.

7.7
Thermomechanical Degradation

7.7.1
Introduction

In polymer science, the importance of chain scission is second only to that of synthesis. Chain scission is in general undesirable, except for some special cases when controlled chain scission can be used to tailor molecules for some specific requirement.

Traditonally chain scission, under the term degradation, is associated with the chemical stability of main chain valence bonds against chemical influences of the environment coupled with that of temperature. Here the activation is purely thermal, associated with bond vibrations within the chain. The same applies to processes during flow, except that here the mechanical loading of the chain raises the energy levels adding to those due to thermal factors alone, thus lowering the barrier to bond dissociation. This will promote chain breakage, or cause such breakage where otherwise its probability would be negligibly small. The major qualitative difference between the two categories lies in the distribution of the bond sites which become activated to the level of breakage. In thermal or chemical degradation they are distributed uniformly, subject only to fluctuations by Boltzmann statistics, while in flow the mechanical addition to the thermal energy is non-uniform leading to stress concentration within the loaded system requiring considerations specially addressed to this issue. The nature of this distribution will depend upon the type of flow-field and whether the chains are isolated or form an entangled network.

7.7.2
Mechanical Scission in Dilute Solutions

Stagnation-point extensional flow-fields can effectively apply a controlled stress to the isolated molecule. Simultaneously, the conformation of the molecule can be monitored. Using elongational flow techniques, both the stress and the temperature applied to the molecules can be varied in order to study their vital influence in the flow induced scission process.

As the strain-rate is increased beyond $\dot{\varepsilon}_c$ along the horizontal plateau in Fig. 7.11 (solid line), the already stretched out chains become increasingly stressed until they rupture. It is then possible to stop the experiment and start again from $\dot{\varepsilon} = 0$, remeasure the new I vs $\dot{\varepsilon}$ curve and derive the resultant molecular weight distribution. Thus the method can both break the chains and characterise the fracture products of the scission.

Odell and Keller [43] found that the fracture behaviour of dilute solutions of PEO and aPS was very similar. As $\dot{\varepsilon}$ was increased beyond $\dot{\varepsilon}_c$ no appreciable fracture occurs until a critical fracture strain-rate $\dot{\varepsilon}_f$ is reached, beyond which the molecules break almost precisely in half. This result has been predicted by a number of authors [75, 76] and might be anticipated from the centrosymmetric nature of the flow-field with respect to the molecule, resulting in a maximum stress at the chain centre. These results were qualitatively explained by applying Stokes' law to a simple stretched out string of beads [43]. The predicted molecular weight dependence of the critical fracture ($\dot{\varepsilon}_f$) based upon such a simple model is

$$\dot{\varepsilon}_f \propto 1/M^2 \tag{7.14}$$

Equation (7.14) has a further important consequence. The critical strain-rates for stretching and fracture both decrease with molecular weight, however, $\dot{\varepsilon}_f$ decreases faster (proportional to M^{-2}) than $\dot{\varepsilon}_c$ (proportional to $M^{-1.5}$). There will be a value of M at which $\dot{\varepsilon}_f = \dot{\varepsilon}_c$ indicating that at a sufficiently large M, the molecules cannot be stretched out without breaking. For aPS this value was estimated to be $M_w = 30 \times 10^6$ and again it was corroborated experimentally (see Fig. 7.13 in [43]).

Stokes' law can also be used to estimate the value of the actual fracture force required to break the chains. This calculation yields values that correspond, within experimental accuracy and the known precision of the parameters involved, to the force needed to break carbon-carbon bonds. One experimental result that remains unexplained by Stokes' law is the precision of chain halving with a very narrow distribution as opposed to the anticipated parabolic distribution. This result was only rationalised when the process of chain breakage was considered to be thermally activated.

7.7.2.1
The TABS Theory

The Thermally Activated Barrier to Scission (TABS) model is based on the assumption that the scission of backbone bonds of a linear polymer is a thermally activated process [43, 79]. A brief description of the model is presented below.

The potential energy U is described as a function of the distance between two covalently bonded carbon atoms, by the well known Morse function:

$$U = U_0 \exp[-2d(r - r_e)] - 2\exp[-d(r - r_e)] \qquad (7.15)$$

where r_e is the distance corresponding to the minimum in the potential energy, d the parameter defining the width of the minimum, and U_0 the dissociation energy.

The breakage of bonds is assumed to be a thermally activated process where the rate is determined by an Arrhenius type equation. The activation energy is a function of the force applied to the individual C–C bond.

It is assumed that bond scission occurs when thermal fluctuations overcome the energy barrier for bond dissociation and that the only role of the elongational flow-induced stress in the chain is to reduce this energy barrier from its equilibrium value. This represents the basis of the Thermally Activated Barrier to Scission (TABS) model.

The relation between the rate of scission (K_0) and the temperature (T) according to this model is given by

$$K_0 \propto \exp[-U_0/(KT) + (a/l)\,\$\,N^2/8] \qquad (7.16)$$

where $\$$ is proportional to strain-rate, N is the number of monomers, a is the stretched bond length and l is the monomer length.

Odell et al. [77] have used the TABS model to predict successfully the precision of chain halving. The model also predicts the width of the distribution of scission around the chain centre. The total scission rate is zero below $\dot{\varepsilon}_f$ (as expected) and then increases very sharply beyond this point. The criticality observed is not matched by the experimental findings, but the model did not take into account the polydispersity of the polymer, which would lead to a broader range in $\dot{\varepsilon}_f$ being covered during the increase in scission rate. Müller [78] has shown that this method gives a scaling law of $\dot{\varepsilon}_f \propto M^{-2.01}$, which is in agreement with previous results from the cross-slot apparatus (Odell and Keller [43]).

The temperature dependence of the scission rate can be predicted by the TABS theory, and Odell et al. [77] have shown that the critical strain rate for fracture should fall only quite slowly with increasing temperature. Narh et al. [79, 80] arrive at good agreement with this prediction, using a 0.02% $M_p = 8 \times 10^6$ aPS/decalin solution in the temperature range 25°C < T < 150°C.

Odell and Taylor [81] have studied the scission of DNA in the opposed jets. The advantage of this system is the absolute monodispersity of the DNA sample, which cannot be achieved using synthetic polymers. The highly accurate technique of pulsed field gel electrophoresis (PFGE) was used for molecular weight characterisation. The experimental distribution of scission products was found to be in very close agreement with the predicted distribution, obtained from the TABS model.

7.7.2.2
Transient Flow Degradation

All the work on chain scission detailed previously has been based on overstretching in a stagnation point flow (or a Quasi-Steady-State-Flow (QSSF)). However, real flow situations exist which are Fast Transient Flows (FTFs), and do not possess a stagnation point, e.g. capillary entrance flows. An FTF produces partial chain extension and poses questions concerning the shape of a partially extended molecule, its flow modifying effects and the possibility of chain fracture.

Ryskin [49] proposed that each molecule would uncoil like a pair of yo-yos connected by a single string. This results in a highly stretched central portion of the chain, with the two ends remaining almost fully coiled. Ryskin states that such behaviour is due to the roughly parabolic distribution of the stretching force over the effective hydrodynamic length of the molecule.

Rabin [82] has calculated that such a molecule could fracture (i.e. the central portion of the chain is sufficiently stressed) at low chain extension in a transient extensional flow, provided the strain rate is high enough. A prediction of the scaling law for the critical fracture strain rate is given:

$$\dot{\varepsilon}_f \propto M^{-1.1} \tag{7.17}$$

This contrasts with the M^{-2} dependence observed in stagnation point flows and expected for extended molecules. Rabin predicted that scission in an FTF would occur mostly at a higher strain rate than stagnation point (QSSF) scission, the molecules would only experience low strains ($\approx$ three times), the scission rates would be high and central scission would be precise.

Nguyen and Kausch [83] have used a piston to drive fluid through a steep contraction, creating a transient extensional flow. Experiments with bisphenol-A-polycarbonate (Nguyen and Kausch [84]) and aPS (Nguyen and Kausch [85]) gave a molecular weight dependence of $\dot{\varepsilon}_f \propto M^{-2}$. Nguyen and Kausch [86] have subsequently reconsidered their strain rate calculations and reported a different scaling law of $\dot{\varepsilon}_f \propto M^{-0.95}$, stressing the close agreement with the Rabin yo-yo model [82].

However, Rabin has predicted that below a certain molecular weight, scission in a transient flow would occur earlier than in a stagnation point flow of the same strain rate [82] (due to a lower power dependence of $\dot{\varepsilon}_f$ upon M in the transient case). Müller et al. [78, 80] have pointed out that the stagnation point flow created by the opposed jets necessarily incorporates a transient flow at its inception and at all points surrounding the stagnation point. Therefore, on the basis of the Rabin theory and the Nguyen and Kausch experiments, there should be a large increase in scission rate at high strain rates, due to transient chain scission. In fact the scission rate reaches a maximum at 30,000 s^{-1}, with the scission rate corresponding to only $\approx 10\%$ of the molecules passing through the jets. From knowledge of the birefringent line width and the jet diameter, Narh et al. [79] calculated that this was equivalent to the proportion of molecular which are highly stretched, and hence concluded that only highly extended chains can suffer scission. This directly contradicts the prediction of scission for partially extended chains.

Nguyen and Kausch [87] have investigated the effect of solvent viscosity on polymer degradation in transient extensional flow. Surprisingly, they observed that the critical strain rate for fracture showed only a weak dependence on solvent viscosity ($\dot{\varepsilon}_f \propto \eta_o^{-0.25}$). This result is unexpected in view of the fundamental predictions of molecular theories of chain dynamcis. The probability of bond scission along the chain was described by a Gaussian distribution function, and highly precise central chain scission was reported (standard deviation, $\sigma = 4\%$ in a good solvent).

Narh et al. [79, 80] have offered a tentative explanation for the transient flow results, based on turbulence being generated in the exit cavity (the conditions of the Nguyen and Kausch experiments make this eventuality likely). Merrill and Horn [88] have shown turbulence induces chain halving in dilute polymer solutions, but high molecular weight polymeric additives act to reduce turbulence. This implies that higher strain rates may be necessary to produce turbulence (and hence fracture) for high molecular weight fractions. This reasoning could explain the reduction in the power law exponent. The near independence of $\dot{\varepsilon}_f$ upon η_o may also be accounted for by turbulence. An increase in solvent viscosity would lead to an increase in the viscous pull on the molecules and, hence, a decrease in $\dot{\varepsilon}_f$. Therefore the molecular stress factor and the turbulence factor have opposite effects on $\dot{\varepsilon}_f$, and could possibly cancel each other out to a large extent.

This model of scission of the partially stretched molecule in a "yo-yo" conformation seemed to agree with the observed molecular weight exponent, but it is hard to explain the very precise halving of the chain a feature still displayed by these experiments on the basis of such a model.

The FTF results of Nguyen and Kausch would predict that below a certain molecular weight, scission would occur in an FTF at a lower strain-rate than in a QSSF of the same strain-rate (because of the reduced power dependence of $\dot{\varepsilon}_f$ upon M as compared to QSSF). This is clearly not a satisfactory situation, since a QSSF necessarily incorporates an FTF at its inception and at all points surrounding the stagnation point.

7.7.2.3
Chain Scission in Simple Shear Flow

Simple shear flow consists of pure shear and rotation. Therefore before the chain is greatly extended it has rotated into a position in which the flow is reversing that extension [89]. Whilst birefringence can be observed in simple shear flow (conventional flow birefringence experiments), it is always small even at high shear-rates.

There appear to be no theoretical reasons to expect degradation in simple shear flows of dilute solutions. The random coil polymer is predicted to adopt an elliptical shape with the principal axes of deformation rotated away from the direction of elongational rate [90] whilst the expansion of the coil is highly restricted [91]. In view of these theoretical considerations it is difficult to envisage how a simple laminar shear flow can cause chain scission in dilute solutions.

Works that claimed to show shear flow degradation used mostly high speed stirring, capillary flow at very high flow-rates or high shear concentric cylinder viscometers. Such experiments probably involve turbulent flow with its concomitant high elongational component. Some experiments have been performed at low Reynolds numbers but it is well documented that dissolved polymers can either induce or suppress flow instabilities and turbulence [64]. Indeed, experiments that have been performed with due attention to the onset of turbulence in capillary flow have confirmed that no shear degradation can be observed in unentangled solutions (dilute) before turbulent flow sets in [92].

There is little conclusive evidence regarding the concentration dependence of degradation. Some authors suggest that the scission-rate increases with increasing polymer concentration [93–95], others that it is independent of concentration [96–100]. A negative concentration dependence (i.e. decrease in scission-rate with increasing concentration) is also often found in the literature [101–104] as well as complex functions of the concentration [105–107]. This is partly due to the nature of shear degradation experiments, where often complex flow-fields and ill defined experimental conditions are used.

If turbulence is causing the chain fracture observed, increasing the polymer concentration can reduce or delay the onset of turbulence and a negative concentration dependence may be expected. Some authors [95] have argued that the true concentration dependence of scission-rate should be positive in shear flows, on the basis of intramolecular entanglements producing a stress concentration build up at the points of physical junction, often invoked in rubber elasticity theory [108]. This treatment, however, may be inadequate to describe the entanglements present in semi-dilute solutions where the overlap density is not as high as in a rubber and where the entanglements are transient in nature.

In summary, it is difficult to form a unified picture of chain degradation in simple shear flow, which is the most commonly studied in the literature of flow-induced scission. From our own background it appears that whenever scission is reported, entanglements or turbulence are likely to be involved.

7.7.3
Degradation in Semi-Dilute Solutions

One would reasonably anticipate marked differences in the degradation behaviour of stretched interacting molecules. The most obvious expectation would be the disappearance of preferential breaking near the centre of the molecule as higher concentrations are approached.

In the semi-dilute regime, the forces required to break the chain are transmitted primarily by valence bonds, i.e. network chains and junctions. It is reasonable to anticipate marked differences in the degradation behaviour of stretched interacting molecules; the most obvious expectation would be the disappearance of preferential breaking near the centre of the molecule as higher concentrations are approached. At concentrations beyond c^+ and at time scales shorter than the disentanglement time of the chains, dramatic increases in the flow resistance are observed (see Sects. 7.5 and 7.6.3).

The scission-rate in elongational flow was found to be independent of concentration in dilute solutions. For semi-dilute solutions the scission-rate was found to increase with concentration [64, 109–111]. The degradation is increasingly randomised as higher concentrations are approached [112]. Perfect random scission would produce a polydispersity index of 2. These results might be expected for a change in degradation mechanism where stress is transferred at junction points in a transient network, instead of the stress being transferred by hydrodynamic interactions of the isolated stretched molecules with the moving solvent molecules. As degradation proceeds the average molecular weight gets much lower; this corresponds to a lower effective degree of entanglement, and we would therefore expect from earlier results a return to non-random (i.e. more central) scission. This could explain the narrowing of the distribution in the final stages of degradation [112]. In order to get perfectly random scission we would have to increase the concentration even more. As a result of the change in degradation mechanism, the relevant parameter that seems to determine the degradation of a transient network is the stress per chain in solution.

The ability of entangled monodisperse polymers to resist degradation seems to be far greater than that of polydisperse polymers. The uniform relaxation times of the monodisperse molecules may resist the flow in concert, whereas the polydisperse molecules get picked off one by one from the longest in the distribution [111].

The information obtained in these works is of potential practical importance. Most flows where polymers are added for hydrodynamic effect are partly elongational in character, and we have shown that central scission can occur in many such flow-fields. The performance of such polymers in providing enhanced flow resistance in porous media, anti-misting effects or drag reduction relies on the stability of the highest molecular weight components.

7.8
Conclusions

It is clear that, for dilute solutions, stagnation point extensional flows are extremely important. It is only for such flows that the conditions for high extension and the strongest non-Newtonian effects pertain. Useful non-Newtonian effects are almost universally accompanied by thermo-mechanical scission of the macromolecules; such scission is also likely to be dominated by effective stagnation points. Components representing such flows deserve the closest examination in any real flow situation, and yet present the most difficult challenge in the numerical modelling of polymer flow.

Over the last twenty years, experimental techniques and observations have been enormously improved. Many researchers have revisited and refined the theoretical approaches and numerical modelling has emerged as a useful adjunct to theory. Nevertheless, the subject remains vital and controversial, with fundamental issues such as the criticality and hysteresis foreseen by de Gennes remaining to be completely resolved.

References

1. Fabula AG, Hoyt JW, Crawford HR (1963) Bulletin APS 8
2. Dauben DL, Menzie DE (1967) J of Petroleum Tech 19:1065
3. James HM, Guth E (1943) J Chem Phys 11:455
4. Flory PJ (1942) J Chem Phys 10:51
5. Peterlin A (1961) Makr Chem 44–46:338
6. De Gennes PG (1974) J Chem Phys 60:5030
7. Hinch EJ (1974) Polymères et lubrification, colloques internationaux du CNRS 233:241
8. Brestkin YuV (1987) Acta Polymerica 38:470
9. Magda JJ, Larson RG, Mackay ME (1988) J Chem Phys 89:2504
10. Wiest JM, Wedgewood LE, Bird RB (1989) J Chem Phys 90:587
11. Fan X-J, Bird RB, Renardy M (1985) J Non-Newtonian Fluid Mech 18:255
12. Liu TW (1989) J Chem Phys 90:5826
13. Darinskii AA, Lyulin AV, Saphiannikova MG (1993) Int J Polymeric Mater 22:15
14. Rallison JM, Hinch EJ (1988) J Non-Newtonian Fluid Mech 29:37
15. Hinch EJ (1994) J Non-Newtonian Fluid Mech 54:209
16. Frank FC, Mackley MR (1976) J Polym Sci Phys Ed 14:1121
17. Zimm BH (1956) J Chem Phys 24:269
18. Frank FC, Keller A, Mackley MR (1971) 12:467
19. Pope DP, Keller A (1978) Colloid and Polymer Sci 255:633
20. Keller A, Odell JA (1985) Colloid and Polymer Sci 263:181
21. Odell JA, Keller A, Müller AJ (1989) Polymers in Aqueous Media (Ed) Glass JE, Advances in Chemistry 223:193
22. Scrivener O, Berner C, Gressely R, Hocquart R, Sellin R, Vlachos NS (1979) Mechanics 5:475
23. Miles MJ, Keller A (1980) Polymer 21:1295
24. Taylor GI (1934) Proc Roy Soc Lond A146:501
25. Cathey CA, Fuller GG (1990) J Non-Newtonian Fluid Mech 34:63
26. Fuller GG, Leal LG (1981) J Polym Sci Polym Phys Ed 19:557
27. Dunlap PN, Leal LG (1975) J Non-Newtonian Fluid Mech 23:5
28. Brestkin YuV, Saddikov IS, Agranova SA, Baranov VG, Frenkel S (1986) Polym Bull 15:147
29. Nguyen TQ, Yu G, Kausch HH (1995) Macromolecules 28:4851
30. Carrington SP, Odell JA (1996) J Non-Newtonian Fluid Mech 67:269
31. Treloar LKG (1975) The physics of rubber elasticity. 3rd Ed. Clarendon Press Oxford
32. Tsvetkov VN, Eskin VE, Frenkel SYa (1971) Structure of macromolecules in solution. Vol 3, C Crane-Robinson (Transl and Ed) National Lending Library for Science and Technology, Boston Spa, Yorkshire, UK
33. Polymer Handbook (1989) Third Edition (Ed) Brandrup I & Immergut EH, Wiley Interscience, USA
34. Tsvetkov VN, Eskin VE, Frenkel SYa (1971) Structure of macromolecules in solution. Vol. 2, Crane-Robinson C (Transl and Ed) National Lending Library

35. Peterlin A (1961) Polymer 2:257
36. Carrington SP, Tatham JP, Sáez AE, Odell JA (1997) Polymer 38:4151
37. Carrington SP (1995) Extensional flow of polymer solutions. PhD Thesis, University of Bristol
38. Frisman EV, Dadivanian AK (1967) J Polym Sci Part C 16:1001
39. Tatham JP (1993) Extensional flow dynamics of macromolecules of different flexibility in solution. PhD Thesis, University of Bristol
40. Odell JA, Keller A, Miles MJ (1985) Polymer 26:1219
41. Farrell CJ, Keller A (1978) Colloid & Polym Sci 256:966
42. Fuller GG, Leal LG (1980) Rheol Acta 19:580
43. Odell JA, Keller A (1986) J Polym Sci, Polym Phys Ed 24:1889
44. Rabin Y, Henyey FS, Pathria RK (1985) in Polymer-Flow Interaction (Ed) Rabin Y, AIP New York
45. Dunlap PN, Leal LG (1987) J Non-Newtonian Fluid Mech 23:5
46. Harlen OG, Hinch EJ, Rallison JM (1992) J Non-Newtonian Fluid Mech 44:229
47. Kuhn W, Grün F (1942) Kolloidzschr 101:48
48. Peterlin A (1966) Pure and Appl Chem 12:63
49. Ryskin G (1987) J Fluid Mech 178:423
50. Mackley MR (1972) PhD Thesis, University of Bristol
51. Schunk PR, de Santos JM, Scriven LE (1990) J Rheol 34:387
52. Schunk PR, Scriven LE (1990) J Rheol 34:1085
53. Mikkelsen KJ, Macosko CW, Fuller GG (1988) Proc Xth Int Congr Rheol, Sydney 2:125
54. Tatham JP, Carrington S, Odell JA, Gamboa AC, Müller AJ, Sáez AE (1995) J Rheol 39:961
55. Larson RG (1990) Rheol Acta 29:371
56. Smith KA, Merrill EW, Peebles LH, Banijamali SH (1974) Polymères et lubrification, Colloques internationaux du CNRS 233:341
57. Menasveta MJ, Hoagland DA (1991) Macromolecules 24:3427
58. Cooper SD, Batchelder DN, Ramalingam P. Submitted to Polymer
59. Larson RG, Perkins TT, Smith DE, Chu S (1997) Phys Rev E 55:1794
60. Carrington SP, Tatham JP, Odell JA, Saez AE (1997) Polymer 38:4595
61. Tatham JP, Keeble LJ (1987) The collapse of long chain polymer molecules. University of Bristol Project 57
62. Adams DN, Garnier T (1990) The collapse of flexible chain macromolecules in solution. University of Bristol Project 59
63. Dyakonova NE, Odell JA, Brestkin YuV, Lyulin AV (1996) J Non-Newtonian Fluid Mech 67:285
64. Müller AJ, Odell JA, Keller A (1988) J Non-Newtonian Fluid Mech 30:99
65. Ng RCY, Leal LG (1993) J Rheol 37:443
66. Special Issue (1988) J Non-Newtonian Fluid Mech 30:1
67. Trouton FT (1906) Proc Roy Soc Lond A77:426
68. Bird RB, Armstrong RC, Hassager O (1977) Dynamics of polymeric liquids. Vol 1, John Wiley & Sons Inc, New York
69. Hassager O (1974) J Chem Phys 60:2111
70. Schweizer T, Mikkelson K, Cathey C, Fuller G (1990) J Non-Newtonian Fluid Mech 35:277
71. Nguyen DA, Sridhar T (1990) J Non-Newtonian Fluid Mech 35:93
72. Laun HM, Hingmann R (1990) J Non-Newtonian Fluid Mech 35:137
73. Muller AJ, Odell JA, Tatham JP (1990) J Non-Newtonian Fluid Mech 35:231
74. Special Volume of (1990) J Non-Newtonian Fluid Mech 35
75. Frenkel J (1944) Acta Physicochim URSS 19:51
76. Harrington RE, Zimm BH (1965) J Phys Chem 69:161
77. Odell JA, Keller A, Rabin Y (1988) J Chem Phys 88:4022
78. Müller AJ (1989) The extensional flow of macromolecules in Solution. PhD Thesis, University of Bristol
79. Narh KA, Odell JA, Müller AJ, Keller A (1990) Polym Commun 31:2

80. Odell JA, Müller AJ, Narh KA, Keller A (1990) Macromolecules 23:3092
81. Odell JA, Taylor MA (1994) Biopolymers 34:1483
82. Rabin Y (1988) J Non-Newtonian Fluid Mech 30:119
83. Nguyen TQ, Kausch HH (1986) Colloid & Polym Sci 264:764
84. Nguyen TQ, Kausch HH (1987) Polym Prep 28:409
85. Nguyen TQ, Kausch HH (1987) Presented at the ACS meeting in New Orleans. Division of Polymer Chemistry, September, 1987). Also commented on by Kausch HH in Polymer Fracture, 2nd. ed, Springer, Berlin
86. Nguyen TQ, Kausch HH (1988) J Non-Newtonian Fluid Mech 30:125
87. Nguyen TQ, Kausch HH (1990) Macromolecules 23:5137
88. Merrill EW, Horn AF (1984) Polym Comm 25:144
89. Frank FC (1970) Proc Roy Soc Lond A319:127
90. Cerf FC (1951) J Chem Phys 48:59
91. Lumley JL (1972) Symposia Mathematica 9:315
92. Nguyen TQ, Kausch HH (1986) Chimica 40:129
93. Rodriguez F, Winding CC (1959) Ind Eng Chem 51:1281
94. Shimada T, Horng PL, Porter RS (1980) J Rheol 24:783
95. Ballauff M, Wolff BA (1988) Adv Polym Sci 85:1
96. Johnson WR, Price CC (1960) J Polym Sci 45:217
97. Minoura Y, Kasuya T, Kawamura S, Nakano A (1967) J Polym Sci A2
98. Nakano A, Minoura Y (1971) J Appl Polym Sci 15:927
99. Abdel-Alim AH, Hamielec AE (1973) J Appl Polym Sci 17:3769
100. Kaverina NI (1956) J Appl Chem USSR 29:1565
101. Porter RS, Johnson JF (1964) J Appl Phys 35:3149
102. Ram A, Kadim A (1970) J Appl Polym Sci 14:2145
103. Fukutomi T, Tsukada M, Takurai II, Noguci T (1972) Polym J 3:717
104. Nakano A, Minoura Y (1975) J Appl Polym Sci 19:2119
105. Goodman J (1957) J Polym Sci 25:325
106. Zakin JL, Hunston DL (1977) Proceedings BHRA Conference on Drag Reduction, Cambridge, Sept C5:41
107. Yu JFS, Zakin JL, Patterson GK (1979) J Appl Polym Sci 23:2493
108. Bueche F (1960) J Appl Polym Sci 4:101
109. Keller A, Müller AJ, Odell JA (1987) Progr Colloid and Polym Sci 75:179
110. Odell JA, Müller AJ, Keller A (1988) Polymer 29:01179
111. Odell JA, Müller AJ, Keller A (1992) Colloid Polym Sci 270:307
112. Müller AJ, Odell JA, Carrington S (1992) 33:2598
113. Narh KA. Private Communication
114. Menasveta MJ, Hoagland DA (1992) Macromolecules 25:7060

CHAPTER 8

Birefringence of Dilute PS Solutions in Abrupt Contraction Flow

Tuan Q. Nguyen, Réza Porouchani and Henning-H. Kausch

Transient elongational flow was created by forcing dilute PS solutions across a narrow contraction. Rheo-optical measurements revealed a localized birefringence zone above a critical strain-rate in the immediate orifice entrance. Birefringence was studied with a polarization-modulation technique for dilute PS solutions (50–500 ppm) as a function of fluid flow rate (1000–38000 s^{-1}), polymer molecular weight (M = 1.93–24 · 10^6) and solvent quality (decalin and 1-methyl-naphthalene). Transient elongational flow is complicated by the presence of local orientation distribution along the different streamlines. To account for this effect, a numerical technique has been devised to compute local birefringence (Δn) from experimental retardation (δ). Orientation correction turned out to be unimportant except at the closest distances from orifice entrance. Results show a steep decrease in birefringence with axial distance. Birefringence profiles determined in a direction perpendicular to the flow show a maximum, not at the center, but in the vicinity of the orifice walls. This effect was explained by means of flow simulation which indicated the presence of extreme strain rates near re-entrant corners. Molecular extension ratio calculated with the Kuhn-Grün theory suggests that polymers may uncoil up to one third of the chain contour length at the approach of capillary entrance. Notable departure from the affine deformation model is, however, observed with multiple birefringence saturation levels changing with flow rate conditions.

8.1
Introduction

Flow through a narrow contraction consitutes one of the simplest and yet, most efficient means to create elongational flow [1]. Besides being of academic interest, this type of flow is ubiquitous in a number of industrial processes such as pumping, filtering, flow through porous media [2] and membranes, extrusion and fiber spinning. In addition, several practical phenomena satisfy the criteria for "transient" converging flow, such as microvortices in turbulent liquids [3, 4] and bubble implosion in cavitational collapse [5]. Because of their technical importance, one can state conservatively that orifice and capillary entrance flows are by far the most widely studied, both by simulation and by experiments [6, 7].

Abrupt contraction flow is characterized by a sharp increase in flow velocity as the fluid approaches the orifice entrance. Owing to this rapid change in extensional strain and strain rate, the terms "transient elongational" or "fast transient" have frequently been associated with this type of flow, although these designations should be reserved for true time-varying situations, such as during the start-up and cessation of flow.

Flow field modeling shows that the polymer transit time in the strong flow region is comparable with, or less than the longest chain relaxation time (τ_1). Thus, one expects that the response of the polymer coil to a converging flow field, in particular with respect to the "coil-stretching" transition and to the degree of chain extension, may be different from that predicted in steady elongational flow [8, 9]. This opinion is shared by a number of authors who have questioned the importance of metastable molecular conformations when polymer-flow interaction takes place far away from equilibrium [10–13]. In spite of this large disparity in the residence times, it transpires from a number of observations that polymer dynamics in abrupt contraction flow and in equilibrium elongational flows are not so very different. As an example, it is now well-established that converging flow provides an effective method of orienting and stretching flexible high MW polymer chains in semi-dilute and concentrated solutions [14, 15]. Non-inertial flow of dilute polymer solutions through short capillaries is frequently reported to exhibit a sudden increase in end-pressure drop when the produt of elongation rate ($\dot{\varepsilon}$) and longest molecular relaxation time (τ_1) is greater than unity [16, 17]. Related phenomena have also been observed in flow through porous media [18, 19], and in the vicinity of a solid obstacle [20, 21]. The appearance of a large extensional viscosity, which occurs at the same critical strain-rate $\dot{\varepsilon}_c \cong 1/\tau_1$ as in certain elongational flows with a stagnation point [22–25], has been generally interpreted as evidence for a flow-induced "coil-stretching" transition with saturated chain extension. These results for abrupt contraction flow can be accommodated within the framework of the "coil-stretching" model, which predicts a large degree of chain extension not only in uniform steady flow, but also in nonhomogeneous elongational flow, provided the macromolecule resides for a sufficiently long time in the "strong" flow region to accumulate the necessary strain [26–28].

However, a number of other experimental findings do not support this simple representation of transient chain uncoiling. Using light scattering, it has been determined that the degree of coil expansion in dilute high MW polyisobutylene barely exceeds a factor of 3–4 in ~ 14:1 abrupt contraction flow [29]. Calculations based on the FENE dumbbells suggest the presence of self entanglements during the fast coil deformation, which may prevent further molecular stretching [30]. Pressure drop measurements in conical sink flows [31] have revealed a sharp increase at a critical flow rate. In these experiments, the calculated degree of chain extension was too small to account for the observed non-Newtonian effects and alternative descriptions such as the presence of "frozen" chains [32], non-uniform extension [12, 33] or kink development [13, 34, 35] are found to be more satisfactory in explaining this phenomenon. Recently, it has been found that the extensional viscosity of dilute polyisobutylene solutions reached a steady-state plateau during liquid filament pull-out at constant extensional rate [36]. Despite the presence of a well-defined stress plateau, detailed analysis of the data indicated again that the dissolved polymer chains could not have stretched by more than half of their full length, presumably as a result of self-entanglements [37]. Degradation experiments performed with dilute PS solutions in abrupt contraction flow demonstrate a distinct propensity for midchain scission [38–44], a feature apparently consistent with highly extended chains [45, 46]. On the other hand, the critical strain-rate for chain fracture ($\dot{\varepsilon}_f$) determined under the same experimental conditions scales with M^{-1} [41]. The inverse relationship with polymer MW is characteristic of partly uncoiled chains [46, 47] but not of highly extended chains, for which a stronger MW dependence ($\dot{\varepsilon}_f$) $\propto M^{-2}$ has been predicted [45], and reported in opposed jets flow [48, 49].

From the aforementioned conflicting results, and contrary to certain assertions, it is evident that the real fate of a flexible polymer chain in an abrupt contraction flow is far from thoroughly understood. Flow birefringence is a valuable technique for the investigation of molecular orientation in simple shear flow and its application to converging flow may help to settle the dispute concerning the different models. A large amount of literature currently exists on the birefringence behavior of flexible polymer chains in flows which possess a stagnation point, such as the four-roll mill [22, 23], the cross-slot [24] or the opposite jets device [23, 50]. However, the stretching behavior of *isolated* macromolecules in abrupt contraction flow has been studied in much less detail. This is somewhat surprising in view of the large number of studies devoted to this type of flow. Flow birefringence measurements in converging flow are traditionally confined to polymer melts [51–54] and concentrated polymer solutions [55–57]. The primary purpose of such work was the determination of the stress distributions in polymer flows, and is limited to the domain of applicability of Brewster's law at low chain orientation.

As has been realized during this study, birefringence measurements in axisymmetric converging flows are much more complex than in stagnation point flows [58]. The first problem was the large variations in optical properties, both in the axial (x) and radial (r) directions. Although pointwise flow birefringence techniques with laser light are well suited to measurements in nonhomogeneous flow fields [59], almost all past experiments have concentrated on steady and transient homogeneous shear flows [60, 61]. Investigations which have determined the spatial dependence of chain orientation by flow birefringence are scarce [54, 62, 63]. The second source of difficulties is the experimental data treatment in axisymmetric flows. Flow birefringence in transmitted light gives only an effective (average) retardation and its transformation into birefringence, necessary for the determination of local polymer orientation, is extremely involved in any 3-dimensional flow [63–66]. For this reason, most birefringence measurements are restricted to slit or wedge flow. The few experiments which have been reported in axisymmetric flows [56, 57, 64] consider only an average value for the birefringence rather than its local distribution and are thus unsuitable for polymer dynamics investigation. To the best of our knowledge, no attempt to convert experimental retardations into birefringence distributions has ever been reported in axisymmetric flow, although notable advances in this direction have been reported [67, 68].

Some years ago, we showed that birefringence could be detected in the capillary entrance flow of dilute PEO and PS solutions [69, 70]. These early experiments should be considered as preliminary since birefringence orientation was not taken into account. In the present study, we endeavour to carry out a more comprehensive investigation using polarization modulated birefringence techniques. A condensed version of the work presented in this chapter has recently been published [71]. Although the primary objective is to achieve a better understanding of the deformation behavior of isolated PS chains in transient elongational flow, the experiments form part of a broader project aimed towards a molecular interpretation of mechanochemical degradation in solution [72, 73]. The subject of flow-induced polymer chain fracture proves to be much more complex than can be accounted for by certain simple models [45, 74]. It will be only marginally treated in Sect. 8.4.8 in conjunction with some birefringence measurements.

8.2
Realization of Abrupt Contraction Flow

8.2.1
Design of the Flow Cell

Unlike stagnation point elongational flow, which necessitates specific experimental arrangements [22–25], "transient elongational flow" (TEF) is readily obtained under a variety of geometries and conditions [42], the simplest means being to force a fluid across a narrow contraction (cf. Introduction).

The set-up used in the present investigation was designed with the dual purpose of flow birefringence and flow visualization measurements (Fig. 8.1). The flow cell was drilled from a block of stainless steel and fitted with three windows: two for birefringence measurements and one at 90° for flow visualization with particle tracing [42, 69]. The windows were made with fused silica with negligible residual birefringence (Hellma Swiss S.A). In the design of the cell, special care was taken to minimize the total cell volume (< 3 cm^3) while keeping the windows sufficiently far from the jet to avoid end effects and preserve flow axial symmetry.

Transient elongational flow was created by pushing the polymer solution across an interchangeable nozzle with different internal diameters (0.8, 0.5 or 0.3 mm). In the present report, only results obtained with the 0.5 mm diameter orifice will be presented. The jet tip was made of synthetic sapphire with a precisely bored circular orifice connected to a 60° tapered channel (Comadur, Le Locle, Switzerland). Tests with an all metallic nozzle (brass or stainless steel)

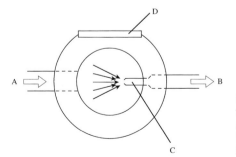

Fig. 8.1. Single jet cell for birefringence, flow visualization and degradation experiments. A: inlet from the syringe pump; B: outlet; C: interchangeable nozzle; D: quartz window for flow visualization

with the orifice machined by electroerosion were less satisfactory owing to micro-irregularities of the orifice lip (Fig. 8.2). These imperfections may lead to unpredictable errors in the entrance flow field as a result of the "re-entrant corner" effect (cf. Sect. 8.2.2). A 500 ml precision syringe pump (ISCO model 500D) was used to transfer the polymer solution at a constant pulseless flow rate from 1 cm^3/min to a maximum of 200 cm^3/min.

The birefringence cell for "stagnation point" flow follows the same design as for transient elongational flow, except that the polymer solution is now pushed through two sapphire jets exactly facing each other. With a system of washers and rotating screws, it was possible to adjust continuously the separation distance between the two jets from 0 to 5000 µm, while keeping the orifice entrances in exact opposition. The gap distance was determined with a microscope to a tolerance better than ± 5 µm.

The flow cell was mounted on a support endowed with three degrees of freedom (x, y and Φ) to adjust the relative position of the flow with respect to the measuring light beam. Two motorized translational stages under computer control allowed variations in the x and y directions with a minimum increment of 1 µm. The azimuthal orientation (Φ) was set manually with a micrometer to align the flow symmetry axis with the x-direction. This step is particularly

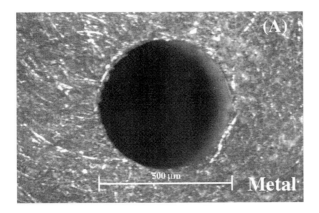

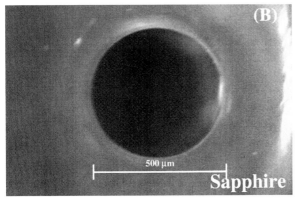

Fig. 8.2a, b. Optical micrographs of the orifice entrance for a nozzle fitted: A with a brass tip; B with a sapphire tip (internal diameter 500 µm)

important in single jet flow and should be performed every time the cell is removed to ensure a symmetric birefringence profile.

Axisymmetric converging geometries are capable of producing stable laminar flow with exceedingly high strain-rates ($> 10^6$ s^{-1} with a 0.3 mm nozzle). However, the three-dimensional character of the flow presents more difficulties to extract local information from experimental data, than is normally the case for a two-dimensional wedge flow (cf. Sect. 8.4.3).

8.2.2
Flow Field Modeling[1]

It is well known that the degree of chain extension depends on the details of the pervading flow field. Hence, birefringence results can only be correctly interpreted if the flow field is known with confidence. It is therefore mandatory that the flow field under the prevalent experimental conditions be properly characterized. In this section, we shall focus exclusively on the results which are the most pertinent to the current investigation.

Owing to its technical importance, abrupt contraction flow is by far the most widely studied flow geometry, both experimentally and theoretically. The hydrodynamics of bidimensional and axisymmetric converging flows was extensively investigated in the 1970s and 1980s, mainly in connection with the die entry flow problem [75–77]. The following features emerge from published results for viscoelastic fluids: at low flow rates (Deborah number $De \ll 1$), the behaviour is essentially Newtonian with a relatively small secondary flow in the corners of the upstream tube. With increasing Deborah number, the secondary flow grows in importance, with the appearance of a multitude of nonlinear transition phenomena roughly classified as vortex growth flow ($De \sim 1$–3), asymmetric flow ($De \sim 5$), rotating flow ($De \sim 10$) and helical flow ($De \sim 10$–15) [75]. These complex viscoelastic motions depend on a number of factors, such as the contraction ratio, the shape of the contraction lip and the fluid properties, and thus remains an active field of research in polymer rheology [54, 78].

To avoid difficulties and uncertainties in modeling viscoelastic flow at high Deborah number, we have always used highly dilute polymer solutions whose rheology shows negligible departures from Newtonian behaviour. The limiting concentration below which viscoelasticity can be ignored is a function of flow geometry, polymer molecular weight and fluid strain-rate [58, 79, 80]. Conditions for the validity of the Newtonian approximation should be checked for each experimental situation.

Flow modeling was carried out on an HP-710 workstation using the POLY-FLOW program [81]. In the simulation, we have tried to match the real flow geometry as closely as possible. In the calculations, fluid compressibility was neglected but viscosity and inertial forces should both be taken into account. Neglect of either element may lead to serious discrepancies with the experimental results as will be evident in the results reported below.

1 Data for Figs. 8.6, 8.7, 8.9, 8.10 and 8.11 are kindly provided by Dr. Debbaut.

8.2.2.1
Streamlines

Streamlines are the most readily discernible features of a flow. Experimentally, streamlines are visualized by mixing decalin with 0.5 μm PVC particles obtained by emulsion polymerization (PVC is insoluble in decalin). Photographs were taken with the same cell and optical bench as for the birefringence measurements after removing all the polarization components. The flow field was illuminated by passing a laser beam through an f = 50 mm cylindrical lens and focussing the produced sheet of light longitudinally along the center flow line. A typical photographic record of tracer particle trajectories is shown in Fig. 8.3 with the laser beam positioned in front of the capillary entrance.

One approach frequently encountered in the literature [31, 46, 82] considers flow in the entrance region to be a uniform converging flow with all the streamlines distributed around a conical contour converging towards a common apex situated inside the capillary entrance. This type of flow is attractive in its simplicity, since the fluid velocity (v) and strain-rate ($\dot{\varepsilon}$) are readily obtained from the flow rate (Q) using the principle of mass conservation:

$$v(s) = Q/[2\pi s^2 (1-\cos\theta_{max})] \tag{8.1}$$

$$\text{and} \quad \dot{\varepsilon} = dv/ds = Q/[\pi s^3(1-\cos\theta_{max})] \tag{8.2}$$

where s is the distance from the apex and θ_{max} the maximum cone opening which may change from $\sim$60° to 90°, depending on the flow conditions (Figs. 8.4, 8.5) [6, 18, 75].

The radial distribution of the velocity angle, defined by the velocity vector relative to the x-axis, is simply given by:

$$\theta(x, r) = \tan^{-1}(r/L_x) \tag{8.3}$$

with L_x = distance separating the cone apex from the axial position x.

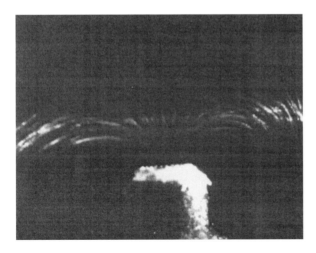

Fig. 8.3. Streak photograph of tracer particles in capillary entrance flow (solvent: decalin at 30 cm³/min, exposure time: 10 s)

The streamlines based on flow field simulation are reported in Figs. 8.4 and 8.5. When viscous forces are predominant, for example at low flow rates (cf. Fig. 8.5 in [7]) or if inertia is neglected (Fig. 8.4), the computed stream functions are almost straight converging lines, as described by the sink flow approximation. Large recirculation zones are also clearly visible in this latter situation. The recirculation vortex shrinks with increasing inertia and good agreement between flow visualization (Fig. 8.3) and the simulations can be obtained only when both viscosity and inertial forces are included into the calculations (Fig. 8.5). Magnification of the inlet region (Fig. 8.5b) reveals that most of the streamlines enter the orifice with a narrow opening angle before diverging rapidly near the edges. The effect is particularly visible when the orientation angle $\psi(x, r)$, defined by the direction of the velocity vector relative to the symmetry axis Ox, is reported as a function of radial position (r) (cf. Fig. 8.10, Sect. 8.2.2.3). The orientation of the velocity vector is always more acute than predicted by the sink flow model, which is only a reasonable approximation at $x < -2r_{0-}$ but is entirely unsatisfactory at any closer distance. These behaviours will be discussed in Sect. 8.2.2.3 in connection with polymer orientation.

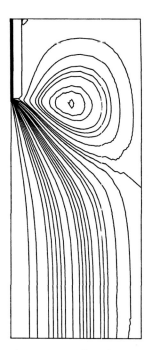

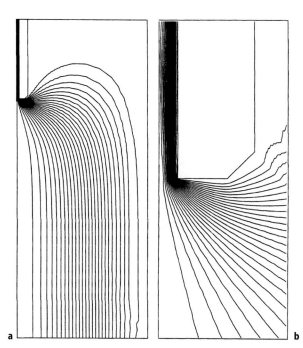

Fig. 8.4. Streamlines in transient elongational flow, computed with the POLYFLOW program without taking into account inertial forces

Fig. 8.5a, b. Streamlines calculated with viscous and inertia forces; **b** is a magnified section of **a** in the entrance region. Calculation conditions: decalin at 100 cm³/min ($\rho_s = 0.880$ kg · m^{-3}; $\eta_s = 2.4$ mPa · s)

8.2.2.2
Velocity Field

To compare results obtained at different flow rates and orifice diameters, it is advantageous to convert data into dimensionless parameters defined by the following relations:

dimensionless axial coordinate	$x' = x/r_o$	(8.4)
dimensionless flow velocity	$v'_x = v_x/v_{av}$	(8.5)
dimensionless strain-rate	$\dot{\varepsilon}'_{xx} = \dot{\varepsilon}_{xx}/(v_{av}/r_o)$	(8.6)

($\dot{\varepsilon}_{xx} = \partial v_x/\partial x$ is the axial strain-rate and v_{av} is the volume average axial fluid velocity at the orifice).

Flow velocity profiles and strain-rate distributions, calculated along the central streamline in terms of the above-defined dimensionless parameters for different flow rates (10, 50, and 200 cm^3/min), are plotted in Fig. 8.6 and Fig. 8.7, respectively. The origin of the coordinates is chosen to be the capillary entrance with negative x-values corresponding to positions in front of the inlet. All the dimensionless flow velocity curves are superposable in front of the entrance with v'_x equal to 1 at the orifice. This coincidence breaks down behind the inlet, in the presence of shearing forces inside the capillary. The flow velocity profile

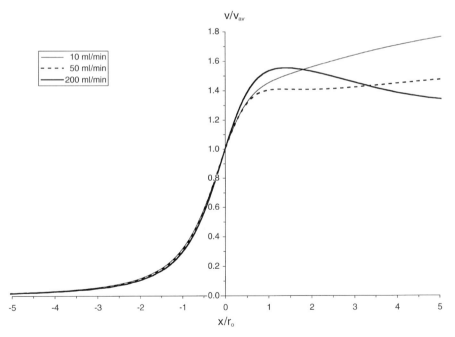

Fig. 8.6. Dimensionless flow velocity distribution (v_x/v_{av}) along the centerline in abrupt contraction flow at the indicated flow rates. A flow rate of 200 cm^3/min corresponds to $v_{av} = 1700$ cm · s^{-1} and a Reynolds number $Re = 3130$ in decalin (2800 in 1-methyl-naphthalene)

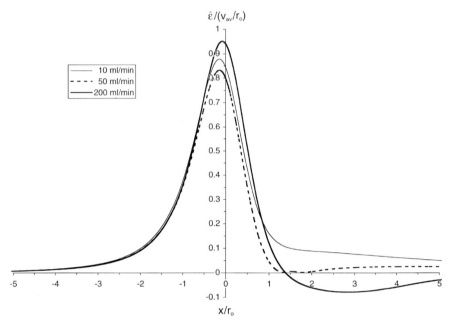

Fig. 8.7. Dimensionless extensional strain-rate calculated along the centerline in abrupt contraction flow at the indicated flow rates from the data of Fig. 8.6

is uniform at the entrance with $v = v_{av}$ (Fig. 8.8). With the formation of a boundary layer, the rate of flow next to the walls gradually decreases until the familiar parabolic Poiseuille profile is fully developed. At this point, $v_x(y = 0) = 2v_{av}$. The minimum traveling distance (x'_{ss}) for the fluid to reach a steady state increases with the Reynolds number (Re) according to the empirical relation [83]:

$$x'_{ss} > 0.06 \, Re \qquad\qquad\qquad (8.7)$$

Flow field calculations indicate that the parabolic profile was actually attained only at $x' \cong 20$ for a flow rate of 10 cm³/min ($Re = 82$ in decalin) and at a distance at least 20 times longer for a flow rate of 200 cm³/min. Changes in the flow field after orifice entrance are irrelevant for the present investigation, but may be important in some chain retraction and degradation phenomena which can occur inside the capillary.

In agreement with previous contraction flow simulations [30, 42], the strain-rate distributions in Fig. 8.7 show a typical spike-like shape with a maximum situated slightly in front of the orifice entrance ($x' \cong -0.012$). As with flow velocities, the dimensionless strain rates are insensitive to the flow rate in front of the entrance. This behavior is anticipated since $\dot{\varepsilon}_{xx}$ is defined as the gradient of v'_x. For the flow rates investigated, the maximum $\dot{\varepsilon}_{xx}$ along the central streamline is 0.83–0.95 (Fig. 8.7). These values are significantly larger than the previously reported value of 0.59 obtained by ignoring the solvent viscosity (inviscid approximation) [30, 42].

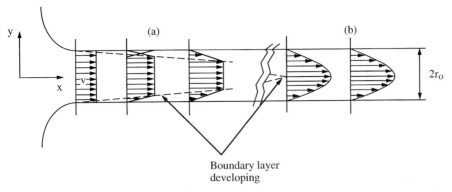

Fig. 8.8. Variation of the velocity profile with axial distance from the inlet in laminar capillary flow. The Poiseuille profile is achieved in (b)

The strain-rate distribution depends on the streamline, with the maximum increasing rapidly at the approach of the orifice lip. This phenomenon, known as the "re-entrant corner" singularity, is responsible for most of the computational difficulties associated with viscoelastic abrupt contraction flows [84]. In the present context, it plays a central role in determining the shape of the birefringence profiles.

It has been shown that the degree of chain extension depends on the first eigenvalue of the rate of deformation tensor $\dot{\varepsilon}_1$, which characterizes the strain-rate along the polymer orientation direction [7, 85, 86]. As with $\dot{\varepsilon}_{xx}$, the parameter $\dot{\varepsilon}_1$ can be made dimensionless by dividing by (v_{av}/r_o). The values thus obtained are plotted as a function of dimensionless curvilinear coordinates ($s' = s/r_o$) for a few selected streamlines in Fig. 8.9.

It should be remembered that theoretical modeling uses sharp re-entrant corners. The corners encountered in a real jet (particularly a metallic one) may have some finite degree of rounding as a result of machining and wear. Therefore, the experimental strain-rate maxima found near the edges may not be as pronounced as shown in Fig. 8.9. The influence of corner rounding on birefringence measurements will be discussed in more detail in Sect. 8.4.4.

8.2.2.3
Molecular Orientation

An important parameter which can be derived from flow field simulation is the average orientation of the polymer chain $\chi(r)$. In a purely elongational flow, polymers are oriented along the strain direction parallel to the velocity vector. As a first-order approximation, it should be possible to equate $\chi(r)$ with the orientation distribution $\psi(r)$ defined in Sect. 8.2.2.1 and shown in Fig. 8.10. Deviation from this simple behavior is expected in presence of vorticity components. A better approach to the problem of chain deformation in a mixed

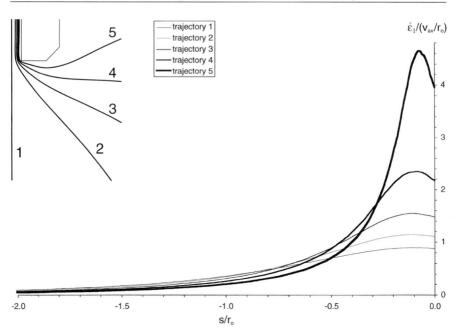

Fig. 8.9. Dimensionless extensional strain-rate ($\dot{\varepsilon}_1'$) calculated along the flow trajectories shown on left at a flow rate of 100 cm^3/min. Streamlines have been selected to encompass 20% of the flow between two consecutive trajectories

shear flow (elongational flow with a shear component) was suggested in [85, 86]. Based on extensive results obtained in the four- and two-roll mill flow, it has been reported that molecular orientation in a flowing system is along the principal eigenvector of the rate of deformation tensor. Polymer orientation plays a crucial role in determining local rheological properties (cf. Sect. 8.4.3) and can be computed from the flow field according to the details given in Appendix B. Results based on these calculations indicate that chains enter the orifice with an almost zero orientation angle, which is significantly different from the direction of the velocity vectors. At some distance from the orifice, the curves gradually collapse to become indistinguishable at $x \geq 2\,r_o$ (Fig. 8.10).

8.2.2.4
Residence Time

The residence time in the high strain-rate region is one of the three main criteria which control the degree of polymer uncoiling [27], the others being the flow character (extensional component > vorticity component) [87] and the fluid strain-rate which must be higher than a critical value $\dot{\varepsilon}_c \sim 1/\tau_1$ [8]. In order to stretch a polymer, the molecular coil must accumulate the require strain, which, in the limit of "no slip" conditions, cannot surpass the degree of extension of the surrounding fluid element.

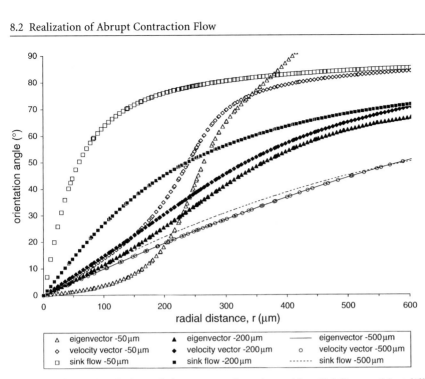

Fig. 8.10. Change in velocity and eigenvector orientations with radial distance (r) at different distances from the inlet ($x = -50$ μm, -200 μm and -500 μm)

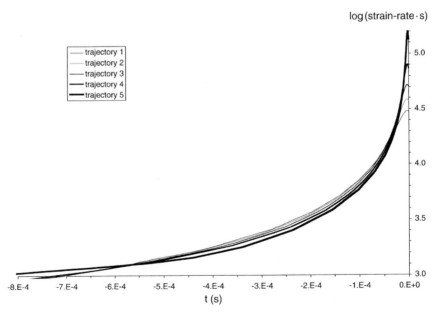

Fig. 8.11. Strain rate as a function of transit time for the different streamlines shown in Fig. 8.9. Flow rate: 100 cm^3/min; t = 0 is taken at the orifice entrance

The strain-rate histories experienced by flowing macromolecules along a few selected streamlines are shown as a function of residence time in Fig. 8.11. The "transient" nature of the flow is evident if one considers that a 3.95×10^6 PS in decalin will start to be deformed only at $\dot{\varepsilon} \cong 5300 \text{ s}^{-1}$ (cf. Fig. 8.47 in Sect. 8.5.1). From Fig. 8.11, this strain-rate corresponds to a residence time of 0.11–0.14 ms before entering the orifice, or only a fraction of the polymer longest relaxation time which is 0.6 ms. During this short transit time, a fluid element can be easily stretched by a factor of 50. Whether or not a polymer chain with 40,000 monomer units is able to deform affinely with the surrounding fluid element on the same time scale is, however, questionable.

8.3
Flow Birefringence Measurements

The rheology of polymeric fluids appears highly complex, and attempts to describe their flow behavior have led to numerous investigations with the purpose of relating macroscopic to molecular chain deformation. Of the experimental methods which have been used to probe the microstructure of flowing liquids, rheo-optical techniques have enjoyed wide acceptance, owing to the relative simplicity of the equipment and the diversity of molecular information which can be obtained. Flow birefringence, in particular, has proven to be a valuable tool for the study of optical, orientational, conformational and hydrodynamic properties of flexible macromolecules in solution.

Flow birefringence occurs when a pure liquid, solution or suspension becomes optically anisotropic in flows with a velocity gradient. Study of this phenomenon has a long history which dates back to its discovery by Maxwell in 1873, in flowing balsam resin. Systematic investigation started with the works of Diesselhorst and Freundlich [88], Vörlander and Walter [89], Raman [90], Signer [91] and Sadron [92] at the beginning of the century. Its occurrence in polymers was first described in [91]. Since then, experimental and theoretical results have become very extensive, and the subject of numerous publications and reviews, some of which are listed in [93–98].

8.3.1
Principles of Optical Rheometry

The general principles of optical rheometry, including the general basis for data analysis, are discussed in a number of excellent monographs and reviews [99–101]. Nevertheless, certain details of the experimental system are worth recalling, as they relate to specific aspects of the birefringence measurements reported in the present paper. An additional impetus for writing this section originates from referees' comments concerning our previous publications in this field. These indicate that the peculiarities of flow birefringence measurement in converging flow can only be correctly grasped if the basic notions of optical rheometry are adequately exposed.

Owing to its versatility, Mueller matrix calculus has become an integral part of the design, data analysis and physical interpretation of polarimetry experiments. The Mueller matrix approach is based on the Stokes vector description of polarized light. The Stokes vector is represented by a four-element column $[s_0, s_1, s_2, s_3]^T$ with dimensions of intensity. The first element (s_0) gives the incident irradiance while the remaining three specify the states of polarization of elliptical light ($+45°$, right-circular and horizontally polarized). Interaction of light with an optical component (or the sample) converts an input Stokes vector (S_0) into an output Stokes vector (S_{out}). Under normal circumstances, the two Stokes vectors are related by a linear transformation which involves multiplication of S_0 by the Mueller matrix (M) representing the optical component:

$$S_{out} = M \cdot S_0 \tag{8.8}$$

For a sequence of N optical elements, the intensity and polarization of the light beam at any position (K) along the optical train can be calculated by repeating the Mueller matrix multiplication the necessary number of times:

$$S_K = M_K \cdot M_{K-1} \cdot \ldots M_2 \cdot M_1 \cdot S_0 \qquad 1 \leq K \leq N \tag{8.9}$$

Mueller matrices for most important optical elements are compiled in a number of references [51, 102]. The outcome of a given experiment can then be determined simply by selecting the appropriate Mueller matrices from the tables and multiplying them together according to Eq. (8.9) using standard rules of matrix algebra.

The method outlined above will be applied to two crossed polarizers, which constitute the simplest means of determining birefringence. It is assumed that the sample has no dichroism, with birefringence oriented at an angle θ, and the two polarizers are ideal with orientation angles α and $\alpha + \pi/2$ respectively, relative to a laboratory frame of reference. Leaving aside the quarter-wave plate in the set-up depicted in Fig. 8.12, the Stokes vector at the detector position is given by

$$S_{out} = M_{P, \alpha + \pi/2} \cdot M_S \cdot M_{P, \alpha} \cdot S_0 \tag{8.10}$$

Since the detector responds only to light intensity, only the first element of the output Stokes vector needs be considered. Denoting I_0 as the incident light intensity, and carrying out the multiplications in Eq. (8.10), one obtains:

$$s_0(out) = (I_0/4) \cdot \sin^2\{2(\theta-\alpha)\} \cdot (1 - \cos\delta) \tag{8.11}$$

It is obvious from Eq. (8.11) that a single intensity measurement would not allow simultaneous determination of δ and θ. A two-step procedure is therefore normally adopted. First, the angle θ is determined by rotating the two polarizers while maintaining their relative orientation, until a sharp minimum is found. At this point, the orientation of the first polarizer (α) is equal to the orientation angle (θ) which is also referred to as the extinction angle. The intensity reaches a maximum on rotating the crossed polarizers by an additional $45°$. This last configuration is used for the determination of the retardation δ. For weakly birefringent samples, such as dilute flowing polymer solutions, $|\delta| \ll 1$ and the intensity is approximately:

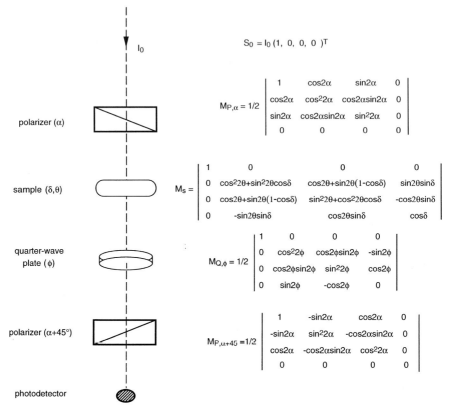

Fig. 8.12. Diagram of the Senarmont compensator

$$I/I_0 \approx \delta^2/16 \tag{8.12}$$

That the signal is proportional to the square of the retardance is an inconvenience which can be removed by inserting a quarter-wave retarder immediately after the sample (Fig. 8.12). The optical axis of this plate is rotated by a small angle ϕ with respect to the first polarizer. Using Mueller matrix calculations, the intensity is found to be

$$I/I_0 \approx \phi \cdot \delta/2 \tag{8.13}$$

The $\lambda/4$ plate can be used as a Senarmont compensator to offset residual window birefringence or to determine the sign of the birefringence.

Optical polarimeters based on retardation compensation have been widely used in a number of applications in which the orientation angle is known a priori. This is the case whenever uniaxial fields, such as electric, magnetic or purely extensional flow fields, are used. In these particular situations, experiments are generally performed by aligning the sample in such a way that its principal orientation direction is at 90° with respect to the polarizers and the retardation obtained from the ratio I/I_0 (Eqs. 8.11 or 8.12). With dark room facilities and careful alignment, the static polarization technique schematized in Fig. 8.12 is capable of resolving retardation to within 10^{-9} radian under quiescent conditions [103]. Its sensitivity decreases drastically, however, under strong flow conditions [58, 72].

Most polymer systems subject to flow are characterized by two unknowns: the birefringence Δn given by the difference of the principal values of the refractive-index tensor in the flow plane, and the orientation angle α of the corresponding axes. Two independent measurements are therefore necessary to determine both the degree of optical anisotropy and the position-dependent orientation angle. This is feasible if δ and α change little with time and position as in steady simple shear flows. Converging flows, on the other hand, are inherently inhomogeneous and it is impractical (if not impossible) to apply the two-step measurement sequence to each position.

8.3.2
Fast Polarization Modulation Technique

Devices such as the "two-color flow birefringence" instrument have been developed to determine simultaneously δ and α in real time, thus extending birefringence measurements to transient flow situations [104]. Another class of polarimeters which has been extensively developed in recent years relies on polarization modulation. The principle is to cause the polarization to oscillate in time, before and/or after the sample section. The electric vector is thus forced to sample cyclically different projections of the refractive index tensor in the plane normal to the direction of light propagation. This renders possible real-time determination of δ and α in a single experiment. In addition, absolute values of δ are obtained without the need for additional calibration. Since the incident intensity is continuously monitored, fluctuations in the laser source are automatically cancelled out, resulting in enhanced sensitivity. The use of a lock-in amplifier further increases the signal-to-noise ratio by rejecting all signals at undesirable frequencies and allows measurements to be made under ambient light conditions. This is a non-negligible advantage given the restrictions imposed by dark room requirements.

It was already known in the late 1950s that modulating the polarization of the probe beam can greatly enhance the precision and sensitivity of flow birefringence experiments [105–108]. This technique only became popular in the mid-1970s, with the availability of precise optical birefringence modulators and affordable lock-in amplifiers. Virtually all modern polarimeters now use polarization-modulation scheme in some part of the optical components [101]. Modulation can be induced either by rotating an optical element with fixed optical properties (polarizer, retarder), or by modulating the optical properties (generally the retardation) of an element with a fixed orientation (photoelastic modulator, Pockels and Faraday cell).

In the present investigation, we used a commercial version of the rotary polarization polarimeter originally described by Fuller and Mikkelsen [109] and distributed by Rheometrics (Rheometrics Optical Analyzer). As detailed in the reference, the use of a rotating half-wave plate offers distinct advantages over the photoelastic modulator, such as simple signal processing and extended spectral response [109]. The main drawback is probe beam broadening as explained in Sect. 8.4.5. The block diagram of the optical train for measuring flow birefringence is depicted schematically in Fig. 8.13 along with the associated Mueller matrices.

Light from the He-Ne laser passes successively through a quartz prism-polarizer oriented at $0°$ with respect to the flow direction, an achromatic half-wave retarder mounted on an air-driven turbine rotating at $\omega/2\pi = 2$ kHz, a birefringence-free condensing lens, the sample, a combination of a quarter-wave plate and a second quartz prism-polarizer oriented at $45°$, and finally arrives at a photodetector.

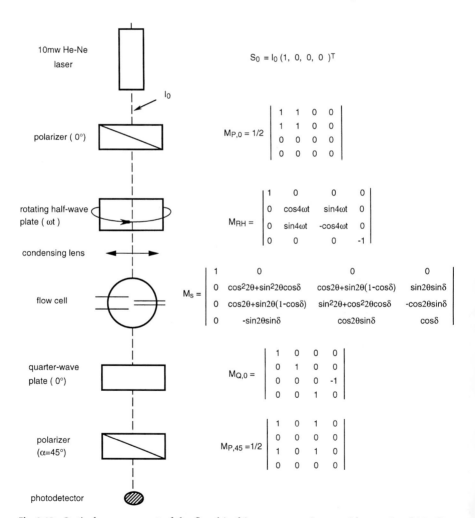

Fig. 8.13. Optical arrangement of the flow birefringence experiment with associated Mueller matrices

Using Mueller matrix calculus, the Stokes vector after the rotating half-wave plate is readily obtained as

$$S_2 = M_{RH} \cdot M_{P,0} \cdot S_0 = (I_0/2) \cdot \begin{vmatrix} 1 \\ \cos(4\omega t) \\ \sin(4\omega t) \\ 0 \end{vmatrix} \tag{8.14}$$

where I_0 is the light intensity before the first polarizer.

Replacing the rotating half-wave plate with a polarizer results in a similar Stokes vector, except that the transmitted intensity and the modulation frequency are now both reduced by half to $I_0/4$ and 2ω, respectively.

The intensity of light, I, at the detector position can be calculated as outlined in Eq. (8.9). If δ and α are the unknown retardance and extinction angle of the sample, respectively, the following equation is obtained:

$$I = I_0/4 \, [1 + R_1 \sin(4\omega t) + R_2 \cos(4\omega t)] \tag{8.15}$$

$$R_1 = -\sin(\delta)\cos(2\alpha) \tag{8.16}$$

$$R_2 = \quad \sin(\delta)\sin(2\alpha) \tag{8.17}$$

Experimentally, the d.c. component ($I_0/4$) is obtained from a low-pass filter which is part of the ROA signal conditioner. The coefficients R_1 and R_2 are detected with a dual-phase digital lock-in amplifier (Stanford Research model 850). Pointwise scanning along the y-direction (with the light beam pointing towards z) gives the transverse $R_1(y)$ and $R_2(y)$ distributions which contain most of the information on the rheo-optical properties of the flowing solution. It can be shown for instance, within the limits of applicability of the stress-optical law, that R_1 depends only on the spatial distribution of normal stresses whereas R_2 is function of the shear stresses [64]. The retardance and extinction angle can be readily computed from R_1 and R_2 using

$$\delta = \sin^{-1}[(R_1^2 + R_2^2)^{0.5}] \tag{8.18}$$

$$\alpha = {}^1/_2 \tan^{-1}(-R_2/R_1) \tag{8.19}$$

The orientation angle, α, is an important experimental parameter which depends on the local flow field. In a homogeneous transient flow, Δn and α change with time following start-up and cessation of flow [60]. In a steady converging flow, Δn and α are time-independent but vary with axial and radial position [71]. The situation is simpler in opposed-jet flow, but only under certain conditions, depending on the separation of the jets and the measuring position. Except for the smallest gaps ($< r_o/2$), polymer chains are oriented along the axial direction ($\alpha = 0°$) only in the immediate proximity of the stagnation point. Close to the orifices, the flow field (and hence the polymer orientation behavior) is very similar to the one encountered in abrupt contraction flow.

Figures 8.14 and 8.15 give some examples of experimental data which have been obtained in abrupt contraction flow conditions.

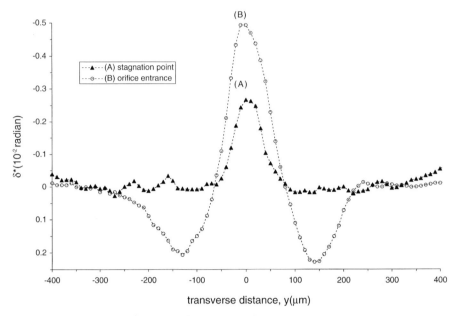

Fig. 8.14. Transverse retardation profiles in opposed-jet flow, at two different positions along the flow symmetry axis: (A) at the stagnation point and (B) at 50 μm from the jet entrance. The center flowline is at $y = 0$. Curvature in the (B) baseline is due to solvent birefringence. Experimental conditions: gap distance 970 μm, PS 3.95×10^6 100 ppm in 1-methyl-naphthalene, flow rate: 200 cm³/min

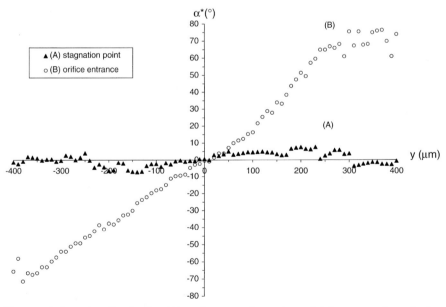

Fig. 8.15. Orientation distribution (A) at the stagnation point and (B) at 50 μm from the jet entrance in opposed-jet flow (experimental conditions as in Fig. 8.14)

8.4
Experimental Results

8.4.1
Experimental Conditions

All the birefringence experiments were conducted at room temperature (295 ± 1 K). Measurements were performed using anionic PS samples dissolved in decalin or in 1-methyl-naphthalene. The polymer solutions are gently stirred for 48 h, then filtered by gravity through a 1-μm PTFE membrane. MWDs were analyzed before and after sample preparation to detect any eventual polymer degradation. Degassing is essential to ensure a good signal-to-noise ratio. Dissolved air tends to generate tiny bubbles along the light path during flow cell loading and deteriorates the measurement. This effect was particularly perceptible in converging flow owing to a weak birefringence combined with long optical path. On-line degassing with a semipermeable membrane (Uniflows, Japan) connected in series to the pump entrance provided the best solution to the problem. However, in view of the large dead volume introduced by this device, the procedure was replaced by helium sparkling. Mechanical vibration during translation of the x-y stage was another source of noise which could be reduced with adequate cushioning.

8.4.2
Material Characterization

8.4.2.1
Chemicals

The PS were narrow molecular weight standards with M_p (MW at peak maximum) equal to 3.95×10^6, 11.6×10^6 and 24×10^6. The polydispersity index, $\overline{M}_w/\overline{M}_n$, increases with polymer MW, from 1.03 for the lowest to 1.30 for the highest MW sample.

The solvents were purified by distillation at reduced pressure prior to the experiments. Decalin is a mixture of 58% *trans*-, 41% *cis*-decahydronaphthalene and 1% tetrahydronaphthalene as determined by gas chromatography. Other important physical, thermodynamic and molecular properties pertinent to the present work are described below.

8.4.2.2
Intrinsic Viscosity of PS Solutions

Intrinsic viscosities were determined by pumping dilute polymer solutions at constant flow rate and recording the pressure drop across a calibrated capillary tube (100 cm length × 0.1 cm diameter). Experiments were conducted at wall

Table 8.1. Selected physical properties of the solvents

	decalin	1-methyl-naphthalene
viscosity (η_S, 22 °C)	2.39 mPa · s	3.02 mPa · s
density (ρ_S, 22 °C)	878.8 kg · m^{-3}	1020 kg · m^{-3}
refractive index (n_D^{20})	1.474	1.614

shear rates < 80 s^{-1}. This condition was selected to avoid shear-thinning with ultra-high MW samples (M above 10^7 dalton).

The following Mark-Houwink relations were established:

PS/decalin (295 K): $[\eta] = 2.34 \cdot 10^{-3} M^{0.53}$ cm$^3 \cdot$ g^{-1} (8.20)

PS/1-methyl-naphthalene (295 K): $[\eta] = 1.87 \cdot 10^{-3} M^{0.70}$ cm$^3 \cdot$ kg^{-1} (8.21)

Some other important physical properties of the solvents are given in Table 8.1.

Gel Permeation Chromatography with viscometric detection were used to determine MWDs of the PS samples. Based on the experimental Mark-Houwink coefficients, it was verified that the MWs given by the manufacturers were accurate to within ± 10%, except for the highest MW sample where errors can surpass 20%.

From the intrinsic viscosities, other important molecular parameters such as the polymer coil dimensions and end-to-end relaxation times could be derived.

8.4.2.3
Molecular Coil Dimensions

From polymer solution theory, the root mean-square end-to-end distance of a "random coil" under quiescent conditions (R_0) is related to the intrinsic viscosity through the Flory-Fox relation [110]:

$$[\eta] \cdot M = \Phi_0 \cdot R_0^3$$ (8.22)

The Flory viscosity constant Φ_0 has a theoretical value of 2.86×10^{23} mol^{-1} in a theta solvent. Depending on the polydispersity of the polymer sample, smaller values of Φ_0 between 2.0×10^{23} and 2.7×10^{23} mol^{-1} have generally been determined experimentally [110, 111]. Ptitsyn and Eizner have extended the Flory-Fox equation to the case of non-theta solvents by replacing the constant Φ_0 with a function $\Phi(e)$ [112]:

$$\Phi(e) = \Phi_0 (1 - 2.63 \, e + 2.86 \, e^2)$$ (8.23)

The parameter e is defined in term of the Mark-Houwink exponent constant (a):

$$e = (2a - 1)/3$$ (8.24)

Using Eqs. (8.20)–(8.24) with $\Phi_0 = 2.5 \times 10^{23}$ mol^{-1}, the following numerical relationships have been obtained:

$$R_0 \text{ (PS/decalin)} = 6.3 \times 10^{-2} \, M^{0.51} \text{ [nm]} \tag{8.25}$$

$$R_0 \text{ (PS/methyl-naphthalene)} = 4.4 \times 10^{-2} \, M^{0.57} \text{ [nm]} \tag{8.26}$$

Equation (8.25) in decalin is in good agreement with the literature values reported for PS under theta-conditions [111]:

$$R_0(\text{theta}) = (6.70 \pm 0.15) \times 10^{-2} \, M^{0.50} \text{ [nm]} \tag{8.27}$$

Denoting P as the degree of polymerization and m_0 as the MW of the repetitive unit, the extended length of PS is given by

$$R_{max} = 0.252 \times P = 0.252 \times M/m_0 \text{ [nm]} \tag{8.28}$$

Together with values for R_0, Eq. (8.28) permits the determination of the maximum extension ratio (R_0/R_{max}) to be used for the calculation of birefringence (Sect. 8.5.2).

8.4.2.4
End-to-End Chain Relaxation Time

The longest intramolecular relaxation time, τ_1, is certainly the single most important parameter in any description of polymer dynamics. Predicted respectively by Rouse [113] and Zimm [114] in the limits of free draining and non-free draining bead-spring chains, τ_1 can be conveniently expressed in the form [115]:

$$\tau_1 = \eta_s[\eta]M/(A_1RT) \tag{8.29}$$

The numerical factor A_1 is 0.822 for the free draining chain [113], 1.184 for the non-free draining chain with pre-averaged hydrodynamic interactions [114, 116] and 0.574 without preaveraging the Oseen tensor [117]. Nevertheless, a number of publications advocate the use of $A_1 = 2.37$ for the non-free draining situation [22, 86, 118]. Since there is some controversy concerning the correct value for A_1, it is preferable to base discussion on experimentally determined quantities. Dynamic light scattering is a technique frequently used for probing the internal motions of flexible polymer chains in solution. From the spectrum of scattered light, different methods of analysis have been applied to estimate τ_1 for comparison with the theoretical predictions [119–121]. In some studies, it has been found that PS at infinite dilution, either in good (benzene) or theta (*trans*-decalin) solvents, behaves according to the Zimm model with $A_1 \cong 1.2$ [115, 119]. Recent investigations [120, 121] have confirmed the value of 1.2 in theta-solvents. In ethyl-benzene, a good solvent for PS, a smaller value of $A_1 = 0.84$ has been obtained [121]. This latter result is qualitatively consistent with an increasing draining effect with the more expanded coils.

Using $A_1 = 1.2$ for decalin and 0.84 for 1-methyl-naphthalene, the following numerical expressions have been derived for τ_1:

$$\tau_1 \text{ (decalin)} = 4.88 \times 10^{-14} \, M^{1.53} \text{ [s]} \tag{8.30}$$

$$\tau_1 \text{ (1-methyl-naphthalene)} = 2.18 \times 10^{-14} \, M^{1.70} \text{ [s]} \tag{8.31}$$

8.4.2.5
Thermodynamic Quality of the Solvents

The thermodynamic quality of the solvent plays the most fundamental role in polymer solution theory and controls the dynamics of chain uncoiling, at least during the initial stages of deformation [122]. Some experiments depend critically on the solvent quality as in "thermodynamically induced shear degradation", in which a few degrees change around the θ-temperature of the solution can influence the degradation rate by several orders of magnitude [123]. The experimentally determined theta-temperature for PS in decalin was 14.8 °C [41]. This value is proportional to the isomer content and should vary between 12 °C and 22 °C, which are the theta-temperatures for the *cis* and *trans* isomer, respectively [124]. At room temperature, the Mark-Houwink exponent for PS in decalin is somewhat larger than the theta-value 0.50. Therefore, it has been argued that decalin should be considered as a "good" solvent rather than as a "theta" solvent for PS at room temperature. The excluded-volume variable (z) can be used to quantify the "quality" of a solvent and settle the dispute. Experimentally, the viscosity expansion factor (α_η) is calculated from the ratio of intrinsic viscosities determined at the experimental and theta-temperature:

$$\alpha_\eta = [\eta]/[\eta]_\theta \tag{8.32}$$

The excluded-volume variable (z) is then estimated from α_η by using one of the available empirical relationships such as [125]

$$\alpha_\eta = (1 + 3.8 \, z + 1.9 \, z^2)^{0.3} \tag{8.33}$$

The excluded-volume variable z is fundamental to polymer solution theory. From 0 under theta-conditions, the value for z increases rapidly with the extent of polymer-solvent interactions. For a 10^7 dalton PS dissolved in decalin and in 1-methyl-naphthalene, values calculated for z are 0.19 and 8.0, respectively. These data indicate that PS coils should behave rather close to the ideal conditions in decalin (marginal solvent), whereas 1-methyl-naphthalene is definitely a good solvent for PS at room temperature.

8.4.2.6
Molecular Weight Distribution

The molecular weight distribution was determined by gel permeation chromatography (GPC) on a Waters 150 CV, equipped with special columns for ultra

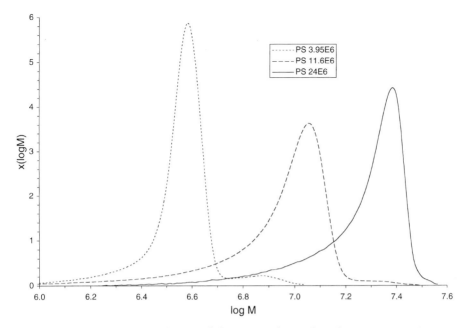

Fig. 8.16. Molecular weight distribution of the PS samples used in the present investigation. Analysis conditions: 3×10^6 Å Ultrastyragel, solvent CH_2Cl_2 at 0.2 ml/min, detection UV at 262 nm

high MW polymers (TSK G7000HXL and Ultrastyragel 10^6 Å) and diode array detection (Kontron model DAD-440). A PC-compatible personal computer was used for data acquisition and analysis. Polymers with MW above 3×10^6 are particularly prone to mechanochemical degradation. To preserve polymer integrity, extreme care should be exercised during the preparation, injection and analysis of the samples. Experimental conditions for a successful GPC characterization in the ultrahigh MW range have been described earlier [126].

8.4.3
Conversion of Retardation into Birefringence

Optical polarimetry measures the retardation (δ) whereas the quantity of interest for conformational studies is the degree of segmental orientation given by the birefringence (Δn). The transmission of polarized light discussed in Sect. 8.3 is based on the assumptions that the birefrigent medium is homogeneous and that the propagation axis is coincident with one of the principal axes of the refractive index tensor. In this case, the retardation δ is linearly related to the birefringence Δn of the sample by

$$\delta = - (2\pi L/\lambda) \cdot \Delta n \tag{8.34}$$

where L is the length of the optical path through the sample and λ, the light wavelength (632.8 nm).

The hypotheses used to derive Eq. (8.32) are nevertheless rarely met in practice and the problem of obtaining local birefringence (Δn) and local orientation (α) from the experimentally determined effective retardation $\delta^*(y)$ and effective extinction angle $\alpha^*(y)$ is non-trivial in any three-dimensional flow, including axisymmetric flows [64]. It is worth noting that real flows are always three-dimensional since genuine two-dimensional flows are not experimentally realizable: boundary layers and confining windows invariably induce variations in flow properties along the optical path [65, 127, 128].

In the section below, we will discuss some of the main difficulties generally encountered in obtaining local birefringence from rheo-optical measurements and the technique which has been used to solve this problem in axisymmetric converging flow.

8.4.3.1
Inhomogeneity of the Birefringence Zone

Most elongational flows are spatially inhomogeneous with large variations in velocities and strain-rates across the flow field. A polymer molecule traveling along a streamline will, therefore, experience different degrees of straining, resulting in a distribution in the degree of orientation. Because of this non-uniform distribution of Δn along the optical path (Fig. 8.17), the local birefringence cannot be determined directly from the retardation measurements as in the case of homogeneous flow.

From a rigorous point of view, the overall optical anisotropy of a medium which exhibits spatial inhomogeneity should be treated with the general formulation based on differential propagation Jones or Mueller matrices [100]. The use of differential propagation matrices requires, however, cumbersome numerical integration. The fact that most of the present studies are limited to the highly dilute regime with maximum recorded retardations of the order of 10^{-2} radian allows significant simplification from the full differential formalism. In the limit of small effective retardance ($\delta^* \ll 1$), it can be shown that the differential propagation Jones (or Mueller) matrix for a composite material is

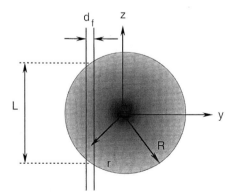

Fig. 8.17. Polarized light propagation through an optically inhomogeneous medium (L = optical path, d_f = probe beam width)

simply the product of the matrices for each of the separate optical elements. In this case, expression for the experimental effective retardation (δ^*) reduces to an integral over the optical path:

$$\delta^*(y) = (2\pi/\lambda) \int\limits_{-\infty}^{+\infty} \Delta n(z)\, dz \qquad (8.35)$$

For radially symmetric phenomena, it is convenient to rewrite Eq. (8.35) in terms of radial coordinates defined as $r = (y^2 + z^2)^{0.5}$:

$$\delta^*(y) = (4\pi/\lambda) \int\limits_{y}^{R} (r^2 - y^2)^{-0.5} \cdot \Delta n(r) \cdot r\, dr \qquad (8.36)$$

In this expression, the integral is taken along a strip at constant y and R is the radius beyond which $\Delta n(r)$ is negligible (Fig. 8.17).

Formally, Eq. (8.36) is a special case of the Volterra integral differential equation which can be inverted analytically to give

$$\Delta n(r) = -(\lambda/2\pi^2) \int\limits_{r}^{R} (r^2 - y^2)^{-0.5} \cdot d(\delta^*(y))/dy)\, dy \qquad (8.37)$$

as was shown by Abel in 1826.

Although Eq. (8.37) provides a formal way to recover $\Delta n(r)$ from $\delta^*(y)$, its practical application is difficult because experimental data are obtained as a numerical set containing noise rather than as an analytical function. For this reason, a numerical method is generally used for the integration [129].

8.4.3.2
Numerical Inverse Abel Transform

The inverse Abel transform is a general technique which should be employed whenever one needs to reconstruct a two-dimensional function from an experimental projection (a line integral). Situations which require Abel inversion are widespread and can be found in fields as diverse as radiotherapy, plasma diagnostics, radar mapping, heat transfer and spectroscopy. Owing to its technical importance, the problem of finding the correct inverse Abel transform remains an active field of research in applied engineering with continuous development of new algorithms in the literature. One of the earliest numerical methods of solving the inverse Abel transform was described over 60 years ago and consists of dividing the optical zone of interest into annuli, with the assumption that optical properties were constant within each annulus [130, 131] (cf. Appendix A). This technique has been recently applied to the specific problem of retardation conversion in opposed-jet flow [132, 133]. The zonal description of the cross-section, and other variants, have one main drawback which has been frequently emphasized in the past [134–136]: the recursive relationship used to solve for the unknowns $\Delta n_i(r)$ (Eq. A10 of the Appendix) is applied starting from the outermost annulus which, unfortunately, also happens to contain the least accurate part of the data. With successive calculations, errors add up and are magnified. Despite this sensitivity to cumulative errors, the above method has the advantage of simplicity and we will adopt it, since orientation angle correction could be readily accommodated. Using simulated data, we found that the procedure could give reliable results providing some precautions are taken, as follows.

8.4.3.2.1
Symmetrization of the Experimental Data

The Abel transform is applicable only to axisymmetric functions whereas experimental retardations are rarely symmetric as a result of noise and flow fluctuations. To improve the *S/N* ratio and minimize random asymmetry, the retardation signals are typically obtained after averaging over 3–5 scans for the weakest birefringence. Owing to the rapid decay of retardation with distance from the entrance in contraction flow (2%/µm), precise flow cell alignment with respect to the measuring direction is primordial to avoid systematic dissymmetry, which cannot be corrected for by repeated scanning. Presence of residual asymmetry in the experimental profiles was assessed in the next step by fitting the data with an even function. (For the sake of completness, it should be mentioned that extension of the Abel transform to non-symmetrical situations has been applied in tomography [137, 138]. In any case, additional measurements along different complementary axial directions are required.)

8.4.3.2.2
Data Smoothing

Since the forward Abel transform is an integration (Eq. 8.36), inversion necessarily involves data differentiation (Eq. 8.37), a process which exacerbates the effect of noise. Reduction of spurious noise, therefore, constitutes a key step in any successful Abel inversion algorithm [134, 139–142]. In the "Nestor-Olsen" method, smoothing was accomplished by a polynomial fit [139–142]. For the present purpose, the set of retardation data was least-square fitted to a symmetric analytical function using a commercial curve-fitting software (Table Curve 2D, Ver. 4.0, from Jandel Scientific) (Fig. 8.18). The function obtained is then inverted according to the procedure described in Appendix A. It should be stressed here that the fitted function is entirely empirical; it depends on the actual experimental conditons and does not imply (as far as this study is concerned) any theoretical or physical meaning.

8.4.3.3
Transmission of Polarized Light at Oblique Incidence

An additional complicating feature of polarimetry in converging flow is the variation of the principal axes of the refractive index tensor with local position.

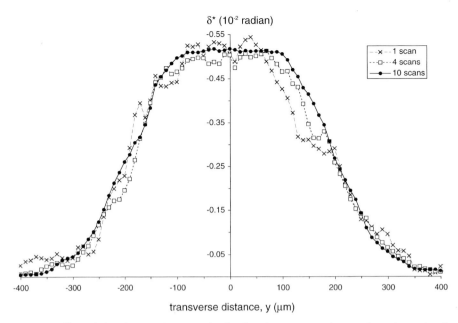

Fig. 8.18. Effect of signal averaging on noise level and asymmetry in transient elongational flow retardation profile (flow rate: 200 cm³/min, 200 ppm PS 3.95 × 10⁶ in 1-methyl-naphthalene, 100 µm from the entrance; the solution was intentionally not degassed to dramatize the effect of signal averaging)

As a consequence, the propagation direction of the polarized beam does not coincide with one of the principal axes of the refractive tensor. The situation pertinent to the present studies is depicted in Fig. 8.19. In the ensuing development, we will follow closely the approach which has been given in [100].

In Fig. 8.19, an ensemble of flexible polymer chains are uniaxially stretched in the direction x of the laboratory frame of reference. Due to the symmetry of the deformed chains, the principal axes of the refractive index tensor will have $\mathbf{n}_1$ aligned along x, while the components $\mathbf{n}_2$ and $\mathbf{n}_3$ of equal amplitudes are contained in the (y, z) plane. The probe beam is directed along the unit vector $\mathbf{e}_3$ forming a polar angle θ with the z-axis, while its projection onto the (x, y) plane defines the azimuthal angle ϕ. Since light is propagating along the $\mathbf{e}_3$ axis, it will only sample material properties in the plane $(\mathbf{e}_1, \mathbf{e}_2)$ which contains the electric displacement vector. In the limit of small birefringence, the refractive index (Δn) and orientation angle (α) measured in the $(\mathbf{e}_1, \mathbf{e}_2)$ plane are given by

$$\Delta n = \Delta n_{12} (\sin^2\phi + \cos^2\phi \cos^2\theta) \tag{8.38}$$

$$\tan 2\alpha = (2 \cos \theta \sin\phi \cos\phi)/(\sin^2\phi - \cos^2\phi \cos^2\theta) \tag{8.39}$$

where $\Delta n_{12} = (n_1 - n_2)$.

The spinning $\lambda/2$ plate of the Rheometrics Optical Analyzer rotates the electric vector of the incident beam at a frequency $4\omega t$ in the plane $(\mathbf{e}_1, \mathbf{e}_2)$ (Eq.

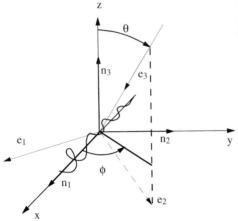

Fig. 8.19. Transmission of polarized light through an homogeneous birefringent material at oblique incidence

8.14). As a result, the azimuthal angle ϕ obeys the same time dependence in the (x, y) plane and should be replaced by $\phi(t) = 4\omega t$ in Eqs. (8.38) and (8.39). With each measurement lasting typically 0.1 s, the signal is averaged over several hundreds cycles, resulting in the following expressions:

$$\overline{\Delta n} = \Delta n_{12} \overline{(\sin^2 4\omega t + \cos^2 4\omega t \cos 2\theta)} = \Delta n_{12} \left(\frac{1}{2} + \frac{1}{2} \cos^2 \theta \right) \tag{8.40}$$

$$\overline{\alpha} = \frac{1}{2} \tan^{-1} \overline{\{(2 \cos \theta \sin 4\omega t \cos 4\omega t)/(\sin^2 4\omega t + \cos^2 4\omega t \cos^2 \theta)\}} \tag{8.41}$$

(the bar denotes a time-average value over a whole number of cycles).

Symmetry considerations show that the mean value of the right-hand term of Eq. (8.41) is zero for each half-cycle. However, because the phase of the lock-in amplifier is synchronized with the orientation of the rotating half-wave retarder, set initially to 0° in a position parallel to the x-axis, the quantity (α) effectively obtained from the measurement is the projection of ϕ in the (e_1, e_2) plane.

When both local orientation and local birefringence are taken into account, the equation for the effective retardation (Eq. 8.36) becomes

$$\delta^*(y) = (2\pi/\lambda) \int_{-\infty}^{+\infty} \Delta n(z) \, dz =$$

$$= (4\pi/\lambda) \int_{y}^{+\infty} (r^2 - y^2)^{-0.5} \cdot \Delta n_{12}(r) \cdot \left[\frac{1}{2} + \frac{1}{2} \cos^2 \theta(r) \right] \cdot r \, dr \tag{8.42}$$

Equation (8.42) contains two unknowns which are $\Delta n_{12}(r)$ and $\theta(r)$. The problem of solving for both quantities solely from the experimental $\delta^*(y)$ and $\alpha^*(y)$ is under-determined and becomes tractable only if some additional information on $\Delta n_{12}(r)$ and/or $\theta(r)$ is available. This is the case, for example, in oriented fibers [143], in Couette flow [144] or along the symmetry axis in opposed-jet flow [58, 129], where Δn and/or θ remain constant along the optical path.

In the present study, flow simulation is applied to obtain an estimate of $\theta(r)$. As mentioned in Sect. 8.2.2.3, chain orientation can be determined from the velocity vector, or better still, from the first eigenvalue of the rate of deformation tensor (Fig. 8.10). Once numerical values for $\theta(r)$ are known, Abel inversion of Eq. (8.42) can be performed according to the numerical procedure detailed in Appendix A. Some typical transverse retardation profiles along with the calculated birefringence data will be given in the following paragraphs (Sects. 8.4.6 and 8.4.7).

8.4.4
Extraneous Birefringence

Quantitative interpretation of the birefringence data for a polymer solution necessitates independent measurements of the birefringence characteristics of the optical windows and of the solvents.

Parasitic retardation from the windows comes from two sources which are compression-induced (during mounting) and pressure-induced birefringence (from extra pressure exerted by the flow). Both of these birefringences were barely discernable from noise level in the present design and were neglected in subsequent calculations.

Solvent birefringence is frequently observed with small asymmetric molecules. Under experimental conditions, decalin does not show any measurable birefringence whereas a weak but reproducible retardation was detected with 1-methyl-naphthalene. The birefringence characteristics of the latter are given in Figs. 8.20–8.22. Over the investigated range, solvent retardation (δ_s) at fixed axial distance from the orifice showed a linear increase with the fluid strain rate (Fig. 8.20) whereas the effective orientation (α_s^*) was independent of this flow parameter and tended asymptotically to a value of 45° at some distance from the center flowline (Fig. 8.21). It is worth noting that δ_s for 1-methyl-naphthalene (as for most investigated solvents) is positive whereas the polymer retardation (δ_p) is negative for PS as a result of the transverse orientation of the phenyl groups with respect to the chain backbone [143].

Variation of local birefringence with flow rate and axial distance was much more gradual in pure solvent than in dilute polymer solutions. The orientation behaviour of pure 1-methyl-naphthalene was also different from that of the polymer, as can be assessed from a comparison of Fig. 8.21 with Fig. 8.31.

Small asymmetric molecules are generally modelled as either rigid ellipsoids [90] or slightly deformable spheres with large rotational diffusion constants [96]. The static theory developed for rigid particles predicts that the main axes of the orientation should correspond to the principal axis of strain of the flow (which is 45° in simple shear flow) at all velocity gradients accessible to experiment. The dynamic theory allows for some supplemental orientation along the direction of the principal eigenvector of the rate of deformation tensor, as with polymer chains. The latter situation may be observed for sufficiently elongated solvent molecules at high strain-rates [145]. As stated in Sect. 8.4.3, the use of a correct orientation distribution is important for the conversion of the effective retardation into the local birefringence. The results obtained are only weakly de-

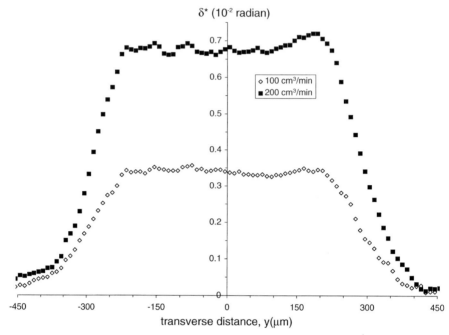

Fig. 8.20. Effective transverse retardation profiles for pure 1-methyl-naphthalene recorded at $x = -50$ μm, at flow rates of 100 and 200 cm^3/min

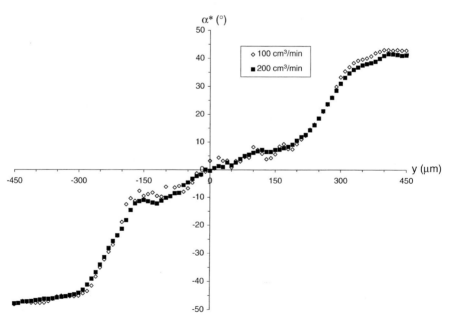

Fig. 8.21. Effective transverse orientation distributions for pure 1-methyl-naphthalene (experimental conditions as for Fig. 8.20)

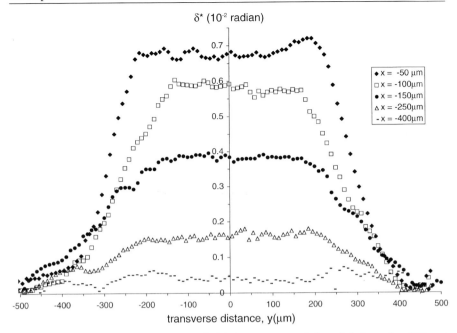

Fig. 8.22. Effective retardation profiles $\delta^*(y)$ in pure 1-methyl-naphthalene, for different axial distances (x), at a flow rate of 200 cm^3/min

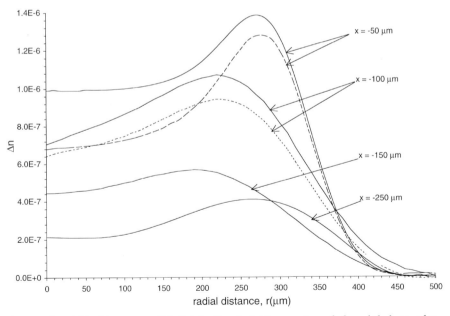

Fig. 8.23. Local birefringence radial distribution $\Delta n(r)$ in pure 1-methyl-naphthalene, calculated with the orientation of the principal axis of strain (——). Calculations with orientation angle of the principal eigenvector of the velocity gradient (------) are also given for $x = -50$ μm and -100 μm

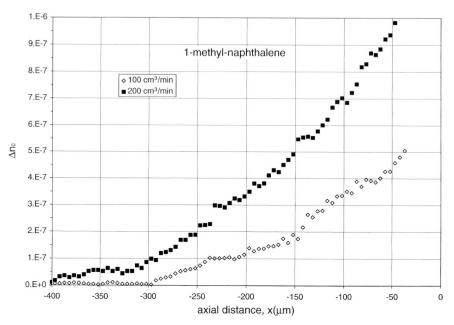

Fig. 8.24. Central local birefringence $\Delta n_0(r)$ in pure 1-methyl-naphthalene as a function of axial distance (x), at flow rates of 100 and 200 cm³/min

pendent on the orientation model. In subsequent calculations, we will assume that the principal axis of strain is determinant for solvent orientation.

The birefringence and orientation of the solution depend on contributions from both the polymer and the solvent. The solvent effects should be taken into account to determine correctly the value and the orientation of the excess flow birefringence introduced by the polymer. For this purpose, the solution is considered as a two-component system in which the solvent is characterized by the birefringence Δn_s and orientation α_s, and the polymer by the corresponding Δn_p and α_p. Assuming that the effects of the different components in a polydisperse system are linearly additive, the following equations known as the Sadron rule [146] have been derived, based on the vectorial representation of the refractive indices [147] (Fig. 8.25)

$$\Delta n^2 = \left(\sum \Delta n_i \sin 2\alpha_i\right)^2 + \left(\sum \Delta n_i \cos 2\alpha_i\right)^2 \tag{8.43}$$

$$\tan 2\alpha = \left(\sum \Delta n_i \sin 2\alpha_i\right)/\left(\sum \Delta n_i \cos 2\alpha_i\right) \tag{8.44}$$

where the summation should be taken over all the components of the mixture.

In simple shear flow where the solvent orientation angle (χ_s) is always 45°, it has been shown that the birefringence from the polymer alone can be obtained by [148]:

$$\Delta n_p^2 = \Delta n^2 + \Delta n_s^2 - 2\,\Delta n_s \Delta n \sin 2\alpha \tag{8.45}$$

$$\cos 2\alpha_p = (\Delta n/\Delta n_p)\cos 2\alpha_p \tag{8.46}$$

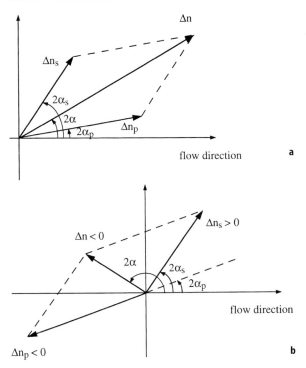

Fig. 8.25a, b. Vectorial representation of solution flow birefringence; **a** with positive Δn_p and Δn_s; **b** with Δn_p and Δn_s opposite in sign ($\Delta n_p < 0$)

where the subscript p refers to the polymer, s to the solvent, and no subscript, to the measured solution property.

In other types of flow Eq. (8.47) should be used:

$$\Delta n^2 = \Delta n_s^2 + \Delta n_p^2 + 2\,\Delta n_s \Delta n_p \cos\,(2\alpha_s - 2\alpha_p) \qquad (8.47)$$

Only the positive root of the quadratic Eq. (8.47) has a physical meaning, giving

$$\Delta n_p = -\,\Delta n_s \cos(2\alpha_s - 2\alpha_p) + [\Delta n_s^2(\cos^2(2\alpha_s - 2\alpha_p) - 1) + \Delta n^2]^{0.5} \qquad (8.48)$$

The excess birefringence can be calculated from the experimental measurements Δn and Δn_s, and from the flow simulation results for $\cos\,(2\alpha_s - 2\alpha_p)$. In the limit of small $(2\alpha_s - 2\alpha_p)$, Eq. (8.47) or (8.48) is practically equivalent to the simple relation $\Delta n \cong \Delta n_s + \Delta n_p$.

Solvent contribution to the total birefringence is substantial when Δn_p and Δn_s are comparable in magnitude. This is the case, for instance, in TCP and in Arochlor solutions [22, 145]. Flow birefringence in pure 1-methyl-naphthalene is comparable with the level attained with the PS solutions in decalin (Figs. 8.28, 8.29). However, owing to the higher degree of chain orientation which can be achieved in a good solvent, Δn_s is typically ~ 2.5 times smaller than Δn_p under prevalent experimental conditions. Solvent correction is thus relatively unimportant for most of the experimental data reported in this work, except at low flow rates and large axial distances as a result of the more gradual decrease of Δn_s (cf. results in Sect. 8.4.7).

8.4.5
Probe Beam Dimensions

The finite cross section of the probe beam has always been a matter of concern in pointwise scanning methods. If the optics are not properly dimensioned, serious distortion of the experimental profile may occur, due to the averaging response over the measuring area. This is particularly important in converging flow, because of the strong spatial dependence of the retardation signal. In the prevailing design, a converging lens of $f = 25$ mm was used to focus the laser beam. The far-field diffraction-limited width at half-height of the beam in the focal region is [149, 150]:

$$d_f = (\lambda/n\pi) \cdot (4f/D) \tag{8.49}$$

whereas the effective length, l_f of the beam waist region is given approximately by

$$l_f = (\lambda/n\pi) \cdot (4f/d)^2 \tag{8.50}$$

with λ = laser wavelength (632.8 nm), D = diameter of the laser beam (0.76 mm at half-width) and n = composite refractive index of the medium ($\cong 1.5$).

The beam diameter was determined by blocking and translating the focused light across a knife edge. Assuming an axisymmetric irradiance distribution for the focused beam, the transmitted light intensity (I_t) is given by

$$I_t(y) = \int_{-\infty}^{y} \Phi(y') \, dy' \tag{8.51}$$

$$\text{with} \quad \Phi(y) = \int_{y}^{R} (r^2 - y^2)^{-0.5} \cdot I(r) \cdot r \, dr \tag{8.52}$$

The derivative of $I_t(y)$ gives $\Phi(y)$ which can then be converted into $I(r)$ with the inverse Abel transform (cf. Eqs. 8.35–8.37). Experimentally, $\Phi(y)$ can be fitted with a Gaussian function both in the absence and in the presence of the rotating half-wave plate (Fig. 8.26). Using the fact that the inverse Abel transform of a Gaussian function is also a Gaussian of identical standard deviation, the half-width of the focused beam is readily obtained from $\Phi(y)$.

Without the rotating half-wave plate, the light intensity profile is consistent with the prediction of Eq. (8.49) with $d_f = 19.8$ µm. Under the measuring conditions ($\lambda/2$ plate + flow), the value obtained is higher by a factor of 1.65 ($d_f = 33$ µm). It was verified that this additional broadening originates exclusively from imperfections of the rotating half-wave plate, since the original beam profile was recovered by replacing the turbine with a photoelastic modulator (PEM) [57, 72]. Using a shorter focal lens can improve lateral

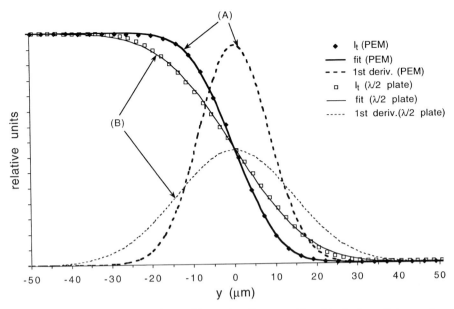

Fig. 8.26. Determination of the focused beam irradiance profile with the laser light passing (A) through a photoelastic modulator, (B) through a rotating half-wave plate. *Data points* are measured values, *continuous lines* are TableCurve fit with an error function, *dotted lines* are first derivatives of the transmitted intensities. Transmitted intensities are scaled to the same value for unobstructed beams; first derivative distributions are normalized to a constant area

resolution. However, the lower limit for the focal distance is $f \cong 15$ mm, to avoid divergence of the laser beam over the birefringence zone which is approximately 0.8 mm (Eq. 8.50).

Signal spreading by a dispersion function is a fundamental problem in science and engineering. Its treatment can be found in any introductory textbook on signal processing [151]. Any correction of signal dispersion starts with the convolution integral:

$$\delta(y) = \int_{-\infty}^{\infty} \delta'(y_0) \cdot G(y, y_0)\, dy_0 \tag{8.53}$$

where $\delta(y)$ is the actual detector response, $\delta'(y_0)$ the hypothetical signal obtained with a probe beam of infinitesimal dimensions, and $G(y, y_0)$ the spreading function.

In the present situation, the kernel $G(y, y_0)$ represents the shape of the probe beam, given by the function $\Phi(y)$ in Eq. (8.52), which can be reasonably assumed to be constant over the scanning distance. In this case, Eq. (8.53) can be replaced by

$$\delta(y) = \int_{-\infty}^{\infty} \delta'(y_0) \cdot G(y - y_0)\, dy_0 \tag{8.54}$$

Equation (8.52) is known as a Fredholm integral equation of the first kind and its inversion permits one to obtain the correct signal without instrumental broadening:

$$\delta'(y) = \int_{-\infty}^{\infty} \delta'(y_0) \cdot G^{-1}(y - y_0) \, dy_0 \qquad (8.55)$$

Once the probe beam profile is known with sufficient confidence, a deconvolution technique, generally based on Laplace transformations, can be applied to recover the correct birefringence distribution according to Eq. (8.55). The effect of probe beam size is most noticeable when its dimensions are comparable to those of the signal profile. This may be the case, for instance, in opposed-jet flow where birefringence zone was narrow, but is only a marginal effect in the present investigation (Fig. 8.27). Although reducing the beam size is still the best solution to obtain a correct signal, application of an adequate deconvolution technique should enable one to recover the original birefringence profile under appropriate conditions with minimal distortion.

In the example shown in Fig. 8.27, a numerical Gaussian deconvolution technique included with the PeakFit package [152] was employed to deconvolute the experimental retardation signal. The effect of signal spreading was barely discernable in the transversal (y) direction, whereas a ~5% error is estimated in the axial (x) direction.

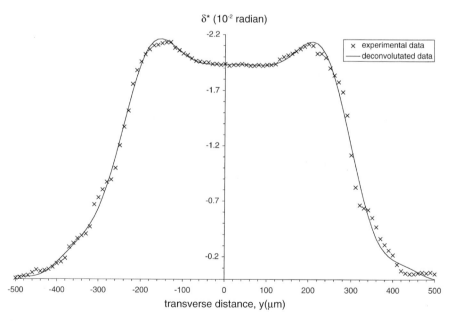

Fig. 8.27. Effect of probe beam size on transverse retardation profile (200 ppm 3.95×10^6 PS in 1-methyl-naphthalene, 200 cm^3/min, $x = -50$ μm)

8.4.6
Birefringence in Dilute Decalin Solutions

8.4.6.1
Optical Micrography

Direct observation and optical micrography were performed with a photographic arrangement based on the Senarmont compensator (Fig. 8.12). Visual techniques are still the most convenient means for a qualitative survey of the birefringence zone. This proves to be useful at the beginning of a series of experiments to assess the conditions of flow symmetry and stability: fluctuations generally mark the departure from Newtonian-like behavior. The optical micrographs given in Fig. 8.28 confirm that the lack of a stagnation point in transient elongational flow drastically changed the birefringence pattern: instead of being localized and constant along the flow direction as in opposed jets flow, the birefringence zone now extends over the whole orifice entrance. It should be obvious, however, from the discussion in Sect. 8.3.1, that static polarimetry whether used with a photodetector, a CCD camera or any other photon recording device, is unsuitable for a quantitative determination of retardations without an a priori knowledge of orientation angle distribution.

8.4.6.2
Quantitative Retardation Measurements

Birefringence was studied with the polarization-modulation technique, as a function of flow rate and polymer molecular weight (1.93×10^6 – 10.2×10^6). Consistent with the micrographs, pointwise scanning along the center flow

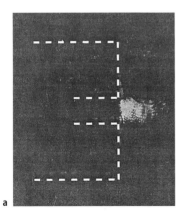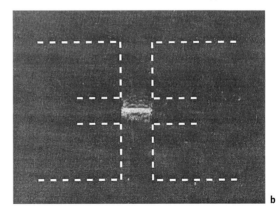

a   b

Fig. 8.28a, b. Optical micrograph of the birefringence zone: **a** in capillary entrance flow; **b** in opposed-jet flow [70] (experimental conditions: 400 ppm 3.95×10^6 PS in decalin, flow rate: 100 cm^3/min in (a) and 200 cm^3/min in (b)

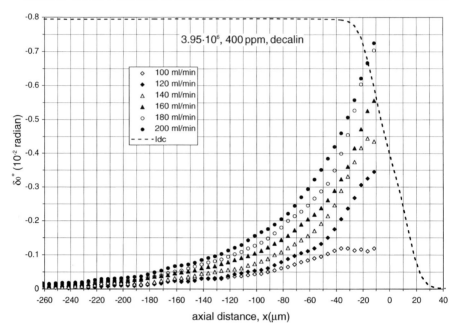

Fig. 8.29. Central retardation (δ_0^*) as a function of axial distance for a 400 ppm 3.95×10^6 PS solution in decalin, at the indicated flow rates. The *dotted line* is I_{dc} plotted in arbitrary units. Experimental points at $x > -40\ \mu m$ should be dismissed for the reasons mentioned in text

direction (x) indicated a rapid decrease of the retardation signal with distance from the orifice (Fig. 8.29). For a proper comparison between different runs, it is imperative to define precisely and reproducibly the position of the orifice entrance. This was accomplished by taking the zero of x at the experimental position where I_{dc} signal reaches 50% of its constant value (Fig. 8.29). With the present set-up, the closest approach to the orifice without occlusion of the laser beam was 40 μm. The present birefringence technique normalizes the experimental signal to I_{dc} (Eq. 8.14), thus allowing measurements to be performed even for a partly blocked beam. However, the results thus obtained are not reproducible owing to scattered light from the sapphire tip (blackening the jet tip could improve the precision of these data, but no efforts have so far been attempted).

Some typical transverse retardation distributions, recorded at different positions from the orifice, are shown in Fig. 8.30. Despite the large changes in retardation, the effective extinction angle distributions do not depend substantially on the axial distance over the investigated range (Fig. 8.31).

The effective retardations, of the order of a few milliradians, are comparable in magnitude to the values obtained in opposed jets flow under similar experimental conditions [41]. However, since the birefringence zone is ten times larger in single jet flow than in opposed jets flow, it can be estimated that birefringence is actually smaller by the same factor. Quantitative values of local birefringence are given in Figs. 8.32 and 8.33 below.

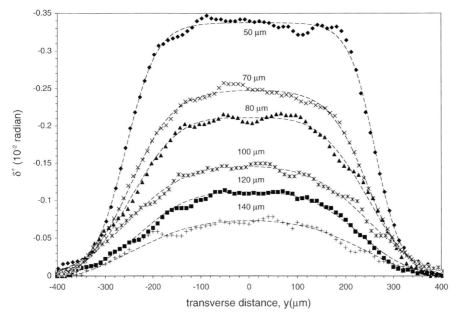

Fig. 8.30. Transverse retardation profiles for a 400 ppm 3.95×10^6 PS solution, scanned at the indicated distances from the inlet. Flow rate $= 200 \text{ cm}^3/\text{min}$. *Solid curves* are fitted results from TableCurve 2D

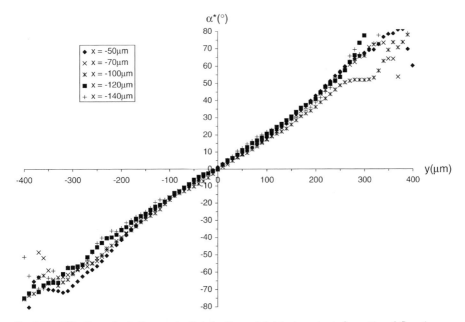

Fig. 8.31. Effective orientation angle distribution $\alpha^*(y)$ in transient elongational flow (same experimental conditions as in Fig. 8.30)

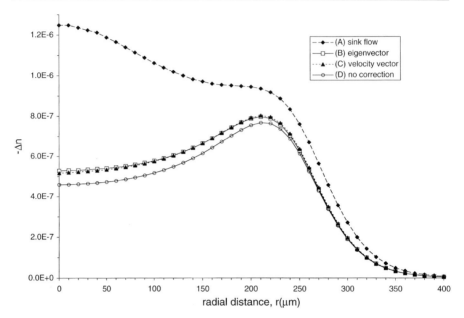

Fig. 8.32. Effect of local orientation angle correction on the radial birefringence distribution (experimental data are from Fig. 8.30 with $x = -50\,\mu m$). (A): corrected with the sink flow approximation; (B): corrected with the orientation of the first eigenvector of the rate of deformation tensor; (C): corrected with the orientation of the velocity vector; (D): without angle correction

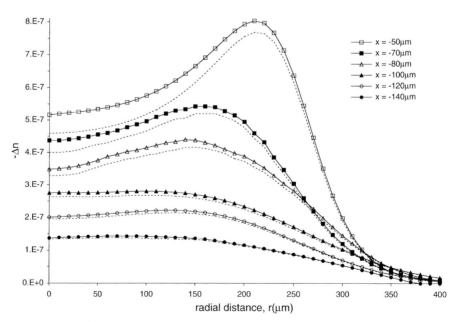

Fig. 8.33. Radial birefringence distribution, $\Delta n(r)$, calculated at the indicated axial positions (———) with the orientation of the first eigenvector of the rate of deformation tensor; (-----) without orientation correction. Experimental data are from Fig. 8.30

8.4.6.3
Radial Birefringence Distribution

Conversion of experimental retardation into birefringence was performed according to the method outlined in Sect. 8.4.3. Some typical birefringence profiles calculated with different orientation correction procedures are shown in Figs. 8.33 and 8.34. Radial birefringence distributions calculated without orientation correction show a broad dip at the center (Fig. 8.33). Correction with the polymer orientation angle parallel either to the velocity vector or to the principal eigenvalue of the rate of deformation tensor slightly increases the birefringence without changing the overall shape of the distribution. Data obtained with the sink flow approximation are also included for comparison, although birefringence obtained with this model is clearly unreliable, owing to overestimation of the polymer orientation angles.

The central minimum in the birefringence distribution diminishes rapidly with increasing distance from the orifice, and disappears completly at $x < -100$ μm (Fig. 8.33).

The presence of maxima near the edges of the capillary entrance was the most striking feature of the radial birefringence distribution. Since the flow field is only purely extensional along the flow symmetry axis, it was intuitively anticipated that the propensity for polymer deformation should diminish rather than increase at the approach of the entrance lip, where shearing is

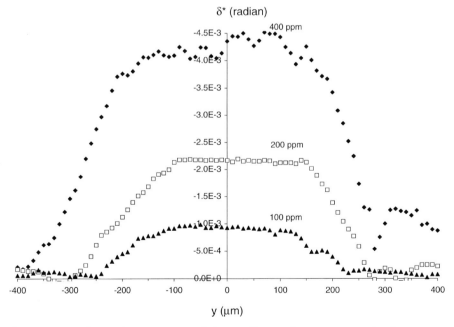

Fig. 8.34. Dependence of transverse retardation profile on polymer concentration for a 3.95×10^6 PS in decalin. Flow rate = 200 cm³/min, $x = -40$ μm

predominant. In the opposed-jet geometry, a central dip in the birefringence profile indicates the occurrence of interchain entanglements, or significant departures from simple Newtonian flow behavior, as a result of the "screening" effect from region of high elongational viscosity [80, 153]. Therefore, the possibility of flow field modification from deforming polymer chains is one plausible explanation for this unanticipated result. Insofar as these effects depend on the relative extension and proximity of the molecular coils, birefringence profiles should change with polymer concentration and fluid strain-rate. To assess for the possibility of non-Newtonian flow effects under the prevailing experimental conditions, retardations were measured over a limited range of polymer concentrations. It was verified that retardations in decalin solutions increased linearly with polymer concentration at least up to the limit of 200 ppm for the 10.2×10^6 PS and 500 ppm for lower MW samples. In addition, it was observed that the shape of the transverse retardation profile $\delta*(y)$ was independent of polymer concentration and flow rate over the investigated ranges (Fig. 8.35). This confirms that the polymer solutions are truly dilute at the experimental concentrations, with a Newtonian-like behavior. The presence of a valley in the birefringence distribution cannot, therefore, be ascribed to the aforementioned "screening" effect.

The role of "re-entrant corner" on flow birefringence in abrupt contraction flow was unknown to the authors at the time of the first publication [71] and it was suggested that the dip in the calculated birefringence distribution may result from an incorrect local orientation assignment. It now seems likely that this suggestion must be abandoned in favor of a more satisfactory explanation provided by detailed flow field calculations. The results of flow field simulations indicate the existence of extreme elongational strain-rates for the outermost streamlines close to orifice entrance lips (Fig. 8.19). These transient strain rates can promote a high degree of chain orientation. Assuming that chains are deformed affinely with the surrounding fluid element above a critical strain-rate, the degree of chain orientation was calculated for the different streamlines, giving a peak birefringence near the orifice walls (cf. Appendix C). Even if quantitative agreement between calculations and measurements is not expected in view of the crudeness of the affine deformation model, the calculated radial birefringence distributions reproduce surprisingly well most of the experimental observations, i.e., the presence of maximum birefringence near the entrance edges for short axial distances, and the relative increase of this maximum with polymer relaxation time. The fact that independent simulation confirms some hitherto unexplained experimental results gives further confidence to the conversion procedure which has been applied to compute radial birefringence distributions from the recorded retardations.

With a tapered inlet, it is expected that change in velocity sustained by the fluid as it passes the corner would be less severe, leading to a more uniform flow field. In contrast to the large number of studies in abrupt contraction flow, a survey of the literature indicates few experiments which consider the importance of finite corner rounding. Only two investigations have dealt explicitly with the role of re-entrant corner on flow birefringence measurements: one in a 4:1 planar contraction [154] and one in a circular hole [63]. From the published results, it is possible to conclude that the effect of lip curvature on the flow characteristics is negligible with Newtonian or viscoleastic fluids at low De, except for the region very close to the corner [76]. The consequences of corner smoothing is more pronounced at high De but seems to depend not only on the curvature radius but also on the nature of the fluid [155–157].

In spite of the complexity of the problem, it would be interesting to be able to minimize the "re-entrant corner" by altering the entrance geometry, for a better resolution of the central birefringence zone. Some preliminary experiments in that direction have recently been undertaken in our laboratory. One main difficulty encountered was experimental: with a rounded corner, the contraction is now situated inside the capillary section. The minimum measuring distance is thus increased by the lip radius of curvature. Since most of the interesting birefringence features occur at a distance $< r_0$, the rounding radius is limited to a fraction of this quantity. The second difficulty concerns birefringence modeling: since rounding imparts an additional length scale for fluid acceleration, and hence on residence time, the orientation dynamics may change and partly compensate for the decrease in fluid strain-rate.

8.4.6.4
Axial Birefringence Distribution

The axial retardation distributions, $\delta_0^*(x)$, obtained by scanning along the central flow direction ($y = 0$), are shown in Fig. 8.36. The effective axial retardation is given by Eq. (8.42) by setting $y = 0$:

$$\delta_0^*(x) = (4\pi/\lambda) \int_0^R \Delta n_{12}(r) \cdot [^1/_2 + {}^1/_2 \cos^2\theta(r)] \, dr \tag{8.56}$$

Knowledge of the central retardation, $\Delta n_0(x)$, is important in verifying some theoretical predictions of flow-induced flexible chain deformation. Formally, $\Delta n_0(x)$ should be calculated from the experimental transverse retardation profile using the technique described in Sect. 8.4.3 and the Appendix. Due to the sharp decay of $\delta_0^*(x)$, the measurement of $\delta^*(y)$ becomes excessively dependent on noise at some distance from the orifice, particularly at low flow rates. Incidentally, calculations with a series of carefully determined $\delta^*(y)$ revealed that $\Delta n_0(x)$ becomes proportional to $\delta_0^*(x)$ at orifice distances greater than 100 μm:

$$\delta_0^*(x) = C(x) \cdot \Delta n_0(x) \tag{8.57}$$

From 6600 radians at $x = -40$ μm, $C(x)$ decreases rapidly to 5700 at $x = -70$ μm, then levels off at 5300 ± 200 for $x \le -100$ μm. The same trend was observed for all the polymer samples investigated. Subsequently, the fitted curve for $C(x)$ is used to determine the central birefringence distribution from $\delta_0^*(x)$, as reported in Sect. 8.5.1 (Figs. 8.39, 8.40, 8.47, 8.48).

8.4.7
Birefringence in Dilute 1-Methyl-Naphthalene Solutions

Whether or not flexible polymer chains may eventually reach a state of full extension (or at least full segmental orientation) in transient elongational flow remains a key issue which necessitates further investigation. In capillary entrance flow, the degree of molecular stretching is largely controlled by the dimensionless parameter $\dot{\varepsilon} \cdot \tau_1$. Thus, the theoretical limit to coil deformation is dictated by experimental conditions, which are the maximum flow rate without turbulence and the availability of well-characterized polymer fractions of sufficiently high MW. The use of a thermodynamically good solvent is another possibility for increasing the degree of molecular stretching, since swollen molecular coils have a higher τ_1 and favor the hydrodynamics gripping action of the accelerating fluid. In addition, it is expected that a more expanded coil is less prone to internal friction and self-entanglement during a fast molecular straining.

In this investigation, 1-methyl-naphthalene was selected as a good solvent for PS for the following reasons:

a) unlike Arochlor or tri-cresyl-phosphate (TCP), 1-methyl-naphthalene is non-toxic and can be redistilled under partial vacuum for further usage;
b) its refractive index closely matches those of PS, thus avoiding any complication which may arise from form birefringence;
c) it has almost the same viscosity as decalin, and most of the difference in the birefringence behavior between these two solvents could be uniquely ascribed to the effect of solvation;
d) the level of solvent birefringence is weak in comparison with the polymer birefringence so that correction can be performed with a minimum of uncertainties (butyl-naphthalene has all the desirable properties of 1-methyl naphthalene, in addition to a higher viscosity [158], but is no longer available commercially).

Some representative flow birefringence results obtained in 1-methyl-naphthalene are presented in Figs. 8.35–8.41. Under identical experimental conditions, the level of retardation which can be reached in 1-methyl-naphthalene is roughly one order of magnitude higher than in decalin. Also, for a given polymer MW, birefringence could be observed at a much lower flow rate as a result of the increase in τ_1. These results amply confirm the higher propensity for flow-induced segmental orientation in polymer chains dissolved in a good solvent.

As in decalin, the effective retardation (Fig. 8.35) and birefringence decreased rapidly with axial distance (Figs. 8.36 and 8.38) and flow rate (Fig. 8.37). It seems from Fig. 8.38 that the central birefringence tends to level off at a short distance from the orifice entrance. To verify this behavior, the birefringence was measured with ultra-high MW polymer solutions (11.6×10^6 and 24×10^6 PS). The results presented in Figs. 8.39 and 8.40 demonstrate clearly that the birefringence saturated at sufficiently high strain rates. The effect is more conspicuous with increasing polymer MW. This interesting behavior could be observed either at constant flow rate, but with increasing strain-rate at

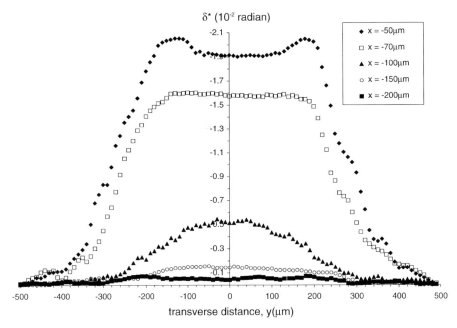

Fig. 8.35. Transverse retardation profiles for a 200 ppm 3.95 × 10⁶ PS solution in 1-methyl-naphthalene, scanned at the indicated distances from the inlet. Flow rate = 200 cm³/min

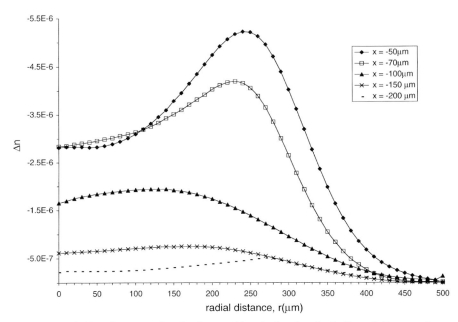

Fig. 8.36. Radial birefringence distribution, $\Delta n(r)$, calculated at the indicated distances from the inlet with the orientation of the first eigenvector of the rate of deformation tensor. (Experimental data from Fig. 8.35)

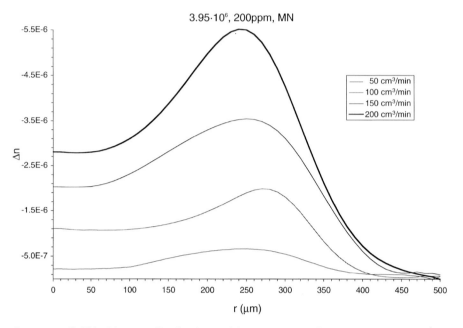

Fig. 8.37. Radial birefringence distribution $\Delta n(r)$ at $x = -50$ μm, for a 200 ppm 3.95×10^6 PS solution in 1-methyl-naphthalene, calculated with the orientation of the first eigenvector of the rate of deformation tensor

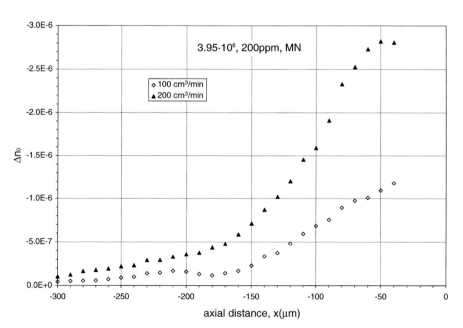

Fig. 8.38. Central local birefringence Δn_0 as a function of axial distance (x), for a 200 ppm 3.95×10^6 PS solution in 1-methyl-naphthalene, at flow rates of 100 and 200 cm^3/min

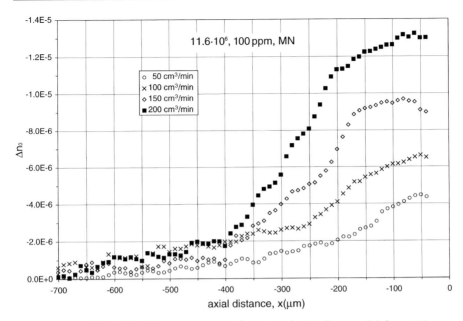

Fig. 8.39. Central local birefringence Δn_0 as a function of axial distance (x) for a 100 ppm 11.6×10^6 PS solution in 1-methyl-naphthalene, at the indicated flow rates. All results are corrected for solvent birefringence

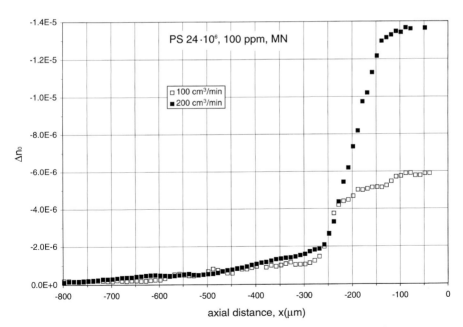

Fig. 8.40. Central local birefringence Δn_0 as a function of axial distance (x) for a 100 ppm 24×10^6 PS solution in 1-methyl-naphthalene, at the indicated flow rates

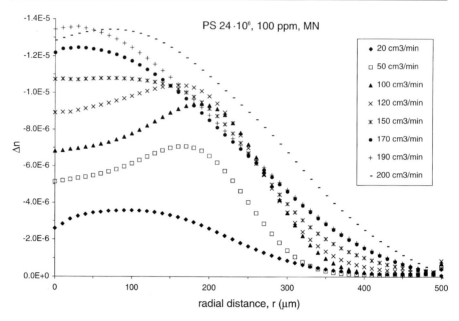

Fig. 8.41. Radial birefringence distribution $\Delta n(r)$ at $x = -50\ \mu m$, for a 100 ppm 24 × 10⁶ PS solution in 1-methyl-naphthalene, at the indicated flow rates

the approach of orifice entrance (Figs. 8.39 and 8.40), or at fixed axial position, with increasing flow rate as shown in Fig. 8.41. Birefringence saturation does not mean, however, that chains have reached a state of maximum orientation since the value for the plateau continued to increase with flow rate, at least for the 11.6 × 10⁶ sample. The possible origins of this plateau will be discussed in a coming section (Sect. 8.5.2).

True saturation seems to be reached with the highest MW polymer (24 × 10⁶ PS). From the computed radial birefringence distributions (Fig. 8.41), true saturation was observed when the birefringence reached a value of $\Delta n_{max} =$ -1.35×10^{-5} for a 100 ppm PS solution.

8.4.8
Flow-Induced Degradation

Chain scission is known to occur in abrupt contraction flow [38–44]. To assess the extent of degradation during birefringence measurement, the polymer solution was collected after a single passage through the orifice and analyzed by GPC. Under the experimental conditions, all samples show evidence of chain fracture above a critical strain rate ($\dot{\varepsilon}_f$) which scales with initial polymer MW as M^{-1} [41]. For example, for a 11.6 × 10⁶ PS in 1-methyl-naphthalene, degradation started at $\dot{\varepsilon}_f \sim 9000\ s^{-1}$ ($Q = 50$ ml/min) to become almost quantitative (> 90%) as the fluid strain rate reached 34,000 s^{-1} ($Q = 180$ ml/min) (Fig. 8.42).

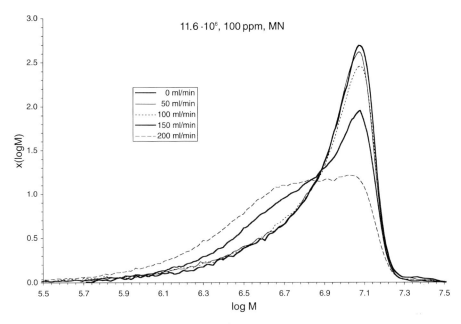

Fig. 8.42. GPC chromatograms of a 11.6×10^6 PS in 1-methyl-naphthalene degraded after a single passage at the indicated flow rates. Analysis conditions: 2×10^6 Å + 10^5 Å Ultra-styragel, solvent CH_2Cl_2 at 0.2 ml/min, detection UV at 262 nm

At this point, a large decrease in flow birefringence is expected since most of the chains are broken into halves and the test fluid should behave as a 5.8×10^6 PS solution. Such a drop, however, was not observed during the first run but only with a recycled solution (Fig. 8.43). This phenomenon could be rationalized if degradation occurred beyond the measurement location (i.e., at $x >$ -40 μm, or inside the capillary section) or if a hysteresis phenomenon, as predicted by de Gennes [8] prevented the strained chains from retracting during their transit time. In any case, the presence of flow-induced degradation seems to have little incidence on the birefringence results.

An important issue which we could not presently answer is the region where degradation effectively took place. It was assumed in previous investigations that degradation started preferentially from the center streamline where flow is purely extensional [43, 159]. In this case, the birefringence recorded at the on-set of polymer degradation ($\dot{\varepsilon}_f$) indicates a modest degree of chain deformation with (R_0/R_{max}) of the order of $10-15\%$. However, results presented elsewhere in this book, suggest that chains could be broken near the orifice lip, either immediately in front of the entrance (re-entrant corner effect) or inside the capillary section, from the action of high wall shear rates on already extended macromolecules [160].

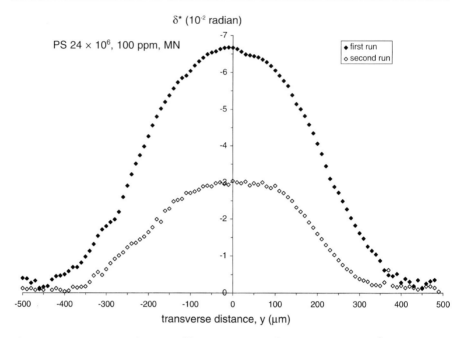

Fig. 8.43. Transverse retardation profiles at $x = -50$ μm for a 100 ppm 24×10^6 PS solution in 1-methyl-naphthalene, during the first and second run, at a flow rate of 200 cm^3/min

8.5
Discussion

Abrupt contraction flow has the peculiarity of being steady in an Eulerian frame of reference, but transient in a Lagrangian frame which moves with the fluid element. A polymer flowing along a streamline will experience subcritical conditions ($\dot{\varepsilon} < 1/\tau_1$) for most of its residence time, and supercritical conditions ($\dot{\varepsilon} > 1/\tau_1$) only during the last $10^{-3} - 10^{-4}$ s before entering the orifice (Fig. 8.11). Because the transit time in the strong flow region is generally less than τ_1, the polymer coil will respond to the transient flow field in a manner different from the steady situation. However, since most of the polymer kinetic models which have been developed refer to steady-state elongational flows, we will first examine the predictions of some of these models when transposed to the abrupt contraction flow situation.

8.5.1
Affine Deformation Model

Studies of flow through or exiting from a narrow orifice was initiated by Trouton in 1906 [1]. Early interest in this flow geometry was connected with fiber-spinning mechanisms [161]. The first orientation measurements in elongational flow were reported by Zabicki and Kedzierska in the late 1950s, for different polymer melts leaving a circular spinneret [162–164]. From bire-fringence data, these authors determined that orientation was insignificant below a strain-rate of 0.5 s^{-1} but increased rapidly with the gradient until satu-ration at about 10 s^{-1}. In these early works, the role of elongation rate could not be separated from cooling effects and the results were explained by the for-mation of polymer threads upon solidification at the exit [162]. Theoretical as-pects of chain orientation in extensional flow started to be developed at that time. Using an ideally elastic necklace model with no or constant hydro-dynamic interaction, Takserman-Krozer [165] and Peterlin [166] predicted an abrupt increase in Δn as the fluid strain-rate approaches a critical value $\dot{\varepsilon}_c = 0.5/\tau_1$. More realistic models of chain deformation in "strong" flow with the inclusion of conformation-dependent hydrodynamic interactions were pro-posed some ten years later, independently by de Gennes [8] and by Hinch [9]. With the weakening of hydrodynamic interactions on increasing coil deforma-tion, the coil-stretching transition sharpens, leading to a sudden passage from a coiled to a quasi-extended conformation at a critical strain-rate $\dot{\varepsilon}_c$.

More specific to the abrupt contraction flow situation, the problem of flexible polymer deformation during passage through conical and cylindrical pores was addressed by Daoudi and Brochard [167]. From their theoretical con-siderations, it was concluded that molecular coils follow affinely the deforma-tion of the local volume element when the modulation frequency (or the strain-rate) is higher than the inverse chain relaxation time, i.e.,

$$\dot{\varepsilon}_c \geq 1/\tau_1 \tag{8.58}$$

In pore entrance flow, the model stipulates that molecular coils remain es-sentially unperturbed up to some distance r_c from the orifice entrance defined by the strain-rate condition

$$\dot{\varepsilon}(r_c) = \dot{\varepsilon}_c \tag{8.59}$$

After the coil-stretching transition ($r < r_c$), chains are extended at the same rate as the surrounding fluid element without adjustment. For a 3-dimensional sink flow, the stretching ratio λ is simply given by the relation (cf. Eq. 8.2)

$$\lambda = R/R_0 = (r_c/r)^2 \tag{8.60}$$

where R is the end-to-end distance of the extended chain.

Essentially the same deformation model was used by Tabor and de Gennes [4] to explain the mechanism of drag reduction in dilute polymer solutions.

A special situation occurs when the pore radius (r_p) is comparable or smal-ler to the molecular hydrodynamic radius. Using scaling arguments, Daoudi and Brochard [167] and de Gennes [26] predicted that, beyond a critical flow rate corresponding to $r_c \cong R_0$, all chains can pass through the pore regardless of the polymer size and pore diameter (Fig. 8.44). This remarkable result is the outcome of two competing effects which exactly compensate each other ac-cording to the affine deformation model: although larger chains must undergo

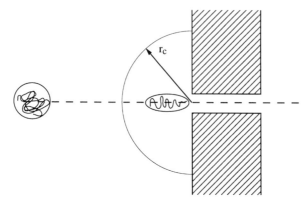

Fig. 8.44. Flexible polymer chain entering a pore

greater deformation to pass the orifice, they begin to deform at a greater distance r_c and thus can reach the same dimensions at the entrance. From this model, the reflection coefficient σ which characterizes the fraction of polymers rejected by the pore, should show a first-order phase transition with flow rate. Experimental observations with highly dilute PS solutions in theta [168] and good solvents [169] confirmed the size independency of the rejection factor, but are entirely inconsistent with the sharp transition prediction at a critical flow rate Q_c (Fig. 8.45).

The affine deformation representation uses a Hookean spring, it does not take into account any of the time-dependent phenomena globally known as internal viscosity, or the dependence of conformational elastic force on the degree of stretching. With all these approximations, it is doubtful that this description may be applied to a real situation. The picture is appealing, however, in its simplicity and we will examine it here as a starting model for further discussion and refinement.

To assess the possibility of effective molecular deformation, the degree of fractional coil extension (β) was computed along the center streamline from

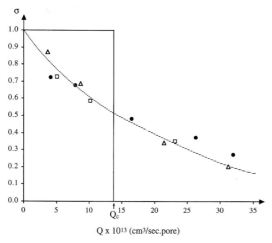

Fig. 8.45. Reflection coefficient (σ) vs flow rate (Q). The *solid line* is theoretical prediction, the *data points with hatched line* are experimental results from [169]. The *symbols* denote different ratio of R_s/r_p (R_s: polymer Stokes radius)

$Q \times 10^{13}$ (cm³/sec.pore)

the velocity components of the flow field calculated with POLYFLOW (Sect. 8.2). The procedure for calculating β is depicted schematically in Fig. 8.46. First, the abscissa x_c is determined for a given $\dot{\varepsilon}_c$ from the strain-rate distribution as shown in Fig. 8.46. From x_c, the corresponding flow velocity ratio v_{obs}/v_c is determined (v_{obs} denotes the flow velocity at the position of observation) and the fractional chain extension calculated:

$$\beta = R/R_{max} = (R_0/R_{max}) \cdot (R/R_0) = (R_0/R_{max}) \cdot (v_{obs}/v_c) \tag{8.61}$$

Here, R and R_{max} denote the end-to-end chain distance at the position of observation and the maximum extended length, respectively.

Once the molecular extension ratio has been obtained, the birefringence can be calculated using the Kuhn-Grün theory [170]:

$$\Delta n = \Delta n_{max} \cdot \{1 - 3\beta/\mathcal{L}^{-1}(\beta)\} \tag{8.62}$$

where Δn_{max} is the birefringence of completely oriented chains, and $\mathcal{L}^{-1}$ the inverse Langevin function.

For computation purposes, Eq. (8.62) could be replaced with the Padé approximant [171]:

$$\Delta n = \Delta n_{max} \cdot \{2\beta^2/(3 - \beta^2)\} \tag{8.63}$$

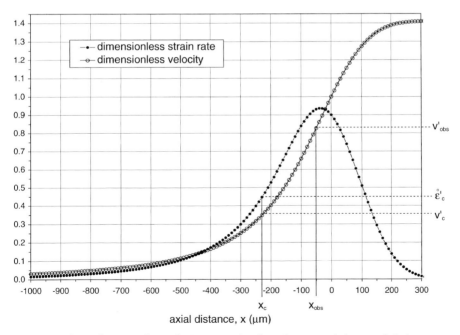

Fig. 8.46. Relation between flow velocity, extensional strain-rate and degree of chain extension in abrupt contraction flow

or by a polynomial approximation [172]:

$$\Delta n \cong \Delta n_{max} \cdot [0.6 \, \beta^2 + 0.2 \, \beta^4 + 0.2 \, \beta^6] \tag{8.64}$$

Either expression is accurate to better than 3%.

Equation 8.62 or the alternatives, Eqs. (8.63) and (8.64), contain two fundamental parameters, Δn_{max} and β, whose exact determination is crucial in the elaboration of a molecular deformation model.

8.5.1.1
Determination of Δn_{max}

The electronic polarizability (α) and the refractive index (n) are related by means of the Lorentz-Lorenz equation [143]:

$$(n^2 - 1)/(n^2 + 2) \cdot (M/\rho) = (4 \, \pi/3) \, \mathcal{N}_A \alpha \tag{8.65}$$

where $\mathcal{N}_A$ is the Avogadro's number, ρ is the polymer density (in $g \cdot cm^{-3}$), M is the polymer molecular weight.

A similar relation, known as the Kuhn-Kuhn equation [173], is applicable for an anisotropic medium with refractive indexes n_1 and n_2 (Δn is the difference and n the arithmetic average of n_1 and n_2):

$$\frac{(n_1^2 - 1)}{(n_1^2 + 2)} - \frac{(n_2^2 - 1)}{(n_2^2 + 2)} \cong \frac{6n}{(n^2 + 2)^2} \cdot \Delta n = (4 \, \pi M/3\rho) \, \mathcal{N}_A (\alpha_1 - \alpha_2) \tag{8.66}$$

The approximation is valid for $\Delta n < 0.3$.

Defining $\Delta \alpha = (\alpha_1 - \alpha_2)$ as the difference in polarizabilities of a subunit parallel and normal to the chain axis respectively, Eq. (8.66) can be rearranged into

$$\frac{\Delta n}{n\rho} = \frac{2 \, \pi}{9} \cdot \frac{(n^2 + 2)^2}{n^2} \cdot \frac{\mathcal{N}_A \Delta \alpha}{M} \tag{8.67}$$

Literature data for the intrinsic birefringence of fully oriented atactic PS chains in the bulk is $\Delta n_{max} = - 195 \cdot 10^{-3}$ [174], corresponding to $\Delta \alpha = - 164 \times 10^{-25}$ cm^3/molecule [148].

The problem of determining the intrinsic birefringence of fully oriented isolated polymer chains is, however, more complex and there is no general consensus as to the exact value Δn_{max} to be used for fully extended PS chains in solution. Based on the Kuhn-Grün calculations of optical anisotropy of Gaussian macromolecules [170], Peterlin has derived the following expressions for the flow birefringence of dilute polymer solutions:
(i) in simple shear flow [175]:

$$\frac{\Delta n}{nc} = \frac{2 \, \pi}{15} \cdot \frac{(n^2 + 2)^2}{n^2} \cdot \frac{\mathcal{N}_A \Delta \alpha}{M} \cdot \frac{1}{R_0^2} \cdot \Gamma \cdot \{((\langle x^2 - y^2 \rangle)^2 + 4 \, \langle xy \rangle^2\}^{0.5} \tag{8.68}$$

(ii) in uniaxial elongational flow [176]:

$$\frac{\Delta n}{nc} = \frac{2\pi}{15} \cdot \frac{(n^2 + 2)^2}{n^2} \cdot \frac{\mathcal{N}_A \Delta \alpha}{M} \cdot \frac{1}{R_0^2} \cdot \Gamma \cdot (\langle x^2 - y^2 \rangle) \tag{8.69}$$

where n is the refractive index of the solution, c the polymer concentration in $g \cdot cm^{-3}$, the brackets $\langle \rangle$ denote averaging over all the configurations of the chain. The factor Γ is a function of the expansion ratio (R_0/R_{max}) and varies between 1 for random coiled and 5/3 for fully extended chains.

Equations (8.68) and (8.69) predict that near complete chain extension, with a resultant saturation in birefringence at Δn_{max}, will occur at strain-rate $\mathring{\gamma} > \sim 100/\tau_1$ in simple shear and $\dot{\varepsilon} > 0.5/\tau_1$ in uniaxial elongational flow. Under these conditions, the terms between brackets in Eqs. (8.68) and (8.69) reduced to $(R_{max})^2$.

In the bead-spring model of a flexible linear chain, R_{max} is replaced by Zb and R_0^2 by Zb^2, Z being the number and b the length of statistical (Kuhn) subunits in the polymer chain. The length of the Kuhn segment is a measure of chain equilibrium rigidity, and is proportional to the characteristic ratio C_∞ according to [177]:

$$b = (C_\infty + 1) \cdot \ell \tag{8.70}$$

with ℓ = average length of the skeletal bonds (0.154 nm for PS).

Denoting P as the degree of polymerization and m_0 as the monomer molecular weight, the number of monomer units per Kuhn subunit is given by

$$\zeta = P/Z = M/(m_0 Z) \tag{8.71}$$

Combining this with $C_\infty = 10.0$ gives a value of $\zeta = 6.7$ for PS [178].

For PS in the saturation limit, Eqs. (8.68) or (8.69) can be rearranged into

$$\frac{\Delta n_{max}}{nc} = \frac{2\pi}{9} \cdot \frac{(n^2 + 2)^2}{n^2} \cdot \frac{\mathcal{N}_A \Delta \alpha}{\zeta \cdot m_0} = 4.88 \cdot 10^{21} \cdot \Delta \alpha \tag{8.72}$$

From birefringence measurements in simple shear flow, it was determined that $\Delta \alpha$ for PS in the matching solvent bromoform lies within the range of -132 to $-159 \cdot 10^{-25}$ cm³/molecule [148, 179]. Although $(\Delta n_{max}/nc)$ could be calculated from the available experimental data, there is always the possibility of measurement errors in the determination of these parameters. The intrinsic segmental anisotropy, $\Delta \alpha$, for instance, may depend on the "specific" effect of the solvent as has been demonstrated for poly(vinyl acetate) [179, 180]. Solvent effects are also present in PS and values for $\Delta \alpha$ are generally ~20% lower in aromatic solvents than in bromoform [158]. For these reasons, some investigators [24, 25] advocate an empirical approach in the determination of Δn_{max}. Assuming that all chains are fully extended at sufficiently high fluid strain-rates, the value of Δn_{max} can be calculated from the saturation birefringence. Some typical values of Δn_{max} estimated by this method are summarized in Table 8.2.

Table 8.2. Experimental determination of saturation flow birefringence

Authors, ref.	Flow geometry	Solvent [a]	$\Delta n_{max}/nc$
Munk-Peterlin [181]	Couette cell	Aroclor	−0.088
Pope-Keller [25]	Opposed jets	p-Xylene	−0.040
Fuller-Leal [22]	Four-roll mill	Polychlorinated-biphenyl	−0.025
Dunlap-Leal [86]	Two-roll mill	Chlorowax Tri-cresyl-phosphate	−0.031
Cathey-Fuller [129]	Opposed jets	Tri-cresyl-phosphate	−0.018
Nguyen et al. [58]	Opposed jets	Decalin 1-Methyl-naphthalene	−0.039 −0.038
Odell-Carrington [50, 133]	Opposed jets	Decalin	−0.076
Nguyen et al. (this work)	Single jet	1-Methyl-naphthalene	−0.084

[a] All experiments were performed at room temperature; the refractive indices of the solvents are in the range 1.5–1.6.

Large variations in saturation birefringence were often noticed within the same group, using the same flow birefringence apparatus, when the solvent [129], polymer MW [129] or concentration were changed [182]. Modifying the gap distance in opposed-jet flow also has a profound influence on Δn_{max} [72, 183]. This peculiar effect will be commented on in the Conclusions section.

By retaining only the largest saturation values from Table 8.2, a range of − (0.076–0.088) cm^3 · g^{-1} is obtained for ($\Delta n_{max}/nc$), in good agreement with values calculated from Eq. (8.72). For reason of self-consistency, we will use ($\Delta n_{max}/nc$) = − 0.084 cm^3 · g^{-1} in the present study.

8.5.1.2
Determination of β

The degree of coil expansion β is directly related to the critical strain-rate $\dot{\varepsilon}_c$. Although $\dot{\varepsilon}_c$ could be determined from the onset of birefringence at low flow rates, the procedure relies on measurements under noisy conditions and is thus highly susceptible to experimental errors. For this reason, it is preferable to use information from the whole birefringence curve with $\dot{\varepsilon}_c$ as a fitting parameter. However, due to the inadequacy of the affine deformation model (cf. Sect. 8.5.2), it is not possible to reproduce the whole birefringence curve with a single $\dot{\varepsilon}_c$. By arbitrary selecting $x \cong -40$ µm as the adjusted region (Figs. 8.47 and 8.48), critical strain-rates of 5300 s^{-1} and 2100 s^{-1} have been obtained for a 3.95 × 10^6 PS dissolved in decalin and in 1-methyl-naphthalene, respectively. Much higher values of $\dot{\varepsilon}_c$ would be determined if the fit was performed on the low strain-rate portion of the curves.

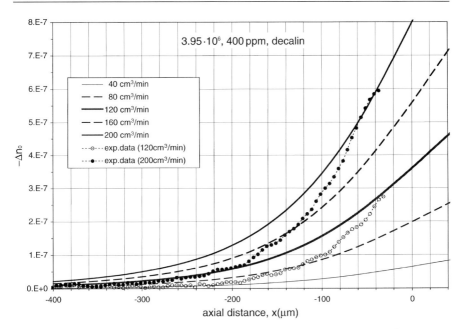

Fig. 8.47. Experimental and calculated axial birefringence distribution, $\Delta n_0(x)$, for a 3.95 × 10^6 PS in decalin, at different flow rates. The affine deformation model was used for the calculations with $\dot{\varepsilon}_c = 5300$ s^{-1}

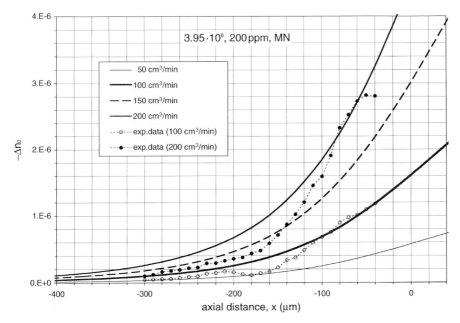

Fig. 8.48. Experimental and calculated axial birefringence distribution, $\Delta n_0(x)$, for a 3.95 × 10^6 PS in 1-methyl-naphthalene, at different flow rates. The affine deformation model was used for the calculations with $\dot{\varepsilon}_c = 2100$ s^{-1}

The coil-stretching theory [8, 184] predicts that $\dot{\varepsilon}_c \cong 0.5/\tau_1$. This relation seems to be confirmed by direct fluorescence visualization of deformed DNA chains [185]. Other polymer dynamics theories suggest, however, different proportionality constants (B) between $\dot{\varepsilon}_c$ and $1/\tau_1$ [10, 186]. The molecular relaxation time for a 3.95×10^6 PS is 0.60×10^{-3} s in decalin, and 3.6×10^{-3} s in 1-methyl-naphthalene (Eqs. 8.30 and 8.31). These values indicate that the constant B in the present flow conditions should be at least 3 in decalin and 8 in 1-methyl-naphthalene (essentially the same conclusions have been reached in opposed-jet flow [58]). A value of B significantly larger than 1 has been actually proposed by Rabin in case kinetic barriers play an important role in the process of coil-stretching transition [10]. In order to reduce these barriers so that the coil-stretching transition may take place during the finite transit time, the strain-rate has to be increased well beyond that required for the equilibrium situation.

The preceding figures reveal that the affine deformation model gives a reasonable description of polymer uncoiling at extension ratios below $\sim 40\%$ (Figs. 8.47 and 8.48). Even within this range, agreement is only qualitative since experimental orientations were systematically below the calculated values. For larger extension ratio reached by the highest MW samples, the birefringence increased rapidly at the approach of the orifice entrance, before reaching a seeming saturation. The value for this plateau increased with flow rate, however, insofar as Δn_{max} was not reached.

8.5.2
Other Polymer Dynamics Models Relevant to Abrupt Contraction Flow

One general criticism of early models of polymer dynamics, including the affine deformation model described in Sect. 8.51, was the application of a linear spring law. In molecular presentation, this connector law is tantamount to a neglect of finite chain extensibility. Although this weakness has long been recognized, its correction leads to nonlinear Langevin equations with inextricable mathematical complexity. The different approximative schemes which have been devised to include *finite extensibility* in a mathematically tractable manner have resulted in a number of *nonlinear elastic* dumbbell (FENE [187], FENE-P [188], FENE-CR [189], FENE-PM [190]) and bead-spring chain models. Apart from avoiding the catastrophic infinite chain extension in strong flow, the use of a nonlinear force law can alter results in some more subtle manner. It has been shown, for example, that application of anharmonic spring dumbbells to the pore entrance flow problem (Fig. 8.44) changed the flow-rate dependence of $\sigma(Q)$ from a first-order to a second-order phase transition, a result more consistent with the experimental findings [3].

The "de Gennes-Hinch" dumbbell with a bead friction coefficient increasing almost linearly with chain extension, predicts an abrupt coil-stretching transition over a very short range of the Deborah number. Only a few experiments, however, have confirmed the sharpness of this transition [22, 50]. Birefringence recorded at the stagnation point in opposed-jet flow [58, 129], and in the two-

and four-roll mill flows [191] showed that beyond $\dot{\varepsilon}_c$, chain orientation continued to increase typically over more than one order of magnitude in fluid strain-rate, before reaching saturation. The breadth of this transition was much greater than could be accounted for by polymer dispersity alone [58]. Birefringence results in opposed-jet flow are discussed elsewhere in this book [50]. The gradual chain orientation reported in the two and four-roll mill experiments, and during flow around a corner, was successfully explained with the FENE-CR model [191–193]. Recent results obtained from this same group have additionally confirmed the adequacy of the Chilcott-Rallison dumbbell model when applied to a transient extensional flow, such as during start-up and cessation of flow in a four-roll mill [193].

The use of an inverse Langevin type spring law, in which the force increases faster than the degree of stretching, would delay chain orientation to higher fluid strain-rates. This model is in qualitative agreement with the current experimental observations and may be used to explain birefringence results up to $\beta \cong 0.4$ for the low MW samples (Figs. 8.47 and 8.48). It cannot, on the other hand, account for the existence of a saturation birefringence specific to each flow rate as pictured in Figs. 8.39 and 8.40 for the highest MW samples.

In a series of three papers, Pincus has developed the "blob" model of stretched polymer chain in presence of different force fields [194]. The difficulties associated with the application of the Pincus "tensile blobs" model to the intermediary stage of chain uncoiling have been reviewed by Rabin et al. [10]. The "trumpet" and "stem-and-flower" models of chain stretching in strong flow based on Pincus blobs was recently proposed by Brochard-Wyart [195, 196]. Since blob sizes vary in inverse proportion to the transmitted frictional forces, one anticipates that large blobs are prevalent at small strain-rates, resulting in small average segmental orientation. At high strain-rates, the blobs shrink and orientation approaches saturation. Although the "blob" model accounts for the apparent molecular slippage at the beginning of chain stretching, it fails to explain the existence of several plateaus in the axial birefringence distributions.

One of the most important experiments specific to converging flows was the pressure drop measurements performed by James and Saringer in conical channels [31]. From the stress generated, these authors infer that isolated flexible polymer chains in converging flow experience strong drag forces well before full chain extension. To explain this observation, King and James suggested that polymer chains start to uncoil up to some degree of stretching, but then "freeze" in partially extended conformations following the formation of intramolecular entanglements [32]. A different picture was proposed by Ryskin with the "yo-yo" model [33]. In a strong flow, the chain is pictured as being stretched taut at the center where frictional stress is maximal, while being curled near the ends. Extra stress is provided by this strained portion of the chain which behaves hydrodynamically as a rigid rod. More recent chain conformation models in transient strong flow were presented by Larson [13] and Kausch and Nguyen [35], based on kink formation observed by Acierno et al. [197], and Hinch and Rallison [11] in their numerical simulations. The "yo-yo" model envisions an affine deformation of the chain ends, and a birefringence which can saturate only at full chain extension. Both of these predictions are at odds with the light-scattering results at the stagnation point in opposed-jet flow [198] and the present birefringence results. In addition, recent fluorescence depolarization experiments in strong flow with anthracene-labelled PS did not show any preferential orientation when the chromophore group is attached at the center [73, 199].

It appears from the present results that the state of segmental orientation in abrupt contraction flow is more governed by the strain-rate (i.e., the frictional hydrodynamic forces) than by the residence time (which determines the accumulated strain). Such behavior is more consistent with the hairpin or kink representations of chain unfolding [11–13, 32, 34, 35] than with any model which depends on affine or pseudoaffine molecular deformation [26, 33, 168, 189]. The presence of well-defined birefringence plateaus at intermediate states of segmental orientation indicates the presence of transient conformations which are sufficiently rigid to resist further stretching by the velocity gradient, in a similar way to the "frozen chains" suggested by James and Saringer [31].

Among the proposed models, the "kinks dynamics" models of Larson [13] and Hinch [34] are the most appealing since they take into account both the kinetics (residence time) and stress (strain-rate) factors. At the present stage, it is difficult to differentiate the "kinks dynamics" from the "frozen chain" model based on intramolecular entanglements. The fact that transient saturation is observed in the good solvent 1-methyl-naphthalene, may lend some credit to the "kinks" model, since intramolecular knots are unlikely to occur for MW below 10^8 [200]. However, Hinch has pointed out that strong flow "does have a component which squashes the molecule onto a line, and from there it is difficult to apply the leverage needed to unroll a loop, even if it is not knotted" [201].

8.6
Prospects and Conclusions

Flow birefringence has been used in this study to probe the orientation of isolated PS chains in an axisymmetric orifice entrance flow. This quantity was found to vary from 0 to ~ 100 %, depending on flow rate and polymer MW. The birefringence recorded in the flow direction showed an almost exponential increase, which may eventually reach saturation, at the approach of orifice entrance. This behavior is in *qualitative* agreement with the prediction of the affine deformation model (cf. Sect. 5.1). This superficial correspondence may have induced some authors to conclude the aforementioned model to be valid in certain pressure drop and flow birefringence experiments [16–21]. However, quantitative analysis of the data has revealed large differences between theoretical predictions and experimental data. The most conspicuous disparity was the presence of several birefringence saturation levels, corresponding to various degrees of partial chain orientation. This peculiar behavior is reminiscent of the results obtained during liquid filament pull-out under quasi-steady elongational flows [35, 36]. Similar behavior is also observed at the stagnation point in opposed-jet flow, in which the level of birefringence seems to saturate at different levels depending on the gap separation distance [72, 183]. This has been claimed to be caused by a "massive screening" when the gap separating the jets is larger than the orifice diameter [202]. Experiments performed in highly dilute PS solutions (10 ppm) reproduced the phenomenon but failed to indicate any non-Newtonian effects [72, 183]. No definite explanation for this behavior could be proposed at this time. Based on recent experimental observation of single DNA chains in cross-slot flow, it is tempting to speculate that the birefringence measured at the stagnation point in opposed-jet flow is the result of an ensemble-averaged orientation of different classes of conformations, which coexist for an extended period of time. The birefringence curve calculated by taking the spatio-temporal average orientation from all the DNA conformations as a function of strain-rate shows a remarkable similarity to the birefringence curves reported in opposed-jet flow [58]. Even the presence of a small notch in the high strain rate region observed in "bulk" birefringence data can be reproduced (cf. Fig. 32 in [185]). It should be recalled at this point that in the context of worm-like chains, Tsvetkov has already drawn attention to the fact that "conformational polydispersity" always occurs in a system of macromolecules, and should be taken into account in birefringence measurements ([147, p. 289]).

Although steady-state conformations of flexible polymer chains in strong flow now seem to be well understood, the kinetics of the transition from the coiled to the extended state remain essentially a matter for conjecture. Most of the problems associated with the limited residence time in the strong flow region and the lifetime of metastable states were formulated by researchers such as Hinch [11, 34], Rabin et al. [10] and Larson [13, 184] more than ten years ago. Modeling metastable conformations requires recognition of the noncrossability of the springs in the bead-spring chain model, a topological problem which, according to Edwards [203], should provide "a splendid challenge to the theorist". Although there has been some attempt to model entanglements by Monte-Carlo simulation [204], it seems that a better comprehension of these non-equilibrium states will continue to rely for some time on experimental findings. In this respect, abrupt contraction flow provides an interesting opportunity to investigate metastable chain conformations under well-defined steady flow conditions.

The present investigation, which is part of a PhD thesis [205], has much room for future improvement. As an extension of the work, investigation of additional variables such as solvent viscosity and chain flexibility may provide supplemental information on the transient states of polymer deformation during the coil-stretching transition. Several ways might be envisaged to increase the precision of the birefringence data, such as use of a more robust Abel inversion technique [134], redesign of the capillary entrance geometry to minimize the re-entrant corner effect, reduction of the probe beam size to allow measurements closer to the orifice and experimentation with a smaller orifice diameter (to increase the strain rate and permit use of lower MW polymers).

Apart from these technical refinements, the problem could be investigated with new or complementary techniques such as:

- fluorescence detection of the degraded polymer (one possibility would be to scavenge the created radicals with a fluorescent-forming probe, such as NTEMPO[2] [206]). This experiment is important to localize the position where bond scission occurred, before the effect of polymer degradation can be assessed;
- the old dictum "seeing is believing" seems to be applicable in science and direct fluorescence visualization of streaming macromolecules, as with DNA, should provide important insight into transient chain conformations. Observing the passage of DNA chains through microscopic pores, for instance, could provide clues about the origins of the discrepancy shown in Fig. 8.45. Direct visualization of PS with MW above 10^7 should also be feasible after grafting the polymer with chromophore groups;
- the synergy offered by combining classical measurement techniques such as pressure drop, spatially resolved light scattering, flow birefringence and flow velocimetry should provide a powerful tool for future investigations.

8.6.1
Final Words

In conclusion to this chapter, we would like to stress that this work is intended to be purely experimental, as have been our preceding publications in this field [48, 58, 71]. Judging from past referee comments, we have the sentiment that the field of polymer dynamics in strong flow is marked by much emotional involvement. We would like to repeat that we did not set out to develop any new theoretical model. Rather, our essential motivation has been to rationalize the observed results with existing models of chain uncoiling in strong flow. Needless to say, any constructive comments which would allow a better comprehension of the phenomena would be warmly welcomed.

Acknowledgments. The authors gratefully acknowledge the assistance of Dr. C.J.G. Plummer in re-reading the manuscript, Dr. B. Debbaut for providing flow simulation data and for helpful discussion, and the Swiss National Science Foundation for continuing financial support of the project.

Appendix A

The experimental scheme for calculating $\Delta n(r)$ from $\delta^*(y)$ is illustrated in Fig. 8.A1. First, the y-axis (and hence r) is divided into a number (N) of equal in-

---

2 Systematic name: 4-(1-naphthoyloxy)-2,2,6,6-tetramethylpiperidine-1-oxyl.

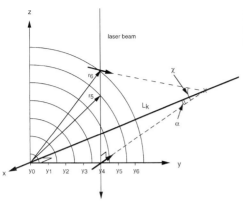

Fig. 8.A1. Relation between the local orientation angle $\chi(r)$ and local extinction angle $\alpha(y, z)$ in converging flow

crements (ℓ) going out from the center, so that

$$y_n = n \cdot \ell \tag{8.A1}$$

$$r_k = (k + 0.5) \cdot \ell \tag{8.A2}$$

where n and k are integers between and including 0 and N. In the present study, ℓ is equal to 10 µm and typically 100 ($= N$) retardation data points are recorded for each transversal scan.

The optical properties of the flowing solution are modeled by approximating the continuous variation of the birefringence and orientation in the z-direction with a series of discrete concentric retarders. Birefringence (Δn_k) is considered to be constant within the region defined by the circles with radii r_k and r_{k-1}. The orientation angle (χ_k) is defined by L_k which is assumed to be constant within each annulus (cf. Fig. 8.A1). Using Pythagore, the variation of χ_k along the light propagation direction (z) is obtained:

$$\chi_k = \cos^{-1}[(L_k^2 + y_n^2)/(L_k^2 + y_n^2 + z^2)]^{1/2} \tag{8.A3}$$

In a uniform sink flow, L_k is a constant which depends only on the axial position (x). In a general situation, L_k is a function of r_k:

$$L_k = r_k/\tan \theta(r_k) \tag{8.A4}$$

The radial orientation distribution $\theta(r)$ is obtained from flow field simulation (Sect. 8.2.2.3). The transmission of light through *each slice* can then be calculated with the formalism for polarized propagation in homogeneous material at an oblique incidence (Eq. 8.42).

The contribution of the k-th annulus to the total effective retardance at the position y_n is given by

$$\delta_k(y_n) = \Delta n_k (2\pi/\lambda) \int_{z_{k-1,n}}^{z_{k,n}} (^1/_2 + ^1/_2 \cos^2 \chi_k) \, dz \tag{8.A5}$$

with $z_{k,n} = [(k + 0.5)^2 - n^2]^{0.5} \cdot \ell$ and $z_{k-1,n} = [(k - 0.5)^2 - n^2]^{0.5} \cdot \ell$.

Using Eq. (8.A3), the above expression can be rewritten as

$$\delta_k(y_n) = \Delta n_k(\pi/\lambda) \int_{z_{k-1,n}}^{z_{k,n}} [1 + (L_k^2 + y_n^2)/(L_k^2 + y_n^2 + z^2)]\, dz \tag{8.A6}$$

which yields after integration

$$\delta_k(y_n) = \Delta n_k(\pi/\lambda) \cdot (z_{k,n} - z_{k-1,n}) + \Delta n_k(\pi/\lambda) \cdot (L_k^2 + y_n^2)^{0.5} \cdot$$

$$\{\tan^{-1}[z_{k,n}/(L_k^2 + y_n^2)^{0.5}] - \tan^{-1}[z_{k-1,n}/(L_k^2 + y_n^2)^{0.5}]\} \tag{8.A7}$$

Summation over δ_k gives the effective total retardance:

$$\delta^*(y_n) = \sum_{k=n}^{N} \delta_k(y_n) \tag{8.A8}$$

It is convenient to rewrite Eq. (8.A8) in a matrix notation:

$$\delta^*(y) = \mathbf{A} \cdot \Delta n(r) \tag{8.A9}$$

where $\mathbf{A}$ is an upper triangular matrix of rank N.

The inversion of $\mathbf{A}$ to obtain $\Delta n(r)$ could be done by successive substitution. From the outermost value $\delta_N(y_N)$, the corresponding Δn_N is calculated from the relation

$$\delta_N(y_N) = \Delta n_N(\pi/\lambda) \cdot \{(N + 0.25)^{0.5} \cdot l +$$

$$(L_N^2 + N^2 l^2)^{0.5} \cdot \tan^{-1}[((N + 0.25)^{0.5}/(L_N^2 + N^2 l^2)^{0.5}]\} \tag{8.A10}$$

The value obtained for Δn_N is then substituted into the equation for $\delta_k(y_{N-1})$ to obtain Δn_{N-1}. The process continues until all the Δn_k are obtained according to the general equation

$$\Delta n_n = 1/A(n, n)\,[\delta^*(y_n) - \sum_{k=n+1}^{N} \Delta n_k \cdot A(k, n)] \tag{8.A11}$$

Appendix B

In cylindrical coordinates (r, θ, z), the velocity and velocity gradient are given by:

$$v = (v_r, v_\theta, v_z) \tag{B1}$$

$$\text{grad } v = (\partial v/\partial r)\, e_r + 1/r\, (\partial v/\partial \theta)\, e_\theta + (\partial v/\partial z)\, e_z. \tag{B2}$$

Under axisymmetric conditions, $\partial/\partial\theta = 0$ and the velocity gradient tensor can be simplified as:

$$G_{ij} = \begin{vmatrix} A & B & 0 \\ C & D & 0 \\ 0 & 0 & E \end{vmatrix} \tag{B3}$$

with $A = \partial v_r / \partial r$
      $B = \partial v_z / \partial r$
      $C = \partial v_r / \partial z$
      $D = \partial v_z / \partial z$
      $E = v_r / r$.

The tensor G_{ij} can be decomposed into a symmetric (e_{ij}) and an antisymmetric (ω_{ij}) part according to:

$$G_{ij} = e_{ij} + \omega_{ij} \tag{B4}$$

with

$$e_{ij} = \begin{vmatrix} A & {}^1/_2(B+C) & 0 \\ {}^1/_2(B+C) & D & 0 \\ 0 & 0 & E \end{vmatrix} \tag{B5}$$

$$\omega_{ij} = \begin{vmatrix} 0 & {}^1/_2(B-C) & 0 \\ {}^1/_2(-B+C) & 0 & 0 \\ 0 & 0 & 0 \end{vmatrix}. \tag{B6}$$

For an incompressible fluid, div $v = 0$ and

$$A + D + E = 0. \tag{B7}$$

The symmetric part corresponds to pure deformation whereas the antisymmetric term corresponds to pure rotation of the fluid element.

The deformation induces rotation with change in length of a vector element δr. There are dilatation without rotation only along the directions which can be determined by diagonalization of the e_{ij} tensor. Mathematically, this is calculated by setting the determinant of the matrix in Eq. B8 to zero[3]:

$$\det \begin{vmatrix} A-\lambda & {}^1/_2(B+C) & 0 \\ {}^1/_2(B+C) & D-\lambda & 0 \\ 0 & 0 & E-\lambda \end{vmatrix} = 0. \tag{B8}$$

The eigenvalues of the matrix are readily obtained as:

$$\begin{aligned} \lambda_1 &= {}^1/_2 \left[(A+D) + ((A-D)^2 + 4(BC)^2)^{1/2} \right] \\ \lambda_2 &= {}^1/_2 \left[(A+D) - ((A-D)^2 + 4(BC)^2)^{1/2} \right] \\ \lambda_3 &= -(A+D). \end{aligned} \tag{B9.}$$

The principal orientation directions, along which there is no flow-induced rotation, are given by the equation:

$$y = (a - \lambda_i) \cdot x/b. \tag{B10}$$

3 Guyon E, Hulin JP, Petit L (1991) Hydrodynamique physique, InterEditions, Paris, p. 108–117.

In Eq. B9, λ_1 has the largest (positive) value and indicates the direction extensional deformation; λ_2 and λ_3 are negative and correspond to compression of
the fluid element.

It is assumed that the polymer chain is oriented in the same way as the fluid
element. This hypothesis is valid only in the Newtonian limit approximation.

Appendix C[4]

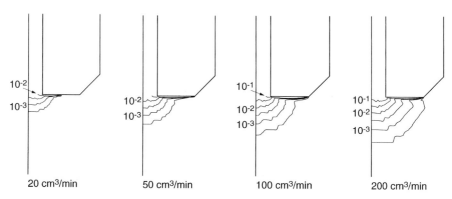

Fig. C1. Contour lines of the birefringence Δn for single jet flow calculated with the affine deformation model at the indicated flow rates, using $\dot\varepsilon_c = 1650\ \mathrm{s}^{-1}$

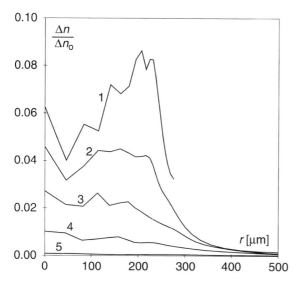

Fig. C2. Radial birefringence
distributions calculated with
the affine deformation model
for a flow rate of 100 cm³/min,
at several distances x from the
jet entrance. $1:x = -20\ \mu\mathrm{m}$;
$2:x = -50\ \mu\mathrm{m}, 3:x = -100\ \mu\mathrm{m}$,
$4:x = -200\ \mu\mathrm{m}, 5:x =$
$-500\ \mu\mathrm{m}$. The peak in the
birefringence distribution
should be interpreted in light
of the stress singularity at the
re-entrant corner (cf. H.K.
Moffat (1964) J. Fluid. Mech.
18:1)

4 Personal communication from B. Debbaut.

List of Symbols and Abbreviations
(symbols which are used only once are explained in the text)

a	Mark-Houwink exponent
C_∞	characteristic ratio
De	Deborah number
M	molecular weight
M_N	Mueller matrix for the Nth optical element
Q	flow rate
r_o	orifice diameter
r_p	pore radius
Re	Reynolds number
R_0	mean-square molecular end-to-end distance
R_{max}	molecular contour length
s	curvilinear coordinates
$\mathbf{S}$	Stokes vector
v	fluid velocity

Greek symbols

α	local orientation
α^*	effective orientation
β	degree of fractional coil extension
χ	chain orientation angle
$\chi(r)$	chain orientation distribution
δ	retardation
δ^*	effective retardation
$\dot{\varepsilon}$	fluid strain rate
$\dot{\varepsilon}_c$	critical strain rate for coil-stretching
$\dot{\varepsilon}_f$	critical strain rate for chain fracture
ϕ	azimuthal angle
Φ_0	Flory viscosity constant
$\dot{\gamma}$	simple shear rate
$[\eta]$	intrinsic viscosity
$\psi(r)$	velocity vector orientation
λ	laser wavelength
θ	polar angle
L^{-1}	inverse Langevin function
n	refractive index
Δn	local birefringence
Δn_{max}	birefringence of completely oriented chain
τ_1	molecular relaxation time

References

1. Trouton FT (1906) Proc Roy Soc London A77:426
2. Müller AJ, Sáez AE (1998) In: Nguyen TQ, Kausch HH (eds) Flexible polymer chain dynamics: theories and experiments. Springer, Berlin Heidelberg New York, Chap XI
3. Jhon MS, Sekhon G, Armstrong R (1993) Adv Chem Phys 66:153
4. Tabor M, de Gennes PG (1986) Europhys Lett 2:519
5. Nguyen TQ, Liang ZQ, Kausch HH (1997) Polymer 38:3783
6. Bird RB, Armstrong RC, Hassager O (1987) Dynamics of polymeric liquids, 2nd edn, John Wiley, New York, vol 1, p 76
7. Keunings R (1989) In: Fundamentals of computer modeling for polymer processing, Tucker CL (ed) Carl Hanser Verlag
8. de Gennes, PG (1974) J Chem Phys 60:5030
9. Hinch EJ (1974) Proc Symposium on Polymer Lubrification; Brest
10. Rabin Y, Henyey FS, Pathria RK (1985) In: Rabin Y (ed) Polymer-flow interaction, AIP conference proceedings, no. 137, American Institute of Physics, New York, p 43
11. Rallison JM, Hinch EJ (1988) J Non-Newtonian Fluid Mech 29:37
12. Wiest JM, Wedgewood LE, Bird RB (1989) J Chem Phys 90:587
13. Larson RG (1990) Rheol Acta 29:371
14. Brestkin YuV, Gotlib YuYa, Klushin LI (1989) Vyskomol Soedin, Ser. A 31:1143
15. Kilian HG, Lagaly G (eds) (1993) Progr Coll Polym Sci 92
16. Hoa NT, Chauvetau G, Gaudu R, Anne-Archard D (1982) CR Acad Sci Paris 294:927
17. Chauvetau G, Moan M, Magueur A (1984) J Non-Newtonian Fluid Mech 16:315
18. James DF, McLaren DR (1975) J Fluid Mech 70:733
19. Haas R, Durst F (1982) Rheol Acta 21:150
20. Ambari A, Deslouis C, Tribollet B (1984) Chem Eng Comm 29:63
21. Cressely R, Hocquart R (1980) Optica Acta 27:699
22. Pope DP, Keller A (1977) Coll & Polym Sci 255:633
23. Fuller GG, Leal LG (1980) Rheol Acta 19:580
24. Srivener O, Berner C, Cressely R, Hocquart R, Sellin R, Vlachos NS (1979) J Non-Newtonian Fluid Mech 5:475
25. Pope DP, Keller A (1978) Coll & Polym Sci 256:751
26. de Gennes PG (1979) Scaling Concepts in Polymer Physics, Cornell University Press, Ithaca, New York, p 191
27. Marucci G (1975) Polym Engin Sci 15:229
28. Tanner RI (1988) Engineering Rheology, Clarendon Press, Oxford, rev ed, p 185
29. Smith KA, Merrill EW, Peebles LH, Banijamali SH (1975) Polymères et lubrication, Colloques Internationaux du CNRS, Wolf C (ed) Paris, vol. 233, p 341
30. Armstrong RC, Gupta SK, Basaran O (1980) Polym Engin Sci 20:466
31. James DF, Saringer JH (1980) J Fluid Mech 97:655
32. King DH, James DF (1983) J Chem Phys 78:4749
33. Ryskin G (1987) J Fluid Mech 178:423
34. Hinch EJ (1994) J Non-Newtonian Fluid Mech 54:209
35. Kausch HH, Nguyen TQ (1992) Proc IUPAC Polymer '91, Melbourne, Australia, Feb 10–15, 1991, p 55–56; Kausch HH, Nguyen TQ. Makromol Chem, Macromol Symp 53:243
36. Tirtaatmadja V, Sridhar T (1993) J Rheol 37:1081
37. James DF, Sridhar T (1995) J Rheol 39:713
38. Merrill EW, Leopairat P (1980) Polym Eng Sci 20:505
39. Horn AF, Merrill EW (1984) Nature 312:140; Merrill EW and Horn AF (1984) Polymer Communications 25:144
40. Nguyen TQ, Kausch HH (1986) Coll & Polym Sci 264:764
41. Nguyen TQ, Kausch HH (1988) J Non-Newtonian Fluid Mech 30:125
42. Nguyen TQ, Kausch HH (1990) Macromolecules 23:5137

43. Nguyen TQ, Kausch HH (1992) Adv Polym Sci 100:73
44. Reese HR, Zimm BH (1990) J Chem Phys 92:2650; in this publication, a more random scission has been reported with DNA molecules although the presence of nicks could not be ruled out with this compound.
45. Frenkel J (1944) Acta Physicochim URSS 19:51
46. Rabin Y (1988) J Non-Newtonian Fluid Mech 30:119
47. Knudsen KD, Hernández Cifre JG, García de la Torre J (1996) Macromolecules 29:3603
48. Odell JA, Muller AJ, Narh KA, Keller A (1990) Macromolecules 23:3093
49. Odell JA, Keller A, Muller AJ (1992) Coll & Polym Sci 270:307
50. Odell JA, Carrington SP (1998) In: Nguyen TQ, Kausch HH (eds) Flexible polymer chain dynamics: theories and experiments. Springer, Berlin Chap VII
51. Janeschitz-Kriegl H (1983) Polymer melt rheology and flow birefringence, Springer, Berlin
52. Han CD, Decker LH (1973) J Appl Polym Sci 17:2329
53. Aldhouse STE, Mackley MR, Moore IPT (1984) J Non-Newtonian Fluid Mech 21:359
54. White SA, Gotsis AD, Baird DG (1987) J Non-Newtonian Fluid Mech 24:121
55. Quinzani LM, Armstrong RC, Brown RA (1994) J Non-Newtonian Fluid Mech 52:1
56. Eisenbrand GD, Goddard JD (1982) J Non-Newtonian Fluid Mech 11:337
57. Baaijens H (1994) Evaluation of constitutive equations of polymer melts and solutions in complex flows, PhD Thesis, Eindhoven University of Technology
58. Nguyen TQ, Yu G, Kausch HH (1995) Macromolecules 28:4851
59. Fuller GG (1990) Ann Rev Fluid Mech 2:387
60. Pearson DS, Kiss AD, Fetters LJ (1989) J Rheol 33:517
61. Osaki K, Bessho N, Kojimoto T, Kurata M (1979) J Rheol 23:457
62. Rajagopalan D, Byars JA, Armstrong RC, Brown RA, Lee JS, Fuller GG (1992) J Rheol 36:1349
63. Galante SR, Frattini PL (1993) J Non-Newtonian Fluid Mech 47:289
64. Li J-M, Burghardt WR (1995) J Rheol 39:743
65. Öttinger HChr (1997) submitted to J Non-Newtonian Fluid Mech
66. McHugh AJ, Mackay ME, Khomani B (1987) J Rheol 31:619
67. McAfee WJ, Pih H (1984) Exp Mech 14:385
68. Funatsu K, Kajiwara T (1988) In: Cheremisinoff N (ed) Encyclopedia of fluid mechanics, vol 7, Gulf Publishing, p 359–405
69. Hunkeler D, Nguyen TQ, Kausch HH (1991) Polym Prep 32:667
70. Hunkeler D, Nguyen TQ, Kausch HH (1996) Polymer 37:4257; Polymer 37:4270
71. Yu G, Nguyen TQ, Kausch HH (1998) J Polym Sci Part B Polym Phys 36:1483
72. Yu G (1996) Experimental investigation of single chain dynamics in elongational flow, PhD Thesis, No. 1523, EPFL, Lausanne, Switzerland
73. Vurpillot P (1997) Dynamique des chaînes Macromoléculaires par Dépolarisation de Fluorescence, PhD Thesis, No. 1647, EPFL, Lausanne, Switzerland
74. Nguyen TQ, Kausch HH (1992) Polymer 33:2611
75. Nguyen H, Boger DV (1979) J Non-Newtonian Fluid Mech 5:353
76. McKinley GH, Raiford WP, Brown RA, Armstrong RC (1991) J Fluid Mech 223:411
77. Feigl K, Öttinger HC (1994) J Rheol 38:847
78. Boger DV (1987) Ann Rev Fluid Mech 19:157
79. Chow A, Keller A, Müller AJ, Odell JA (1988) Macromolecules 21:250
80. Harlen OG, Hinch EJ, Rallison JM (1992) J Non-Newtonian Fluid Mech 44:229
81. POLYFLOW (1996) Version 3.5.0, Polyflow S.A., Louvain-la-Neuve, Belgium
82. Padmanabhan M, Macosko CW (1997) Rheol Acta 36:144, and references therein
83. Schlichting H (1968) Boundary layer theory, 6th edition, McGraw-Hill, New York
84. Keunings R (1986) J Non-Newtonian Fluid Mech 20:209
85. Fuller GG, Leal LG (1981) J Polym Sci Polym Phys Ed 19:557
86. Dunlap PN, Leal LG (1987) J Non-Newtonian Fluid Mech 23:5

87. Giesekus H (1962) Rheol Acta 2:122
88. Diesselhorst H, Freundlich H (1915) Physik Z 16:419
89. Vorländer D, Walter R (1925) Z Physik Chem 118:1
90. Raman CV, Krishnan KJ (1928) Phil Mag 5:769
91. Signer R (1930) Z Physik Chem (A) 150:247
92. Sadron Ch (1936) J Phys Radium 7:263
93. Kuhn W, Kuhn H, Buchner P (1951) Ergeb Exakt Naturwiss 25:1
94. Cerf R (1959) Adv Polym Sci 1:383
95. Janeschitz-Kriegl H (1969) Adv Polym Sci 6:170
96. Peterlin A, Munk P (1972) In: Weissberger A, Rossiter B (eds) Physical methods of
    chemistry, Interscience, New York, p 271
97. Tsvetkov VN (1964) In: Ke B (ed) Newer methods of polymer characterization,
    Interscience, New York, p 563
98. Scheraga HA, Signer R (1960) In: Weissberger A (ed) Physical methods of organic
    chemistry, vol. 1, New York, p 238
99. Azzam RMA, Bashara NM (1977) Ellipsometry and polarized light, Amsterdam, North-
    Holland
100. Fuller GG (1995) Optical rheometry of complex fluids, Oxford University Press, New
     York, Chap 2
101. Shindo Y (1995) Opt Engin 34:3369
102. Schellmann JA (1987) In: Samori B, Thulstrup EW (eds) Polarized spectroscopy and
     ordered systems, Kluwer, Boston, p 231–274
103. Mayer M, Sturm J, Weill G (1993) Biopolymers 33:1359
104. Chow AW, Fuller GG (1984) J Rheol 28:23
105. Zimm BH (1958) Rev Sci Instrum 29:360
106. Wayland II (1959) CR Acad Sci Ser B 249.1228
107. Wayland H (1960) CR Acad Sci Ser B 250:688
108. Pen'kov SN, Stepanenko BZ (1963) Opt Spektrosk 14:156
109. Fuller GG, Mikkelsen KJ (1989) J Rheol 33:761
110. Fujita H (1990) Polymer solutions, Elsevier, Amsterdam etc., p 58
111. Kurata M, Tsunashima Y (1989) In: Brandrup J, Immergut EH (eds) Polymer handbook,
     3rd ed., John Wiley & Sons, New York, Toronto, p VII–38
112. Ptitsyn OB, Eizner YuE (1960) Sov Phys Tech Phys 4:1020
113. Rouse PE (1953) J Chem Phys 21:1272
114. Zimm BH (1956) J Chem Phys 24:269
115. Nemoto N, Makita Y, Tsunashima Y, Kurata M (1984) Macromolecules 17:425
116. Benmouna M, Akcasu AZ (1978) Macromolecules 11:1187
117. Bixon M, Zwanzig R (1978 J Chem Phys 68:1890
118. Peterlin A (1963) J Chem Phys 39:224
119. Tsunashima Y, Nemoto N, Kurata M (1983) Macromolecules 16:584
120. Bhatt M, Jamieson AM (1989) Macromolecules 22:2724
121. Nicolai T, Brown W, Johnsen R (1989) Macromolecules 22:2795
122. Doi M, Edwards SF (1986) The theory of polymer dynamics, Clarendon Press, Oxford,
     p 65
123. Ballauff M, Wolff BA (1984) Macromolecules 17:209
124. Elias HG (1989) In: Brandrup J, Immergut EH (cds) Polymer handbook, 3rd ed, John
     Wiley & Sons, New York, Toronto, p VII–213
125. Barrett AJ (1984) Macromolecules 17:1561
126. Nguyen TQ, Kausch HH (1991) Proceedings intl GPC symposium '91, San Francisco, CA,
     USA, Oct 13–16, pp 373–397
127. Burghardt WR, Fuller GG (1989) J Rheol 33:771
128. Galante SR, Frattini PL (1991) J Rheol 35:1551
129. Cathey CA, Fuller GG (1990) J Non-Newtonian Fluid Mech 34:63
130. Schardin H (1933) Zeits f Instrumenten 53:396

131. Maecker H (1953) Z Physik 136:119
132. Tatham JP (1995) Extensional flow dynamics of macromolecules of different flexibility in solution, PhD thesis, University of Bristol
133. Carrington SP, Tatham JP, Sáez AE, Odell JA (1996) Polymer 38:4151
134. Hansen EW, Law Phaih-Lan (1985) J Opt Soc Am 2:510
135. Dasch CJ (1992) Appl Opt 31:1146
136. Cormack AM (1963) J Appl Phys 34:2722
137. Glasser J, Chapelle J, Boettner JC (1978) Appl Opt 17:3750
138. Anton M, Weisen H, Dutch MJ, von der Linden W, Buhlmann F, Chavan R, Marletaz B, Marmillod P, Paris P (1996) Plasma Phys Control Fusion 38:1849
139. Nestor OH, Olsen HN (1960) SIAM Review 2:200
140. Edels H, Hearne K, Young A (1962) J Math Phys 41:62
141. Barr WL (1962) J Opt Soc Amer 52:885
142. Cremers CJ, Birkebak RC (1966) Appl Optics 5:1057
143. Gurnee EF (1954) J Appl Phys 25:1232
144. Bossart J, Öttinger HChr (1995) Macromolecules 28:5852
145. Geffroy E, Leal LG (1990) J Non-Newtonian Fluid Mech 35:361
146. Sadron C (1936) J Phys Radium 7:381
147. Tsvetkov VN (1989) Rigid-chain polymers: hydrodynamic and optical properties in solution, (transl. from Russion by Korolyova EA), Consultants Bureau, New York (etc) p 305
148. Philippoff W (1963) Proc Intern Congress on Rheology 4 (2):343
149. Chu B (1991) Laser light scattering: basic principles and practice, 2nd ed, Boston (etc) Academic Press, p 158
150. O'Shea DC (1985) Elements of modern optical design, New York (etc) Wiley, p 230
151. Morse PM, Feshback H (1953) Methods of theoretical physics, McGraw-Hill, New York, Part I, Sect. 8.4, p 464
152. Peakfit™ v4.0 (1995) Jandel Scientific Software, Erkrath, Germany
153. Chow A, Keller A, Müller AJ, Odell JA (1988) Macromolecules 21:250
154. Baird DG, Read MD, Reddy JN (1988) J Rheol 32:621
155. Boger DV, Crochet MJ, Keiller RA (1992) J Non-Newtonian Fluid Mech 44:267
156. Walters KW, Webster MF (1982) Philos Trans R Soc London Ser A 208:199
157. Evans RE, Walters KW (1989) J Non-Newtonian Fluid Mech 32:95
158. Munk P, Poupetová D, Bohdanecky M, Dillingerová M (1967) J Polym Sci C 16:1125
159. Nguyen TQ, Kausch HH (1989) Makromol Chem 190:1389
160. Laso M, Picasso M, Öttinger HC (1998) In: Nguyen TQ, Kausch HH (eds) Flexible polymer chain dynamics: theories and experiments, Springer, Berlin Chap V
161. Hill JW, Cuculo JA Rev Macromol Chem and references therein
162. Zabicki A, Kedzierska K (1959) J Appl Polym Scie 2:14; 6:111, 361 (1962)
163. Zabicki A, Kedzierska K (1960) Kolloid Z 171:51
164. Zabicki A (1959) J Appl Polym Sci 2:24
165. Takserman-Krozer R (1963) J Polym Sci A1:2477, 2487; C16:2855 (1967)
166. Peterlin A (1966) J Polym Sci B4:287
167. Daoudi S, Brochard F (1978) Macromolecules 11:751
168. Long TD, Anderson JL (1984) J Polym Sci Polym Phys Ed 22:1261
169. Long TD, Anderson JL (1985) J Polym Sci Polym Phys Ed 23:191
170. Kuhn W, Grün F (1942) Kolloid Z 101:248
171. Cohen A (1991) Rheol Acta 30:270
172. Treloar LRG (1958) The physics of rubber elasticity, 2nd edition, Oxford University Press, New York
173. Kuhn W, Kuhn H (1943) Helv Chim Acta 26:1394
174. Seferis JC. Polymer Handbook, 2nd edition, VI–451
175. Peterlin A (1961) Polymer 2:257
176. Peterlin A (1966) Pure Appl Chem 12:563

177. Flory PJ (1969) Statistical mechanics of chain molecules, Interscience, New York, p 10–18
178. Aharoni SM (1983) Macromolecules 16:1722
179. Tsvetkov VN, Andreeva LN Polymer Handbook 2nd edition, VII–4577
180. Frisman EV, Dadivanyan AK (1967) J Polym Sci C16:1001
181. Munk P, Peterlin A (1970) Rheol Acta 9:294
182. Ng RC-Y, Leal LG (1993) J Rheol 37:443
183. Yu G, Nguyen TQ, Kausch HH, unpublished results
184. Larson RG, Magda JJ (1989) Macromolecules 22:3004
185. Perkins TT, Smith DE, Chu S (1998) In: Nguyen TQ, Kausch HH (eds) Flexible polymer chain dynamics: theories and experiments. Springer, Berlin, Chap X
186. Kobe JM, Wiest JM (1993) J Rheol 37:947
187. Warner HR (1972) Ind Eng Chem Fundam 11:379
188. Peterlin A (1966) J Polym Sci Polym Lett B4:287
189. Chilcott MD, Rallison JM (1988) J Non-Newtonian Fluid Mech 29:381
190. Wedgewood LE, Ostrov DN, Bird RB (1991) J Non-Newtonian Fluid Mech 40:119
191. Singh P, Leal LG (1994) J Rheol 38:485
192. Singh P, Leal LG (1995) J Non-Newtonian Fluid Mech 58:279
193. Harrison GM, Remmelgas J, Leal LG (1998) J Rheol 42:0000
194. Pincus P (1976) Macromolecules 9:386; Macromolecules 10:210 (1977); Picot C, Duplessix R, Decker D, Benoit H, Boue F, Cotton JP, Daoud M, Farnoux B, Jannink G, Nierlich M, de Vries AJ, Pincus P (1977) Macromolecules 10:436
195. Brochard-Wyart F, Buguin A (1998) In: Nguyen TQ, Kausch HH (eds) Flexible polymer chain dynamics: theories and experiments. Springer, Berlin, Chap II
196. Brochard-Wyart F (1993) Eur Phys Lett 23:105; Eur Phys Lett 30:387 (1995)
197. Acierno D, Titomanlio G, Marucci G (1974) J Polym Sci Polym Phys Ed 12:2177
198. Menesveta MJ, Hoagland DA (1991) Macromolecules 24:3427
199. Nguyen TQ, Vurpillot P, Kausch HH (1998) Bulletin of the APS 43 (1): 936
200. Brochard F, de Gennes PG (1977) Macromolecules 10:1157
201. Hinch EJ (1985) In: Rabin Y (ed) Polymer-flow interaction, AIP conference proceedings, no. 137, American Institute of Physics, New York, p 59
202. Carrington SP, Odell JA (1996) J Non-Newtonian Fluid Mech 67:269; Odell JA (1995) private communication
203. Edwards SF (1983) Faraday Symp Chem Soc 18:10
204. Hess-Bellwald K, Oberson C (1997) Internal report, Chair of Calculus and Topology, Dept of Mathematics, EPFL
205. Porouchani R. PhD thesis, EPFL, work to be completed in 1999
206. Moad G, Shipp DA, Smith TA, Solomon DH (1997) Macromolecules 30:7627

CHAPTER 9

The Hydrodynamics of a DNA Molecule in a Flow Field

R. G. Larson, T. T. Perkins, D. E. Smith, and S. Chu

9.1
Introduction

The behavior of dilute flexible polymer molecules in flowing liquids remains controversial, despite a long history of experimental and theoretical study. The simplest theory, introduced by Kuhn [1] some 60 years ago, treats the polymer as an elastic "dumbbell" in which an elastic spring connects two "beads" onto which are lumped the viscous drag forces that in reality act along the entire chain. In the simplest version of the dumbbell model, the drag force F^d on each bead is given by Stokes law, $F^d = \zeta k_B TV$, where V is the velocity of the solvent relative to that of the bead, and the drag coefficient $\zeta k_B T$ is independent of the deformation of the molecule.

More sophisticated theories account for changes in hydrodynamic drag induced by chain deformation. Peterlin [2], followed by de Gennes [3] and by Hinch [4], have argued that as a polymer molecule is stretched, hydrodynamic interactions between different parts of the chain are weakened, and the effective drag coefficient for the chain increases. By analogy with the well established hydrodynamics of undeformed or slightly deformed macromolecules, in which the drag on a polymer coil scales with polymer size the same way as it does for flow around an impenetrable sphere, de Gennes modeled the distorted polymer molecule by an impenetrable cylinder. This leads to the prediction that the drag coefficient should increase nearly linearly with chain extension, until at high extension the drag on the chain approaches the "free draining" limit. In the free-draining limit, the drag coefficient of the chain is proportional to the polymer molecule's contour length L (or molecular weight M), while in the undistorted, coiled state, where hydrodynamic interaction is dominant, the drag coefficient scales with the square root of L in the simplest case of a "theta" solvent, and with a slightly stronger power law for solvents of better quality than that of a "theta" solvent. When it is applied to a shearing flow, the cylinder model predicts that the shear viscosity increases with shear rate, a prediction at odds with most experimental data [5, 6]. A more refined version of this model by Brochard-Wyart [7, 8] partitions the chain into a series of "blobs" inside of which hydrodynamic interaction dominates, but outside of which free draining prevails.

In other theoretical approaches, the deformation of the polymer molecule is assumed to be limited by "internal viscosity" [9] or "straining inefficiency"

[10–12]. These factors introduce shear thinning in shear flows, and slow the rate at which the chain is deformed in transient flows. Others have proposed that "self entanglements" might prevent or delay a full extension of polymer molecules, even in strong flows [13, 14]. These concepts are of unknown importance.

A major reason for the persistence of uncertainties regarding the behavior of isolated polymer molecules in flow fields is that only measurements of collective phenomena, such as birefringence, stress, or scattering, have been available; models are needed to infer the behavior of the individual molecules from the collective behavior. Such models invariably rely on assumptions of questionable validity. For example, the solutions are assumed to be dilute enough that the molecules do not interact with each other, even when high elongations make them more prone to overlap at the minimum concentrations required to obtain adequate signal.

Recently, however, a direct method of studying the dynamics of a single DNA polymer molecule in a flowing solvent has been developed. In this method, Perkins et al. [15, 16] tether a small polystyrene sphere to one end of a DNA molecule; the sphere is held by a nonintrusive laser optical trap in a position away from any solid surfaces. The tethered DNA molecule is then subjected to a uniform flow field, which stretches it in the flow direction. The sphere to which the DNA molecule is attached is too small (0.3–1.0 μm) to affect significantly the flow field around the extended DNA. The DNA molecule is fluorescently labeled so that it can be observed directly by optical microscopy. Its steady and fluctuating dimensions are then measured for various flow rates, solvent viscosities, and molecular lengths. The DNA molecule is long enough to be considered flexible, and the data collected from the study of it make possible the direct and rigorous testing of theories for polymer dynamics in flow.

In addition to its ability to be rendered visible, DNA also has the advantage that its elastic properties in solution have been well characterized. Smith et al. [17] attached an individual DNA molecule to a magnetic sphere at one end and a flat surface at the other, and stretched it by a combination of hydrodynamic and magnetic forces. They found that the equilibrium curve of force (F) vs extension (r) for DNA is remarkably well described by the worm-like chain model, which obeys the approximate formula [18, 19]

$$\frac{F(r)\,A}{k_B T} = \frac{1}{4}\left(1 - \frac{r}{L}\right)^{-2} - \frac{1}{4} + \frac{r}{L} \tag{9.1}$$

Here r is the distance separating the ends of the molecule, L is the molecular contour length, and A is the molecular persistence length, which describes the chain stiffness. Bustamante et al. [19] found that for their unlabeled DNA molecules, $A \approx 0.053$ μm.

Perkins et al. [16] found that the velocity-extension data obtained on fluorescently labeled DNA with L ranging from 22 to 84 μm can be collapsed to a single curve when plotted as fractional chain extension (r/L) vs flow velocity times length to a power m (VL^m), with $m = 0.54 \pm 0.05$. The exponent 0.54 is in the range expected for dominant hydrodynamic interaction, namely $m =$

0.5–0.6. This scaling behavior is not surprising for small chain extensions. However, for high extensions, one would expect from the "cylinder" model that this scaling would fail, and an exponent on L closer to unity would be required to collapse the data. Thus, the experiments appear to show, paradoxically, that even at high chain extensions, the transition from "dominant hydrodynamic interaction" to "free draining" does not occur. While these results are consistent with the lack of shear thickening observed in viscosity measurements, they appear to be in conflict with the theoretical expectations discussed above. Also puzzling is the observation by Perkins et al. that the velocity-extension curve has almost the same shape as the force-extension curve measured by Smith et al. using the magnetic bead attached to the free end of the DNA.

To help resolve the questions raised by the study of Perkins et al., in Sect. 9.2 we attempt to make a more precise estimation of the change in the drag coefficient that a DNA chain of a given length should experience as it is stretched out from an undisturbed coil to a fully extended filament. Then, in Sect. 9.3, we describe some new experiments with DNA molecules longer than those studied by Perkins et al., to look for predicted changes in hydrodynamic behavior for longer molecules. In Sect. 9.4 we describe a Monte Carlo hydrodynamic model for DNA molecules in constant velocity flows; the predictions of this model are presented in Sect. 9.5. In Sect. 9.6 the conclusions are discussed. A condensed presentation of this study can be found in Larson et al. [20].

9.2
Calculation of the Drag Coefficient

We consider two simple limits in which analytic expressions for the drag coefficient are available: that of an undistorted coil, and that of a fully extended strand. The hydrodynamic drag coefficient for the undisturbed coil (i.e., from the Zimm theory) is [21, Eq. (4.60)]:

$$\zeta_{coil} k_B T = \frac{3}{8} (6\pi^3)^{1/2} \eta_s R = 5.11 R \eta_s = \frac{k_B T}{D} \tag{9.2}$$

where η_s is the solvent viscosity, and R is the root-mean-square size of the undisturbed molecule, $R = \sqrt{6} R_G$, where R_G is the molecule's radius of gyration, and D is the center of mass diffusivity of the molecule. Equation (9.2) applies to a "theta" solvent; for a "good" solvent, the same formula applies, except for a minor (3%) change in the prefactor [21, Eq. (4.84)]. If the chain could be completely stretched out into a straight rod of length L, the translational drag coefficient would be given by (Doi and Edwards, Eq. 8.20):

$$\zeta_{rod} k_B T = \frac{2\pi L \eta_s}{\ell n(L/d)} = \frac{6.28 L \eta_s}{\ell n(L/d)} \tag{9.3}$$

where d is the diameter of the rod, i.e., the diameter of a DNA molecule, which is about 2 nm [22, 23].

Notice that ζ_{coil} is proportional to R, while ζ_{rod} is proportional to L. For a theta solvent, $R \propto L^{1/2}$, while for "good solvent" conditions, $R \propto L^\nu$, with $\nu \approx 0.6$. In either case, for long enough molecules, ζ_{rod} should be much higher than ζ_{coil} and the effective drag coefficient should dramatically increase as the molecule is stretched out in the flow [3].

From the above, the values of the ratio ζ_{rod}/ζ_{coil} can be estimated for the DNA molecules studied by Perkins et al. [16]. According to Eq. (9.2), ζ_{coil} can be obtained from the center of mass diffusivity D. For a single DNA molecule, Smith et al. [24] found that D scales as $L^{-\nu}$ over the range $L = 2.0 - 140$ µm, with $\nu \approx 0.611 \pm 0.016$, implying that the DNA coils are somewhat expanded relative to theta conditions. For $L = 83.8$ µm, which is the longest DNA molecule used in the experiments of Perkins et al. [16], we find by interpolating between values tabulated by Smith et al. that $\zeta_{coil} = 1/D = 5.4$ sec$(\mu m)^{-2}$. From Eq. (9.3), on the other hand, $\zeta_{rod} = 11.5$ sec$(\mu m)^{-2}$ for $L = 83.8$ µm, so that $\zeta_{rod}/\zeta_{coil} = 2.1$. These, and estimates of the coefficients ζ_{coil} and ζ_{rod} for DNA molecules of other lengths, are listed in Table 9.1.

As shown in Table 9.1, over the range of lengths $L = 22.4 - 151$ µm, the drag coefficient in the fully extended state is predicted to be only 1.7 to 2.6 times that in the coiled state! The main reason for this surprisingly small increase in the drag coefficient is the presence of the logarithmic dividing factor, $\ell n(L/d) = 9 - 11$ in Eq. (9.3); this factor is large because of DNA's enormous aspect ratio $L/d = 10,000 - 75,000$. Physically, this means the following: in the coiled state, the slender DNA molecule wanders through space, blocking easy flow paths for the solvent; hence, total drag on the coil is equivalent to that on an impenetrable ball, 2 - 4 µm in diameter, around which solvent must flow. In the fully extended state, the DNA at first blush looks even more formidable: it is a 20 - 150 µm long fiber. However, on closer look, this fiber turns out to be a scrawny filament only 0.002 µm wide. Thus, the drag on the long-but-thin filament is not much higher than it is for the impenetrable ball.

For typical synthetic polymer molecules, the situation is similar. For example, for polystyrene of molecular weight $M = 10^6$ in a theta solvent, $R^2 = C_\infty n \ell^2$, where C_∞ is the "characteristic ratio", which is around 10.0 for polystyrene [25], n is the number of backbone bonds, $n = 2 \times 10^4$, and $\ell = 1.54$ Å is the length of a backbone carbon-carbon bond. Thus, we have $R = 0.070$ µm. The fully extended length L of the chain can be estimated to be around $0.8 n \ell = 2.5$ µm, and the chain diameter d is roughly 1.0 nm. Thus, from

Table 9.1. Drag coefficients for DNA

L (µm)	$\zeta_{coil} = 1/D$ sec$(\mu m)^{-2}$	ζ_{rod} sec$(\mu m)^{-2}$	ζ_{rod}/ζ_{coil}
22.4	2.1	3.5	1.7
44.0	3.6	6.4	1.8
67.2	4.8	9.4	1.9
83.8	5.4	11.5	2.1
89.0	5.6	12.1	2.2
151	7.5	19.6	2.6

Eq. (9.2), $\zeta_{coil} = 0.35 \; \eta_s/k_B T \; \sec(\mu m)^{-2}$, while from Eq. (9.3), $\zeta_{rod} = 1.55 \; \eta_s/k_B T$ $\sec(\mu m)^{-2}$, only around four times ζ_{coil}. An even smaller ratio of ζ_{rod}/ζ_{coil} would be obtained in a good, rather than a theta, solvent. In general, from Eqs. (9.2) and (9.3), the ratio ζ_{rod}/ζ_{coil} is $1.23 \, L/(\ell n(L/d)R) \approx 0.1 \, L/R$. Thus, small or modest increases (a factor of two or less) in drag coefficient under flow can be expected for molecules that can be stretched to around 20 times their undisturbed coil size R. A factor of ten or more increase would only occur for very long molecules that can be stretched out to at least 100 times R. For most synthetic polymers, then, no more than modest changes in hydrodynamic drag coefficient with increasing chain extension should be expected.

9.3
Stretching Experiments with Longer DNA Molecules

Although the observable changes in the hydrodynamic drag coefficient are predicted to be modest, in principle they should be measurable if the DNA chains are made long enough. Thus, we describe here new experiments of the type presented by Perkins et al. [16], except that we use longer strands of DNA, up to 151 μm in length. The extension-velocity curves for all chains with lengths ranging from 22 to 151 μm can once again be approximately superposed onto those for shorter chains; see Fig. 9.1. However, Fig. 9.2 shows that the scaling factor λ required to bring the data of Fig. 9.1 into superposition obeys a power law with $m \approx 0.75 \pm 0.02$ when the chain length exceeds 40 μm. The change in apparent exponent from 0.54 (for shorter chains) to 0.75 shows that for long chains there is a measurable departure from the regime of dominant hydrodynamic interaction. Departures from dominant hydrodynamic interaction might also occur for chain lengths less than 40 μm at high chain extensions where they are hard to detect because of the flatness of the extension-force curve.

Although the data in Fig. 9.1 seem to superimpose to within experimental error, for chains long enough that the regime of dominant hydrodynamic interaction begins to break down, the curves should in principle change shape as the chain length L increases. This change in shape is evidently not large enough to stand out above the scatter in Fig. 9.1. A breakdown of dominant hydrodynamic interaction is, however, indicated in Fig. 9.3, where the fractional extension is plotted against V/D, the velocity divided by the molecule's center-of-mass diffusivity, for three "representative" DNA molecules of lengths $L = 44, 89$, and 151 μm. These molecules, whose scaling parameters λ are marked by open squares in Fig. 9.2, were chosen because for them $\lambda(L)$ lies almost exactly on the best fit line in Fig. 9.2. In the "Zimm limit" of dominant hydrodynamic interaction (where the drag coefficient ζ_{tot} equals $1/D$ and scales as L^{ν}), the curve of force vs extension for all three chain lengths would collapse onto a single line. Although at small extensions data for all three chains do appear to collapse onto a single line, at higher extensions ($x/L > 0.3$), data for the length $L = 151$ μm lie clearly to the left of data for the shorter chain with $L = 44$ μm. Thus, the theoretical prediction that the effective drag coefficient of a polymer

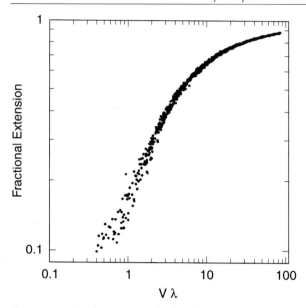

Fig. 9.1. Fractional extension of tethered DNA molecules with contour lengths ranging from 22.4 to 151 μm, vs flow velocity scaled by the shift parameter λ. Each set of data of extension vs velocity can be fitted by Eq. (9.1), which is the analytic expression of Marko and Siggia [28] for a worm-like chain, when the velocity is assumed to be proportional to the force. By fitting each data set to this equation, the contour length (L) and a constant (c) which multiplies the force are determined. The velocity rescaling parameter (λ), plotted in Fig. 9.2, is $\lambda \equiv c(92)/c$, where $c(92)$ is the value of c for one arbitrarily chosen data set, with $L = 92$ μm

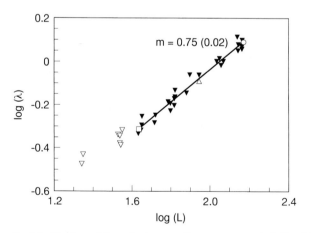

Fig. 9.2. The log of the velocity rescaling parameter λ, defined in the caption of Fig. 9.1, as a function of the log of the DNA contour length. The symbols ($\triangledown$) are for data with L less than 40 μm; these data were excluded from a least squares linear fit to the rest of the data, which yielded a line with slope $m = 0.75 \pm 0.02$. The three open symbols ($\square$, $\triangle$, $\bigcirc$) represent data sets that lie very close to this line, and were chosen as representative data sets for comparison to theory in Fig. 9.3

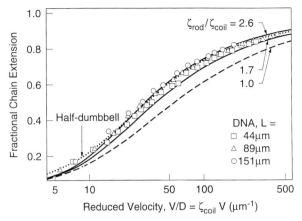

Fig. 9.3. Average fractional chain extension plotted against V/D, the ratio of flow velocity to diffusivity for "representative" DNA chains of length 44 µm (□), 89 µm (△), and 151 µm (○). The *dotted line* is the prediction of the "half-dumbbell model;" i.e., it is the force-extension curve for the worm-like chain with a reduced force $F/k_B T$ equal to 0.73 $\zeta_{coil} V$ applied to the chain end. The *solid lines* are the predictions of the simulation with $h = 1.3$ and 4.0, and the *dashed line* is the prediction for $h = 0$

molecule should increase with increasing chain deformation is confirmed, and there is evidence of a breakdown of "Zimm" behavior at high chain extensions.

However, the change in the power-law exponent from around 0.54 to around 0.75 and the change in shape in the velocity-extension curve as the DNA length increases might be partially due to hydrodynamic interactions of the DNA molecule with the wall of the cell in which the DNA is contained. The DNA molecule is kept around 15 µm from the top and 60 µm from the bottom of the cell. One can estimate from an "image-force" analysis that the walls tend to suppress hydrodynamic interactions between parts of the chain separated from each other by more than twice the distance to the wall (i.e., 30 µm). Although we shall show later that the observed changes in hydrodynamic interaction for $L > 40$ µm are close to those predicted theoretically in the absence of wall interactions, we cannot at this time rule out an influence of the wall on the results reported here for long chains.

Figure 9.3 also shows that the shape of the velocity-extension curve for DNA molecules is almost the same as the force-extension curve for the worm-like chain, given by the dotted line. Therefore, the behavior of the tethered DNA chain in the flow is equivalent to a "half-dumbbell," that is a nonlinear spring with a single fictitious bead at its free end. To obtain the fit shown, the drag coefficient ζ_{bead} for this fictitious bead is set equal to 0.73 times the drag coefficient $\zeta_{coil} = 1/D$ of the undisturbed coil, and the persistence length A is set equal to the value $A = 0.053$ µm measured by Bustamante et al. [19] for unlabeled DNA. The persistence length of the DNA used in our experiments may be somewhat higher than this, due to possible stiffening of the chain by the fluorescent label. According to Eq. (9.1), an increase in "A" merely rescales the abscissa of Fig. 9.3; thus if "A" is larger than 0.053 μm, an equally good fit to the data would be obtained by multi-plying ζ_{coil} by a quantity somewhat larger than 0.73.

9.4
Hydrodynamic Model for DNA

To help understand these results, we develop here a simulation method for predicting the effects of drag, and elastic and Brownian forces on a DNA molecule in a simple uniform flow field, and compare the simulation results to the experimental data of Perkins et al. [16] and the new data reported here.

In a detailed microscopic model, the DNA would be described as a long flexible cylinder, which curls up and fluctuates under the action of Brownian motion. The chain lengths in the experiments of Perkins et al. ranged from 22 to 84 μm, which are 400–2000 times the persistence length, $A = 0.05$ μm. Therefore, in any discretization of the chain that accounts for the detailed con-formation of such a chain's backbone at the length scale of the persistence length, the chain would need to be subdivided into thousands of computational segments, each of size ~ 0.01 μm or less. The Brownian time-scale t_B of a seg-ment of size "a" (the time to diffuse a distance equal to its diameter, "a") is roughly $t_B \sim \eta_s a^3 / k_B T$ where η_s is the solvent viscosity. This gives a Brownian time of less than 10^{-8} s at room temperature for $\eta_s = 1$ cP. This very short Brownian time sets the maximum duration of a time step in a simulation of a chain discretized as finely as this. To obtain well behaved averages of steady-state properties, the simulation must continue for much longer ($\gtrsim 100 \times$ longer) than the longest relaxation time of the chain, which is several seconds in the ex-periments of interest. Thus, $\sim 10^{11}$–10^{12} time steps, or even more, would be required for a single simulation of a chain composed of thousands of small sub-elements. Such simulations are impractical, even with today's supercomputers.

However, the limit of optical resolution in the experiments of Perkins et al. [16] is around 1 μm; hence, all the optically measurable properties of the chain can, we believe, be extracted from a suitable coarse-grained model, whose smal-lest elements are around 1 μm in size. Fortunately this length scale is much longer than the persistence length of the chain, but still much smaller than the chain's contour length. Thus, an appropriate coarse-grained description of the chain is one in which the smallest element is a flexible sub-molecule that is itself many persistence lengths long.

We therefore divide the chain into N "beads," with N as large as 80, distribut-ed uniformly along the chain backbone, and connected together sequentially by "springs." The fully extended length of each spring is $\ell = L/N$. One end of the chain is held fixed at a tethering point at the origin. We number the beads and springs sequentially, with spring 1 being attached to the tether point at one end and to bead 1 at the other. Bead i is then a contour distance Li/N along the chain backbone away from the tether point, and bead N is attached to the free end of the chain. We define the Cartesian bead coordinates such that x_i is bead i's distance downstream from the tether point in the flow direction, and y_i and z_i are distances from the tether point in directions normal to x.

9.4.1
Elastic Spring Force

The elastic spring force F_i^s for each spring is obtained from Eq. (9.1) by replacing L with the length of the submolecule ℓ:

$$\frac{F_i^s A_{eff}}{k_B T} = \frac{1}{4}\left(1 - \frac{r_i}{\ell}\right)^{-2} - \frac{1}{4} + \frac{r_i}{\ell} \tag{9.4}$$

where r_i is the separation distance between bead $i - 1$ and bead i. That is, $r_i^2 \equiv (x_i - x_{i-1})^2 + (y_i - y_{i-1})^2 + (z_i - z_{i-1})^2$ for $i \geq 2$, and $r_1^2 \equiv x_1^2 + y_1^2 + z_1^2$.

The introduction of beads into the worm-like chain increases the molecule's flexibility, because the beads act like "free hinges" which do not transmit a bending moment. The added flexibility would become large if the spacing between beads was comparable to the persistence length of the chain, but because the chain is around 1000 times the persistence length, the spacing between beads is large enough that the chain's flexibility is only slightly increased, even when there are as many as 80 beads in the chain. It is shown in the Appendix that the increase in flexibility of the chain caused by addition of the beads can be effectively offset by a modest increase in the value of the persistence length used in the simulations. Thus, for $L = 67.2$ μm, the behavior of the worm-like chain with $A = 0.053$ μm is recovered in a 40-bead chain by increasing the persistence length to $A_{eff} = 0.0610$ μm, and in an 80-bead chain by choosing $A_{eff} = 0.0700$ μm. In what follows, the simulation method is shown to give the same results for $N = 80$, $A_{eff} = 0.0700$ μm as it does for $N = 40$, $A_{eff} = 0.0630$. Thus, using 40–80 beads provides a fine enough discretization of the chain's contour for our purpose.

9.4.2
Drag Force

The drag force on bead i is given by $F_1^d = \zeta_i k_B T V$, where V is the imposed solvent flow velocity, which is assumed to be uniform and constant in the x direction, corresponding to conditions in the experiments of Perkins et al. Here $\zeta_i k_B T$ is an effective drag coefficient for bead i. In the simplest case of uniform drag, all ζ_is would be equal, $\zeta_i = \zeta$, and the total drag on the molecule would then be divided equally among the beads: $\zeta = \zeta_{tot}/N$, where $\zeta_{tot} k_B T$ is the effective drag coefficient for the whole chain, which should scale roughly as $\zeta_{tot} \propto L^{0.5}$ if hydrodynamic interaction is dominant.

To account for variations in the hydrodynamic drag coefficients that occur when the macromolecule is stretched out in the flow, we use the average separations between beads to compute the effective hydrodynamic interactions between them. The average bead positions $\langle x_i \rangle$ and drag coefficients ζ_i are then calculated self-consistently. In principle, since the hydrodynamic interactions are sensitive to chain configuration, fluctuations in the chain conformation produced by Brownian motion should be reflected in fluctuations in hydrodynamic interactions, but such fluctuation effects are neglected here.

Using the Oseen theory to describe the hydrodynamic interactions between each pair of beads [5], the drag force on bead i is given by

$$F_i^d = V\zeta k_B T - \frac{h}{N} \sum_{j \neq i} \Omega_{ij} F_j^d \tag{9.5}$$

where Ω_{ij} accounts for the hydrodynamic interaction between beads i and j; it falls off as the inverse of the distance between beads i and j. The coefficient h in Eq. (9.5) is a dimensionless pre-factor that accounts for the detailed size and shape of the interacting units along the chain backbone; we will estimate h using the properties of DNA molecules shortly. Since we are using a pre-averaging approach, we let Ω_{ij} depend inversely on the average separation between beads i and j. To keep Ω_{ij} from diverging, we introduce a cut-off length $x_{cut} = R$, where $R \equiv \sqrt{2AL}$ is roughly the root-mean-square separation of the chain ends in the absence of flow. Thus

$$\Omega_{ij} \equiv \begin{cases} 1, & |\langle x_i \rangle - \langle x_j \rangle| \leq R \\ R/|\langle x_i \rangle - \langle x_j \rangle|, & |\langle x_i \rangle - \langle x_j \rangle| \geq R \end{cases} \tag{9.6}$$

where $\langle x_i \rangle$ denotes the average bead position. We do not attempt to account for any variations in hydrodynamic interactions that might occur on length scales smaller than the average radius of the unextended coil. Notice the factor of $1/N$ introduced into Eq. (9.5) to offset the growth of the sum due to an increasing number of terms. Hence, as N increases, a continuous limit is approached in which Ω_{ij} approaches a constant at large N, if i and j are varied with N to keep $\tilde{i} \equiv i/N$ and $\tilde{j} \equiv j/N$ constant.

The drag force on each bead depends on the average x positions of all beads, and these positions depend on the drag forces. Hence, the solution must be obtained iteratively. We update the drag forces on a bead every 10^4–10^5 time steps. This procedure gives converged results for both the drag and the averaged bead positions.

The choice $x_{cut} \approx R$ can be justified as follows. One expects the effective hydrodynamic drag coefficient ζ to begin to deviate from its no-flow value ζ_{coil} when the coil dimension exceeds its no-flow size, R. Because of the neglect of fluctuations in drag, however, in the simulations ζ begins to change as soon as $\langle x_N \rangle$, the average stretch of the chain, exceeds x_{cut}. Since $\langle x_N \rangle \to 0$ as $V \to 0$, if we choose $x_{cut} \ll R$, ζ will begin to depart from ζ_{coil} at velocities below those required to stretch the chain beyond its random-flight dimension R. Hence, we should choose $x_{cut} \approx R$. Fortunately, the simulation results are insensitive to x_{cut} at least over the range $R/2 \leq x_{cut} \leq 2R$; see the Appendix.

The combination of the spring and the drag force can be expressed as the gradient of a bead potential, W_i:

$$F_i = F_i^s + F_{i+1}^s + F_i^d = \nabla_i W_i \tag{9.7}$$

where $\nabla_i \equiv (\partial/\partial x_i, \partial/\partial y_i, \partial/\partial z_i)$, $\underline{F}_i$ is a vector force acting on bead i with magnitude $|\underline{F}_i|$, and

$$\frac{W_i}{k_B T} \equiv \frac{\ell}{A} \sum_{j=0}^{j=1} \left\{ \frac{1}{4}\left(1 - \frac{r_{i+j}}{\ell}\right)^{-1} - \frac{1}{4}\left(\frac{r_{i+j}}{\ell}\right) + \frac{1}{2}\left(\frac{r_{i+j}}{\ell}\right)^2 \right\} - x_i \zeta_i V \tag{9.8}$$

The sum over j from 0 to 1 accounts for the two springs attached to each bead except the end bead. For bead N, there is only one spring and only $j = 0$ should be taken in the sum. A similar consideration applies to the two spring forces in Eq. (9.7). The "bead drag coefficient" is $\zeta_i \equiv F_i^d/V$, where F_i^d is given by Eq. (9.5); thus, ζ_i includes the contributions from both terms on the right-hand-side of Eq. (9.5), and is deformation dependent. The deformation dependence imputed to ζ_i is a convenient way of accounting for the changes in flow field, and hence of the drag force, that are brought about by changes in molecular conformation.

9.4.3
Brownian Motion

Brownian motion causes fluctuations in all directions about the position of minimum potential, and we shall account for it by a standard Metropolis Monte Carlo method. In this method, a bead, say bead i, is chosen at random, and its position is displaced slightly in a direction and by a distance determined by selection of two random numbers, one to determine whether the displacement will be in the x, y, or z direction, and the second to determine the sign and magnitude of the displacement. The displacement of a bead in the chosen direction is determined by a random number uniformly distributed over the interval $(-\ell/10, \ell/10)$, where $\ell = L/N$ is the maximum possible extension of the spring. If the randomly chosen bead displacement would stretch a spring beyond full extension, the bead move is rejected, and the simulation proceeds to the next time step. After choosing a potential bead displacement, the change $\Delta W_i \equiv W_i^f - W_i^i$ in the potential energy is computed, where W_i^i is the potential energy before the displacement and W_i^f is the potential energy after. If $\Delta W_i \leq 0$, the move is accepted. If $\Delta W_i > 0$, the move increases the bead's energy, and so is accepted only with probability $P = \exp(-\Delta W_i/k_B T)$. Then another potential move is attempted using the same procedure. We find that the number of such attempted moves (or "time steps") required for accurate average properties to be computed is of order $10^4 \times N^3$. Thus, runs with $N = 40$ ($N = 80$) beads require around 10^9 (10^{10}) time steps.

9.5
Simulation Results

Figure 9.4a shows typical simulated chain configurations for various magnitudes of the drag force $\zeta_{tot} V$ for a chain with $N = 80$, and $h = 0$. One can observe qualitatively that the chain stretches out as expected when the drag increases,

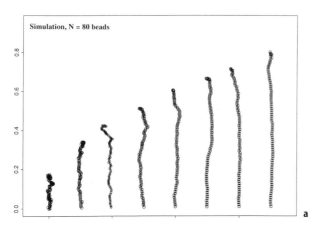

Fig. 9.4. a Typical simulated chain configurations for $N = 80$, $h = 0$, and $\zeta_{tot}V = 12$, 24, 36, 60, 96, 144, 198, and 400 sec/μm^2. **b** Images of experimental DNA molecules at velocities of 1, 2, 3, 4, 5, 7, 10, 12, 15, 20, 30, 49, and 50 μm/s, from Perkins et al. [16]

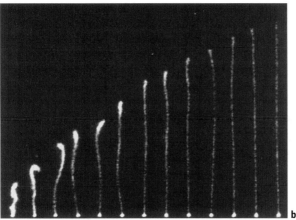

that beads tend to bunch up near the free end, and that the degree of sideways fluctuation in the chain seems to decrease as full extension is approached. These configurations are similar to images of DNA molecules at various flow rates published by Perkins et al. [16]; see Fig. 9.4b.

Figure 9.5 plots the average chain extension $\langle x_{max}\rangle/L$ vs total drag force $\zeta_{tot}V$ for $N = 40$ and $N = 80$, with A_{eff} chosen as described in Sect. 9.4.1, and $h = 0$. The quantity x_{max} is the distance from the tether point of the bead furthest downstream at any given instant. Because of Brownian motion, the end bead is not always furthest downstream; thus $\langle x_{max}\rangle$ is slightly greater than $\langle x_N\rangle$. We plot $\langle x_{max}\rangle$ because it corresponds to the experimentally measured quantity. Figure 9.5 shows that there are only small differences between the results for $N = 40$ and those for $N = 80$. It also shows that the shape of the simulated extension $\langle x_{max}\rangle/L$) vs velocity curve is similar to the extension (r) vs force (F/kT) curve for the worm-like chain with a force applied at the chain end, Eq. (9.1).

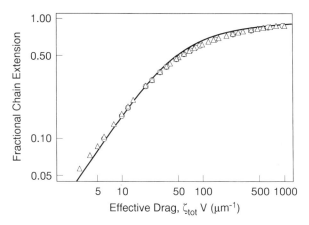

Fig. 9.5. Average fractional chain extension $\langle x_{max} \rangle /L$ vs total drag $\zeta_{tot} V$ for $h = 0$ and $N = 40$ ($\triangle$) and $N = 80$ ($\bigcirc$). The *line* is the fractional extension predicted by the worm-like chain model, Eq. (9.1), with a reduced force F/kT equal to 0.47 $\zeta_{tot} V$ applied to the end of the chain only

Simulations for long ($L \geq 34$ µm) molecules show that for fixed hydrodynamic interaction parameter h, the curves of fractional extension vs drag force are independent of L. This result is to be expected for highly flexible chains ($L \gg A$). Thus, in the model, the effect of chain length enters only through the chain-length dependence of the hydrodynamic interaction parameter h. We find that ζ_{tot}, defined as $\sum_i \zeta_i$, varies from a low-stretch value ζ_{coil} to a higher value ζ_{rod} when the chain's length approaches full extension; the values ζ_{coil} and ζ_{rod} depend on the value of h. For $h = 0, 1.3$, and 4.0 we find $\zeta_{rod}/\zeta_{coil} \approx 1, 1.8$, and 2.6, respectively. As discussed in Sect. 9.2, and shown in Table 9.1, the latter two values of ζ_{rod}/ζ_{coil} correspond roughly to DNA lengths of $L = 44$ µm and 151 µm, respectively. The values bracket the values of L for which the extension-velocity curves can be rescaled using the power-law exponent of 0.75; see Fig. 9.2. We thus include in Fig. 9.3 curves of predicted fractional extension vs drag V/D for $h = 1.3$ and 4.0, to show the changes predicted in this curve when L is increased from 44 to 151 µm; since the parameters of the model were obtained independently, the solid curves in Fig. 9.3 are obtained without adjustable parameters. Note in Fig. 9.3 that with increasing h, the curve shifts to the left and steepens; the steepening occurs because for larger h, the frictional grip of the solvent increases to a greater degree when the chain is extended. The shifting of the curve with increasing h (or L) agrees well with the experimental curves, except for a small offset. If, as we suspect, the persistence length of labeled DNA molecules is somewhat larger (about 15 % larger) than the value $A = 0.053$ µm for unlabeled DNA, then the theoretical lines would almost perfectly bracket the experimental data for molecules of length 44–151 µm. As mentioned earlier, the dashed line in Fig. 9.3 shows that the shape of the velocity-extension curve is very close to that of the force-extension curve for the worm-like chain with force only at the free end (i.e., the half-dumbbell).

The average distribution of bead mass as a function of distance downstream from the tethering point is plotted for various values of V in Fig. 9.6, for $N = 80$ and $L = 67.2$ µm. The method by which the data in Fig. 9.6 were obtained is

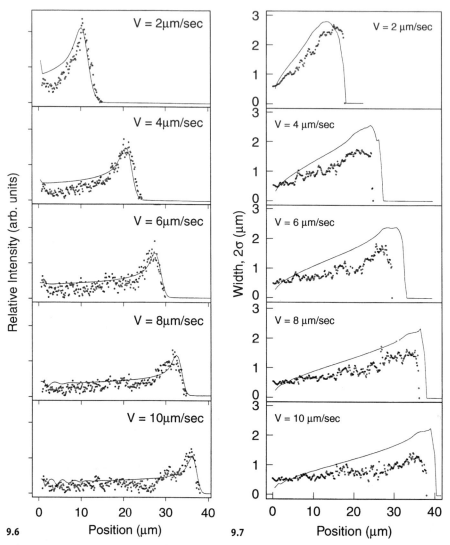

Fig. 9.6. Computed distribution of bead mass as a function of position downstream of the tether point for $N = 80$ and $L = 67.2$ μm at the values of V shown (*lines*), compared to the measured distribution of DNA mass for $L = 67.2$ μm (*symbols*). The values of the parameter $\zeta_{coil} = 4.8$ s (μm)$^{-2}$ is obtained from the diffusivity measurements of Smith et al. [24], and $h = 1.5$ is chosen so that the ratio ζ_{rod}/ζ_{coil} agrees with the theoretical estimate in Table 9.1

Fig. 9.7. Computed width of transverse mass distribution 2σ as a function of position downstream of the anchoring point for the conditions given in Fig. 9.6

described in Perkins et al. [16]. Calculations analogous to those in Sect. 9.2 show that the appropriate value of h for this chain length is around 1.5, and simulations for $L = 67.2\,\mu m$ were carried out for this value of h. From the measurements of Smith et al. [24], the value of $\zeta_{coil} = 1/D$ is around 4.8 s/μm^2 for $L = 67.2\,\mu m$. The simulated distributions were obtained by counting repeatedly ($\sim 10^5$ times) the number of beads in bins of width 0.5 μm, and averaging together the results. The agreement between experimental and simulated mass density (with no adjustable parameters) is excellent; the tendency of the DNA mass to collect near the free end of the chain is correctly predicted.

The relative lateral spread of the bead mass can similarly be computed by compiling statistics for the distributions of y and z coordinates for beads residing in each of the bins described above. The standard deviation 2σ is plotted in Fig. 9.7 as a function of distance downstream of the tether point, for various V. To improve the statistics somewhat, σ is taken as the square root of the variance averaged over the two equivalent directions y and z: $\sigma^2 \equiv (\sigma_y + \sigma_z)/2$. The statistics become poor for 2σ at the farthest positions from the tether point where there is little chain density. For positions within 3.5 μm of the largest positions for which data are reported in Fig. 9.7, the chain density is small (see Fig. 9.6), and chain statistics sparse. Thus, to reduce the noise level, in this range of positions, σ was averaged over a range of positions up to 1 μm upstream and downstream of the nominal x position. Qualitative agreement is obtained in Fig. 9.7 between the shape of the computed profile of σ and that experimentally measured; in particular, the cone-shaped mass distribution is reproduced. As the velocity increases, and the chain becomes more extended, the potential wells in which the beads reside become steeper, and lateral Brownian excursions are suppressed, so that 2σ decreases; however, the predicted decrease in σ is not as great as is seen in the experiments. As discussed in Perkins et al. [16], the longitudinal Brownian excursions are also suppressed as the chain extension increases, in qualitative agreement with the model. Results obtained for $N = 40$ beads (not shown) are in good agreement with those obtained for $N = 80$ in all cases. Thus, the deviations between predicted and measured lateral excursions are probably not caused by any artificial increase in flexibility caused by introduction of beads into the worm-like chain.

The success of the hydrodynamic model suggests that it can be used to analyze the details of the distribution of drag along the DNA molecule, and its dependence on velocity. Figure 9.8 shows the predicted drag coefficient ζ_i as a function of bead number i for different velocities V (or equivalently, different chain extensions r/L) for $L = 67.2\,\mu m$, $h = 1.5$. The drag is greater on the beads at the chain ends than on those in the center of the molecule, because the ends are farthest from the center of mass of the molecule, and hence have the weakest hydrodynamic interactions with other beads. This is especially true of beads near the tether point, since the spacing between beads is greatest at this end of the molecule. The limiting results for $V \to \infty$ are shown by a dotted line. Figure 9.9 shows the reduced total drag ζ_{tot}/ζ_{coil} as a function of the chain extension $\langle x_{max}\rangle/L$ for $h = 1.3$ and 4.0.

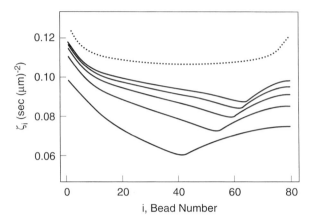

Fig. 9.8. Drag coefficient ζ_i as a function of bead number i for $N = 80$, $L = 67.2$ μm, $h = 1.5$, at velocities V (*from bottom to top*) of 2, 4, 6, 8, and 10 μm s^{-1}, corresponding, respectively, to average chain stretchings $\langle x_{max}\rangle/L$ of 0.172, 0.325, 0.428, 0.499, and 0.552. The *dashed line* corresponds to the high-velocity limit of $\langle x_{max}\rangle/L \to 1.0$

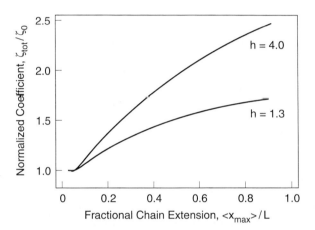

Fig. 9.9. Normalized drag coefficient ζ_{tot}/ζ_{coil} as a function of chain stretching $\langle x_{max}\rangle/L$ for $N = 40$, $L = 67.2$ μm, $h = 1.3$ and 4.0, corresponding to molecular lengths of 44 and 151 μm

9.6
Discussion

The predictions of the coarse-grained hydrodynamic model are, for the most part, in quantitative agreement with the flow experiments on long DNA molecules. The degree of stretch as a function of velocity, and the distribution of mass along the stretched chain, are in nearly quantitative agreement with the measured properties, with no adjustable parameters. The lateral dispersion of the mass distribution is qualitatively predicted; however, the width does not decrease as much with increasing chain extension as is observed in the experiments.

The relationship between flow velocity and chain extension is very similar to that between force and chain extension when the force is applied only at the free end of the molecule. This implies that the gross features of the DNA experiments can be described by the simplest of models: a finitely-extensible elastic dumbbell. In particular, at steady-state, the degree of stretching of the molecule as a function of flow rate can be predicted rather well by assuming, as in the dumbbell model, that all the drag is concentrated on the free end of the molecule, and that the drag coefficient of this bead is independent of the molecular deformation. Not surprisingly, to obtain quantitative agreement between extension-velocity curve of the dumbbell model and that of the more complete multi-bead model, the drag lumped onto the end of the molecule must be reduced somewhat (around 30%) below the total drag distributed along the multi-bead chain.

The surprising accuracy of the dumbbell model in describing the steady-state velocity-extension curve is due to a fortuitous cancellation of effects: the distribution of drag forces along the chain modestly reduces the slope of the curve of fractional extension vs velocity, while the increase in effective drag coefficient with chain extension modestly steepens it. For DNA molecules of length around 40–150 μm, these two effects cancel out, and the steepness of the velocity-extension is close to that of the simple dumbbell model with all drag force lumped onto the chain end and with no deformation dependence of the drag coefficient. The accuracy of the crude dumbbell model is no doubt also enhanced by the shape of the observed mass distribution, with its high concentration of monomer, and hence of drag force, near the chain end. Remarkably, this distribution of DNA mass is in qualitative agreement with that first proposed by Frenkel [26] some 50 years ago for a polymer molecule in an extensional flow. He wrote that "the central portion of the molecule becomes straightened out along the direction of flow…, while the two end portions remain curled in the usual way." A similar idea, called the "yo-yo" model, has been put forward by Ryskin [27]. In the DNA experiments, there is only one "end portion" of the molecule, and the tethering point in the DNA experiments is analogous to the "central portion" of a molecule in a flow gradient such as that of an extensional flow. Otherwise, Frenkel's description of a polymer conformation in a flow field aptly describes the observed DNA conformations.

The expected change in the hydrodynamic scaling law from "dominant hydrodynamic interaction" towards "free draining" is observed on stretching DNA molecules whose contour lengths are long, of the order of 40 μm or longer. For shorter chains, the increase in the effective hydrodynamic drag coefficient ζ_{tot} as the chain is extended is small, especially since part of it occurs when the chain is already so highly stretched that the shape of the curve of chain extension vs velocity is no longer very sensitive to ζ. Even for 40 μm $\leq L$ ≤ 150 μm, the hydrodynamic drag coefficient is only very weakly dependent on the molecular stretch. Using a simplified analysis, Marko and Siggia [28] have recently reached a similar conclusion. If the highest molecular weight of the chain were further increased 10- or 100-fold, more dramatic effects should be observable, but for the molecular weight range of typical synthetic polymers, especially those used commercially, the neglect of the conformation dependence of the drag coefficient in dilute solutions seems justified.

A similar conclusion can be drawn from calculations for shear and extensional flows of untethered molecules described by multiple beads-and-springs models in which deformation-dependent hydrodynamic interactions are accounted for [2, 6, 29 – 33]. These calculations have yielded predictions of steady-state rheological properties in shear and extensional flows that are qualitatively similar to those of the simple dumbbell model, and show that the effective drag coefficient acting on the polymer molecule is fairly insensitive to the deformation of the chain [32]. As mentioned in the Introduction, rheological and rheo-optical experiments on dilute solutions are usually in qualitative accord with the dumbbell model, although, as expected, experiments conducted on extremely high molecular-weight samples do show a slight shear thickening effect that might be attributable to a weak conformation-dependent drag coefficient [34].

Thus, experimental studies of the hydrodynamics of single DNA molecules in a simple flow conform to the classical picture of the dynamics and rheology of flexible polymer molecules, in which three forces – the elastic spring force, hydrodynamic drag with a nearly constant drag coefficient, and the Brownian force – provide a complete physical model, at least in the steady-state experiments described here.

Appendix: Model Validation

We validate our Monte Carlo model using two test. In the first test, we compute the mean square end-to-end separation in the x direction, $\langle x_N^2 \rangle_0$, between the chain end (bead N) and the tether point, for various N, where the subscript "0" denotes the absence of flow. An analytic value for $\langle x_N^2 \rangle_0$ can be estimated by assuming that the end bead resides in a harmonic well, as described in Perkins et al. [16]. When flow is absent, this gives

$$\langle x_N^2 \rangle_0 = \frac{2LA}{3} \tag{9 A1}$$

(see Eq. (4) in Perkins et al., with $x = 0$). In the second test, we compute via simulation the average positions of the N beads in the chain in the limit of a flow weak enough that the spring force law can be linearized, $F_s(r) \approx Hr$, where H is the linear spring constant, given by the classical formula

$$H = \frac{k_B T}{\langle x_N^2 \rangle_0} \tag{9 A2}$$

When a chain with linear spring constant H is divided into N sub-chains, the spring constant of each sub-chain is HN. In a weak flow, the last, Nth, bead is pulled away from the N-1st bead by a drag force ζV; hence the average extension of the Nth spring is

$$\langle x_N - x_{N-1} \rangle = \frac{\zeta V}{HN} \tag{9 A3}$$

The N-1st bead feels an average force ζV from the Nth spring, as well as a flow-induced drag force ζV acting on bead $N-1$. (We take $h = 0$ here.) Hence the extension of the N-1st spring should be $\langle x_{N-1} - x_{N-2}\rangle = 2\zeta V/HN$. And, in general, the extension of the ith spring is

$$\langle x_i - x_{i-1}\rangle = \frac{(N - i + 1)\zeta V}{HN} \tag{9 A4}$$

Thus, for a small enough velocity, the stretching of the chain should decrease linearly with i/N, where i is the bead number. We find, for chains of $N = 10, 20, 40$, and 80 beads, that the spring stretch is indeed linear in i/N; see Fig. 9 A1. The lines through the points in Fig. 9 A1 are least-square fits to the points for each value of N. The slopes of these lines are inversely proportional to the

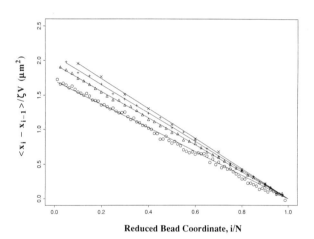

Fig. 9 A1. Reduced average stretch of springs, $\langle x_{N-i} - x_{N-i-1}\rangle/\zeta V$, vs reduced spring number, i/N, with numbering starting from the tether point, for chains with $N = 10 (\times)$, $20 (+)$, $40 (\triangle)$, and $80 (\bigcirc)$ springs. Here $\zeta V = 0.02$ for $N = 80$, $\zeta VN = 0.05$ μm^{-1} for $N = 40$ and 20, and $\zeta V = 0.10$ for $N = 10$. In this and subsequent figures, $L = 67.2$, $A = 0.053$, $h = 0$, except where noted

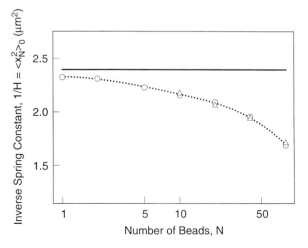

Fig. 9 A2. The *symbols* ($\triangle$) show the effective inverse spring constant,

$$\frac{1}{H} = \frac{\langle x_{N-i} - x_{N-i-1}\rangle \cdot N}{((N - i + 1)\,V\zeta)},$$

as a function of the number of beads in the chain, obtained from the slopes of the lines in Fig. 9 A1. The *symbols* ($\bigcirc$) give the inverse spring constant $1/H = \langle x_N^2\rangle_0$ from the mean-square positions of the end bead $\langle x_N^2\rangle_0$ in simulations without flow. The horizontal *solid line* is from the approximation $H \approx 2LA/3$

spring constant H, according to Eq. (9 A4). Thus, the decrease in the slope with increasing N implies that the effective spring constant H increases with increasing number of beads N. The inverse spring constants $1/H$ for $N = 10, 20, 40$, and 80 are plotted in Fig. 9 A2, along with $1/H$ extracted from fluctuations in the position of bead N in the absence of flow using Eq. (9 A2), for chains containing $N = 1 - 80$ beads. Notice that the spring constants obtained from the no-flow calculations are equal to those from weak flows, to within computational error (which is of order the size of the symbols). Both methods of obtaining H show that it increases with increasing N. The horizontal line in Fig. 9 A2 is the approximate analytic prediction given by Eqs. (9 A1) and (9 A2).

A modest deviation between the simulated values of H for small N and those obtained from the analytic theory using the harmonic approximation is to be expected, since the analytic theory is approximate. The increase in H with increasing N can also be explained rather easily. The worm-like chain has stiffness uniformly distributed along its length. In our beads-and-springs approximation to it, the beads act like free "joints", which add flexibility to the chain. As the chain becomes more flexible with larger N, it requires more force to extend it, and it therefore has a higher effective spring constant H. If we were to add thousands of beads to the chain, each spring would be shorter than the persistence length, $A \approx 0.05$ μm, and Brownian motion would then stretch each spring to nearly its full length. The elasticity of such a chain would therefore approach that of a freely-jointed chain, and would depend only on how many springs the chain is partitioned into, and not on the nominal persistence length A.

The elasticity of our model bead-spring chain can be probed directly by applying a constant force F to the end bead of the chain only, and computing the average stretch of the whole chain, $\langle x_N \rangle$. The relationship between applied force and chain extension, $\langle x_N \rangle$ is plotted in Fig. 9 A3, for $N = 1, 10, 40$, and 80, along with the result expected for a worm-like chain, namely Eq. (9.1), for $A = 0.053$ and $L = 67.2$ μm. Figure 9 A3 shows that if the only bead is the one on the chain end, the elastic properties of the chain are essentially identical to those expected for a worm-like chain. As more beads are added to the chain, however,

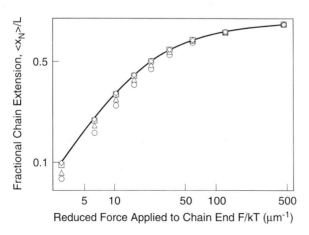

Fig. 9 A3. Average fractional chain extension $\langle x_N \rangle / L$ vs reduced force F/kT applied to the end of chains with $N = 1$ (diamonds), $10 (\square)$, $40 (\triangle)$, and $80 (\bigcirc)$ beads, and $A = 0.053$. The *line* is the prediction in Eq. (9.1), for the worm-like chain

deviations from this behavior occur, and the curve of extension vs force shifts to the right as N increases. Thus, elasticity of the chain as a whole, with a force applied only to the chain end, is most faithfully reproduced by the bead-spring model when N is small.

Of course, our interest is in describing the behavior of DNA under a flow field; in that case the forces are applied along the entire chain contour, and not just at the chain end. For the flow problem, N needs to be large, so that the distribution of the drag force can approach the continuous limit. One can show analytically, in the limit of $A/\ell \rightarrow 0$, that the discretization error in the computed average chain extension produced by replacing the continuous drag force by a force distributed among N beads, is exactly $1/N$. Thus, to reduce this error to 2.5% or less, N must be at least 40. But for $N \geq 40$, the chain elasticity departs significantly from that of the original worm-like chain model; see Fig. 9A2.

A simple way to counteract the increased chain flexibility caused by the introduction of beads into the chain is to increase the value of the persistence length A somewhat, to bring the relationship between $\langle x_N \rangle$ and F back toward that of a pure worm-like chain with $A = 0.053$ and $L = 67.2$ µm. Figure 9A4 shows that the behavior of the worm-like chain with $A = 0.053$ is nearly recovered for $N = 40$ when A is raised to $A_{eff} = 0.0610$, or for $N = 80$ with $A_{eff} = 0.0700$. Notice, however, that the predicted value of $\langle x_N \rangle$ using this method is slightly too low at small force, and slightly too high for large force. These deviations are unavoidable, even when the best-fit A_{eff} is chosen, because the shape of the force-extension curve changes gradually as N increases. For very large $N (N > 1000)$, the force-extension curve would approach that of a freely jointed chain. For $N < 80$, however, and $L > 40$ µm, the fully-extended length of spring ℓ is still a factor of ten or so greater than the persistence length A, and deviations in the shape of the curve from that of the worm-like chain theory remain small enough that they can be corrected satisfactorily simply by adjusting A. The suitability of this procedure is further demonstrated in Sect. 9.5, where the results for chains with $N = 40$ ($A_{eff} = 0.0610$) are shown to be almost identical to those for $N = 80$ ($A_{eff} = 0.0700$).

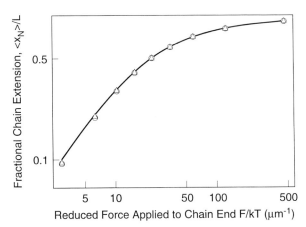

Fig. 9A4. The same as Fig. 9A3 for $N = 40$ and 80, but using adjusted values of the persistence length, $A_{eff} = 0.0610$ for $N = 40 (\triangle)$, and $A_{eff} = 0.0700$ for $N = 80 (\bigcirc)$

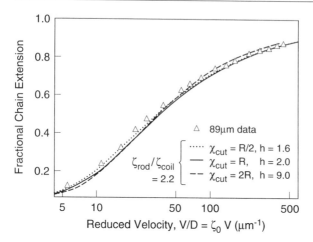

Fig. 9A5. Average fractional chain extension x/L plotted against V/D, the ratio of flow velocity to diffusivity for a DNA of length 89 μm ($\triangle$). The *lines* are the predictions of the simulation using $x_{cut} = R/2, R$, and $2R$, with h chosen so that $\zeta_{rod}/\zeta_{coil} = 2.2$, as estimated in Table 9.1 for DNA of this length $L = 89$ μm

The above discussion focused on a chain of length $L = 67.2$ μm, which is the length of a molecule whose hydrodynamic properties were studied in detail by Perkins et al. [16]. However, for a fixed value of h, the simulations show that the relationship between force and relative extension r/L is virtually independent of molecular length, for almost the full range of molecular lengths studied experimentally, i.e., for $L \geq 30$ μm $\gg A$.

In Sect. 9.4.2, a cut-off distance $x_{cut} \approx R$ was introduced to keep the Oseen expression for the hydrodynamic interactions from becoming singular; see Eqn. (9.6). While physical reasoning tells us that this is the correct order of magnitude for x_{cut}, we would like to assure ourselves that the simulations are not too sensitive to this choice. To do this, simulations were carried out with $x_{cut} = 2R$ and $R/2$. In Fig. 9A5, the results obtained for x/L vs V/D are compared to those for the choice $x_{cut} = R$. For all three choices of x_{cut}, the ratio of drag coefficients ζ_{rod}/ζ_{coil} is fixed at the value 2.2 estimated to be appropriate for $L = 89$; see Table 9.1. Figure 9A5 shows that the predictions for x/L vs V/D are indeed insensitive to the value of x_{cut} over this range; all three choices agree almost equally well with the experimental curve.

List of Symbols and Abbreviations

A molecular persistence length
A_{eff} "effective" molecular persistence length used in simulations
D polymer translational diffusivity
d diameter of polymer chain
F force
F^d drag force
F^s elastic "spring" force
F_i force on bead i
H spring constant

h	hydrodynamic-interaction parameter; see Eq. (9.5)
k_B	Boltzmann's constant
L	contour length of polymer molecule
N	number of beads in a bead-spring chain
R	polymer root-mean-square end-to-end distance
r	extension of DNA molecule
T	temperature
$\underset{\sim}{V}$	velocity
W_i	bead potential for bead i
x	position coordinate in flow direction
x_{cut}	cut-off distance for hydrodynamic interactions
x_{max}	x position of bead furthest downstream
y	position coordinate orthogonal to the flow direction
z	position coordinate orthogonal to the flow direction

Greek

η_s	solvent viscosity
ζ	translational drag coefficient
ζ_{bead}	drag coefficient for a bead of a bead-spring chain
ζ_{coil}	drag coefficient in the coil state
ζ_i	drag coefficient for bead i
ζ_{rod}	drag coefficient in the fully extended state
ζ_{tot}	total drag coefficient summed over all beads
η_s	solvent viscosity
λ	horizontal scaling factor required to superpose data of Fig. 9.1
Ω_{ij}	Oseen hydrodynamic interaction between beads i and j

Other Symbols

$\langle \cdot \rangle$	ensemble average
$\langle \cdot \rangle_0$	ensemble average in the absence of flow

References

1. Kuhn W (1934) Kolloid-Z 68:2
2. Peterlin A (1966) Pure Appl Chem 12:563
3. de Gennes PG (1974) J Chem Phys 60:5030
4. Hinch EJ (1974) Proc Symp Polym Lubrification, Brest
5. Bird RB, Curtiss CF, Armstrong RC, Hassager O (1987) Dynamics of polymeric liquids, 2nd edn, vol 2. Wiley, New York
6. Larson RG (1988) Constitutive equations for polymer melts and solutions. Butterworths, New York
7. Brochard-Wyart F (1993) Europhys Lett 23:105
8. Broachard-Wyart F (1995) Europhys Lett 30:387

9. Cerf RJ (1969) J Chim Phys 66:479
10. Hinch EJ (1977) Phys Fluids 20:S22
11. Phan Thien N, Manero O, Leal LG (1984) Rheol Acta 23:151
12. Dunlap PN, Leal LG (1987) J Non-Newt Fluid Mech 23:5
13. Armstrong RC, Gupta SK, Basaran O (1980) Polym Engng Sci 20:466
14. King DH, James DF (1983) J Chem Phys 78:4749
15. Perkins T, Smith D, Chu S (1994) Science 264:819
16. Perkins T, Smith D, Larson RG, Chu S (1995) Science 268:83
17. Smith SB, Finzi L, Bustamante C (1992) Science 258:1122
18. Vologodskii A (1994) Macromolecules 27:5623
19. Bustamante C, Marko JF, Siggia ED, Smith S (1994) Science 265:1599
20. Larson RG, Perkins T, Smith D, Chu S (1997) Phys Rev E 55:1794
21. Doi M, Edwards SF (1986) The theory of polymer dynamics. Oxford Press, New York
22. Eimer W, Pecora R (1991) J Chem Phys 94:2324
23. Pecora R (1991) Science 251:893
24. Smith DE, Perkins TT, Chu S (1995) Macromolecules 29:1372
25. Flory PJ (1969) Statistical mechanics of chain molecules. Carl Hanser Verlag, New York
26. Frenkel J (1944) Acta Physicochim URSS 19:51–76
27. Ryskin G (1987) J Fluid Mech 178:423
28. Marko JF, Siggia ED (1995) Macromolecules 28:8759
29. Fixman M (1966) J Chem Phys 45:785, 793
30. Öttinger HC (1987) J Chem Phys 86:3731
31. Magda JJ, Larson RG, Mackay ME (1988) J Chem Phys 89:2504
32. Larson RG, Magda JJ (1989) Macromolecules 22:3004
33. Kishbaugh AJ, McHugh AJ (1990) J Non-Newt Fluid Mech 34:181
34. Magda JJ, Lee C-S, Muller SJ, Larson RG (1993) Macromolecules 26:1696

Single Polymers in Elongational Flows: Dynamic, Steady-State, and Population-Averaged Properties

T. T. Perkins, D. E. Smith, and S. Chu

10.1
Introduction

The behavior of dilute polymers in an elongational flow has been an outstanding problem in polymer science for several decades [1–3]. In these flows, a velocity gradient along the direction of flow can stretch polymers far from equilibrium. Extended polymers exert a force back on the solvent leading to the important, non-Newtonian properties of dilute polymer solutions such as viscosity enhancement and turbulent drag reduction.

A homogeneous elongational flow is defined by a linear velocity gradient along the direction of flow such that $v_y = \dot{\varepsilon} y$ where $\dot{\varepsilon} \equiv \partial v_y / \partial y$, the strain rate, is constant. Theory suggests that the onset of polymer stretching occurs at a critical velocity gradient or strain rate $\dot{\varepsilon}_c$ of

$$\dot{\varepsilon}_c \approx \frac{0.5}{\tau_1} \tag{10.1}$$

where τ_1 is the longest relaxation time of the polymer [4]. For $\dot{\varepsilon} < \dot{\varepsilon}_c$, the molecules are in a "coiled" state. But as $\dot{\varepsilon}$ is increased above $\dot{\varepsilon}_c$, the hydrodynamic force exerted across the polymer just exceeds the linear portion of the polymer's entropic elasticity and the polymer stretches until its non-linear elasticity limits the further extension of this "stretched" state.

The simplest model that captures this coil-stretch transition is a dumbbell model with a nonlinear spring. In the dumbbell model, two beads of radius R_{bead} are connected by a spring. The two beads represent the coupling between the fluid and the polymer while the spring represents the entropic elasticity of the molecule. A nonlinear spring is necessary to account for the finite extensibility of the polymer since a polymer cannot extend beyond its contour length L. Neglecting Brownian motion and assuming steady-state, the two forces on each bead separated by an extension x are given by

$$F_{hydro} = 6\pi\eta R_{bead} v_{fluid} \tag{10.2}$$

$$= 6\pi\eta R_{bead} \dot{\varepsilon} \left(\frac{x}{2}\right)$$

$$F_{spring} = f(x/L)$$

$$\cong k_{spring} x + O(x^3) \quad (x \ll L) \tag{10.3}$$

where F_{spring} is approximately linear for small extensions. Thus, when $3 \pi \eta R_{bead} \dot{\varepsilon}$ exceeds k_{spring}, the molecule undergoes a transition from a coiled to a stretched state. De Gennes predicted that this "coil-stretch transition" would be sharpened by an increase in the hydrodynamic drag of the stretched state relative to the drag of the coiled state [1].

10.2
Previous Experimental Work

In many types of elongational flows, such as flow through a pipette tip or a contraction in a pipe, the time a polymer interacts with the elongational flow field is limited. Also, such flow fields are elongational but not homogeneous. The classical technique to simplify the experimental situation is to use a stagnation point flow (Fig. 10.1). Such a flow geometry increases the polymer's interaction or residency time t_{res} in the velocity gradient and it also generates a homogeneous velocity gradient. For polymers whose molecular trajectories approach the stagnation point, t_{res} diverges and measurements are made on those molecules near the stagnation point.

Birefringence is a commonly used technique for inferring the degree of polymer deformation in this geometry [5–9]. For example, in dilute polystyrene solutions, Keller and Odell reported a rapid increase in the birefringence for $\dot{\varepsilon}$ above $\dot{\varepsilon}_c$ followed by a saturation [5]. Such saturation was interpreted as an indication that the polymers had reached equilibrium in a highly extended state.

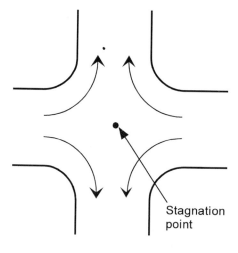

Stagnation
point

Fig. 10.1. Crossed slot flow to increase interaction time of the polymers in the elongational flow

In addition to these birefringence experiments, an analysis of the molecular weight of the polymers after being subject to an elongational flow showed that, at sufficiently high $\dot{\varepsilon}$, chains are fractured in this flow and such fracture occurs preferentially at the center of the chain [10–12]. This result further supports the hypothesis that the polymers reached full extension because the tension within the chain is highest at the center of the polymer only if the chain is extended.

An alternative method to probe the deformation of a polymer is to measure the radius of gyration R_G by light scattering [13–15]. To make a direct comparison between light scattering and birefringence, Menasveta and Hoagland performed both types of measurements on the same experimental apparatus [13]. Their birefringence results showed the same saturation with increasing $\dot{\varepsilon}$ as seen previously. Similarly, their light scattering measurements showed increase in R_G at the same $\dot{\varepsilon}_c$ as their birefringence data, followed by a saturation. However, the saturating value of R_G was only twice the equilibrium size implying that the saturation in birefringence did not represent a highly extended state.

Both of these measurement techniques have an inherent disadvantage: the state of the polymer must be inferred from an indirect measurement of an optical property not directly related to the polymer's conformation. Furthermore, the signal is averaged over molecules having a broad range of t_{res}, inherent in stagnation point flows. And the dynamics of individual polymers are hidden within an average over a macroscopic number of molecules. Finally, recent birefringence experiments show the same saturation at the stagnation point but when the probe region is moved towards an outlet, the birefringence increased [16]. Thus, the origin and meaning of the saturation in birefringence at the stagnation point is uncertain.

Many rheological effects also remain unexplained. In 1980, James and Saringer measured a pressure drop of a dilute polymer solution in a converging flow that was significantly higher than predicted by simple models [17]. Recently, Tirtaatmadja and Sridhar measured extensional viscosities η_E in filament stretching experiments, which were several thousand times higher than the shear viscosities [18]. At large deformations η_E saturated, suggesting again that the polymers were fully extended. However, the measured stress was significantly lower than expected for fully extended polymers and the plateau in η_E occurred before the polymers could have become fully extended [19]. Taken together, these results imply that full extension had not actually been achieved. The stress relaxation after the exponential deformation was stopped contained both a strain-rate independent, "elastic" and a strain-rate dependent, "dissipative" component. The molecular origin of the dissipative component was uncertain [20, 21].

These examples of seemingly contradictory results indicate that, even after a tremendous amount of work, the deformation of polymers in elongational flows is still not understood [19, 22].

10.3
Experimental Technique

In this chapter we report the direct visualization of individual polymers in an elongational flow. A shorter summary article of this work has appeared in *Science* [23]. Additional studies have also recently been reported [D.E. Smith, S. Chu, Science 281, 1335 (1998)]. We fluorescently labeled DNA molecules and imaged them in a standard optical microscope using a low light video camera (Fig. 10.2). By positioning the field of view of the camera at the stagnation point, we can image the unwinding dynamics of single polymers as illustrated in Fig. 10.3. From this visual data, we measure the conformation and extension x of each molecule as a function of strain rate $\dot{\varepsilon}$ and residency time t_{res} in the applied velocity gradient.

Imaging single molecules eliminates four important experimental limitations of classical techniques. First, direct imaging yields the full conformation of individual polymers whereas birefringence measures average local orientation of the chain segments. Second, by tracking single polymers near the stagnation point, we know t_{res} of each polymer, eliminating the distribution of t_{res} normally found in stagnation point flows. Third, by working with single, isolated molecules, we eliminate polymer-polymer interactions and polymer induced alterations of the flow field. Finally, the inherent uniformity in size of lambda bacteriophage DNA (λ-DNA, $L_{strained} \cong 21-22$ μm) eliminates the complications due to polydispersity and enables accurate calculation of ensemble averages.

10.3.1
Direct Imaging of Single Molecules as a Measurement Technique

As outlined above, direct observation of individual molecules provides a simple and unique method for testing models of polymer dynamics. Visualization of single molecules is generally done with video enhanced fluorescence microscopy. Such measurements require an intensified camera and a high numerical aperture (NA ≥ 1.2) microscope objective.

In this chapter, we focus on the deformation of polymers in flow. But in general, optical tweezers [24] are used to manipulate micron-sized microspheres

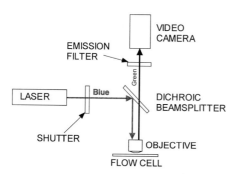

Fig. 10.2. Schematic of optical layout for fluorescence imaging

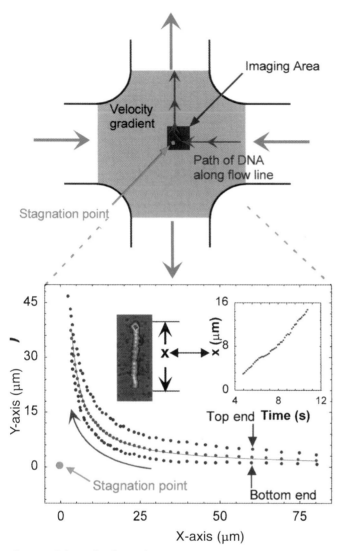

Fig. 10.3. Schematic of experiment

[25–32] although some work is done with magnetic microspheres [33] or electric fields [34–38]. The susceptibility of the DNA to the applied electromagnetic field from the laser is too small to be directly manipulated using optical tweezers. Therefore polystyrene microspheres (~1 µm) are attached to one or both ends of the molecules. These spheres are the handles by which the molecule is manipulated.

Direct observation and manipulation of single molecules is not limited to fluorescently labeled DNA or to polymers. These techniques are making important contributions in systems as diverse as colloidal suspension [29, 39] and molecular motors [30–32]. For instance, Crocker and Grier measured the interaction potential between isolated pairs of charge colloidal spheres [29] while Finer *et al.* [30] measured the force and step size of a single myosin molecule on an actin filament, the fundamental interaction that leads to muscle contraction.

10.3.2
DNA as a Model Polymer

There has been considerable recent work of activity based around visualizing and manipulating single DNA molecules. DNA acts as a scaled up version of the smaller, more flexible synthetic polymers. Locally, it is quite stiff with a persistence length p of 50 nm [40] and has a hydrodynamic diameter of 2 nm [41, 42]. But, for sufficiently long chains, the polymers act like a flexible chain and the entropic and hydrodynamic effects dominate. Within this limit, many properties can be described by three variables: the contour length of the chain, the persistence length, and the quality of the solvent, a description of monomer-solvent interaction [43, 44]. For the results presented in this work, L/p is ~ 400 and the DNA is in a good solvent [45].

In bulk experiments, the polymer physics of DNA has been extensively studied and shown to agree with classical polymer theory [10 12, 16 50]. Initially, the interpretation of single DNA molecule research relied upon this universality of polymer dynamics when $L/p \gg 1$ [26, 27, 43, 44]. However, the interpretation of single DNA molecule data as a representation of polymers in general is now grounded in quantitative agreement between experiment and theory [51–55]. For instance, Smith *et al.* measured non-linear elasticity of single DNA molecules [33] while Vologodoski [56] and Marko *et al.* [52, 53] showed that this data corresponded to the elasticity of a worm-like chain. Marko *et al.* [53] provided a simple analytical approximation to this nonlinear elasticity that is a function only of the fractional extension and persistence length:

$$F_{MS}\left(\frac{x}{L}\right) = \left(\frac{k_b T}{p}\right) \cdot \left\{\frac{x}{L} + \frac{1}{4}\left(1 - \frac{x}{L}\right)^{-2} - \frac{1}{4}\right\} \tag{10.4}$$

where x is the end-to-end extension. Quantitative agreement has also been established between a bead-spring model with self-consistent hydrodynamics developed by Larson et al. [51] and our previous measurements on stretching of a tethered polymer in a uniform flow [28].

DNA molecules have three unique advantages when compared with standard synthetic molecules: length, uniformity, and ease of manipulation.

DNA molecules longer than 100 µm can be routinely manipulated. In this experiment, we use λ-bacteriophage DNA consisting of 48,502 base pairs and having a length of $L = 16.3$ µm for an unstained molecule. Stained λ-DNA is 21–22 µm long due to the intercalation of the dye molecules into the double helix structure [28]. With polymers this long, both its extension and general features of its internal conformation are observable using standard video enhanced microscopy [27].

Second, DNA is inherently monodisperse. Enzymatic replication guarantees that each copy of the λ-DNA is exactly the same size. For longer molecules, concatamers of λ-DNA can be made or alternative bacteriophages (T2, T4, T5) can be used. However, for the experiments described here, we have found commercial sources of T2, T4, or T5 were not adequately monodisperse. Presumably, this polydispersity was caused by shearing and we did not have an independent way to measure the length of each molecule. So, given the diversity of the dynamics that will be discussed below, we choose to limit our study to the dynamics of λ-DNA. Shorter lengths of DNA can be generated using restriction enzymes. Flexibility is perhaps DNA's biggest advantage for studying polymer dynamics over another important bio-polymer, actin. Nevertheless, actin offers the advantage that is very stiff ($p = 8-20$ µm) and its polymer properties have been extensively studied at the single molecule level [57].

In addition, we utilize a number of advances in molecular biology. DNA molecules can be "cut and pasted" together and purified from contaminating protein and DNA on a routine basis. DNA can be stained with highly efficient fluorescent dyes. Such bright dyes enable single molecules to be visualized [58, 59]. Finally, the ends of DNA can be functionalized to facilitate the coupling of DNA molecules to polystyrene beads. Typically this coupling is done via a protein-ligand bond.

10.3.3
Microscope and Imaging

We imaged single DNA molecules using video enhanced fluorescence microscopy (Fig. 10.2). Our microscope is a Zeiss Axioplan with a c-Apochromat 100X NA 1.2 objective. This water immersion, long working distance (~170 µm) objective corrects for spherical aberration deep into water, allowing sharp, bright images of molecules deep within the flow cell.

The DNA was stained with YOYO-1 (Molecular Probes, Inc.), a fluorescent dye that emits in the green portion of the spectrum when excited by 488 nm light from an argon-ion laser. YOYO-1 offers excellent signal-to-noise but it tends to photobleach rapidly and, if care is not taken, there will be photo-induced fragmentation of the DNA in a few seconds. To alleviate this photobleaching, we enzymatically scavenged oxygen from the solution (see Sect. 10.3.8).

Excitation with a laser leads to speckle in the image plane due to the coherence of the laser light. This coherence was greatly reduced by focusing the light onto a spinning, ground glass disk. To couple the light into the microscope efficiently, we used the normal illumination path except the disk took the place of the excitation lamp.

We used a long pass emission filter (515 EFLP, Omega Optical) to increase the transmitted light while blocking the excitation light. A long pass filter allows more transmitted light than the more common notch filter, which is only necessary if several different fluorophores are being used.

The flow cell was positioned such that the stagnation point was centered vertically on the screen and was 15 μm from the left-hand-side. This arrangement maximized the number of molecules with a large t_{res}.

Images were recorded by a silicon intensified target (S.I.T.) camera (Hamamatsu C2400-08), processed by an image processor (Hamamatsu Argus 10), and then recorded onto S-VHS videotape.

Due to lag in the phosphor of a S.I.T. camera, rapidly moving objects appear to have comet-like tails. The decay time is about 3 video frames (33 ms per video frame). By using an intensified CCD camera this lag may be reduced to $^1/_{30}$ sec. To overcome this limitation of the S.I.T. camera, we stroboscopically illuminated the DNA using a mechanical shutter (Uniblitz) to block the laser. The shutter was synchronized to the camera so it opened 4 ms before the start of a frame. A synch pulse was also recorded onto the audio track of the video-tape. Exposure times varied from 10 to 33 ms because, for the highest $\dot{\varepsilon}$, the images started to blur due to motion during the exposure time. At $\dot{\varepsilon} = 0.86$ s^{-1}, the shutter was run at 7.5 Hz which corresponded to an accumulated fluid strain of $\dot{\varepsilon}\Delta t \cong 0.1$. At lower strain-rates, we continued to illuminate strobo-scopically to prevent photobleaching of the dye. Thus we took data at approximately fixed intervals of the accumulated fluid strain ($\dot{\varepsilon} t_{res} \cong 0.1$).

By using the timing information stored on the audio portion of the videotape, only the fully illuminated images were transferred to an optical disk recorder (Panasonic, TQ-2028F). This video data was then digitized by a Data Translation Quick Capture board in sequences of approximately 7500 frames.

10.3.4
Flow Cell Design

As stated earlier, a homogeneous elongational flow is defined by a linear velocity gradient along the direction of flow such that $v_y = \dot{\varepsilon} y$. The strain rate $\dot{\varepsilon}$ is constant but the residency time t_{res} of the polymers in these flows is limited. To increase t_{res}, we used a standard, cross-slot flow geometry (Fig. 10.1). This generated a planar homogeneous extensional flow where $\dot{\varepsilon}_x = -\dot{\varepsilon}$, $\dot{\varepsilon}_y = \dot{\varepsilon}$ and $\dot{\varepsilon}_z = 0$. Note, this is slightly different to an axial extensional flow generated by an opposing jet flow where $\dot{\varepsilon}_x = -\dot{\varepsilon}/2$, $\dot{\varepsilon}_y = \dot{\varepsilon}$ and $\dot{\varepsilon}_z = -\dot{\varepsilon}/2$.

The main design consideration was to ensure that we studied dynamics of polymers unwinding from equilibrium. That is, we wanted to avoid any pre-deformation of the polymer prior to entering the elongational flow. To eliminate pre-deformation of the molecules caused by velocity gradient at the entrance to the flow cell, we used long channels (25 mm) on the inlet. This increased the transit time for a molecule down the inlet channel and allowed any molecule deformed upon entering the flow cell sufficient time ($> 8\tau_1$) to relax back to an equilibrium configuration. Another way to avoid predeformation is to turn on the flow suddenly [D. E. Smith, S. Chu, Science 281, 1335 (1998)].

In a preliminary experiment with a 50-μm deep channel, we imaged polymers in the center of the flow as they moved down the inlet channel. Some polymers were in highly extended states ($\sim 15\,\mu m$) along the flow direction ($\hat{x}$). To clarify the origin of this deformation, we pulsed the pump and imaged only molecules that started within the inlet channel. We still found some polymers in highly extended states.

We hypothesize that the shear rate, $\partial v_x/\partial z$, is strong enough in a 50-μm deep flow cell to cause significant pre-deformation even for molecules at the center of the parabolic fluid flow. A characterization of the strength of shear rate is given by a dimensionless shear rate or a Weissenberg number ($Wi = \dot{\gamma}\tau_1$) where $\dot{\gamma} = \partial v_x/\partial z$ – the velocity gradient perpendicular to the flow direction. Polymers start to deform when $\dot{\gamma}\tau_1 \geq 1$. To estimate the dimensionless shear rate, we note that the velocity profile as a function of depth is parabolic at low Reynolds number ($\sim 10^{-4}$). The shear rate in the flow is then

$$\frac{\partial v_x(z)}{\partial z} = \dot{\gamma} = 2\,\frac{v_{inlet}}{z_{center}^2}\,(z - z_{center}) \tag{10.5}$$

where v_{inlet} is the velocity at the center of flow in the $\hat{x}$ direction and z_{center} is the distance from the top surface to the middle of the flow. Although a naive interpretation of this formula would indicate that $\dot{\gamma} = 0$ at $z = z_{center}$, DNA molecules have a finite size ($R_g = 0.7\,\mu m$) which leads to some shearing even for molecules at $z = z_{center}$. Further, the microscope has a small but finite depth of field ($\sim 1\,\mu m$). Thus, an in-focus molecule nominally at the center of the flow will experience the shear rate approximately equal to a molecule displaced by $1\,\mu m$ from z_{center}. Thus, we estimate $\dot{\gamma}\tau_1 = 8$ for our preliminary experiment, which supports our hypothesis that shearing was the origin of the observed pre-deformation. Further experimentation at $Wi = 1$ showed little to no pre-deformation.

In order to prevent pre-deformations due to shearing, the flow cell was redesigned to minimize the shear experienced by the polymers along the inlet channel. Inspection of Eq. (10.5) shows this can be achieved by increasing z_{center} or decreasing v_{inlet}. By increasing the depth of the flow cell, z_{center} is increased. This is limited by the working distance (170 μm) of the high numerical aperture, water immersion objective. It is important to be able to image several tens of microns past z_{center} to be able to measure accurately z_{center} (see Sect. 10.3.7). By narrowing the width of the flow cell, v_{inlet} is reduced. The narrower channel leads to a smaller region over which there is an elongational flow. Thus, a smaller v_{inlet} is needed to achieve the same strain rate. This is limited by shear flow arising from flow gradients in the $\hat{y}$ direction which, until now, has been neglected.

Our goal was to measure polymers to extensional flow up to $\dot{\varepsilon}\tau_1 = 3.5$ while keeping $\dot{\gamma}\tau_1 < 0.5$. To achieve this goal within these above constraints, we choose the depth of the flow cell to be 220 μm and the width of the inlet channel to be 650 μm (Fig. 10.4). The corners of the crossed slot were rounded with a radius of curvature of 325 μm.

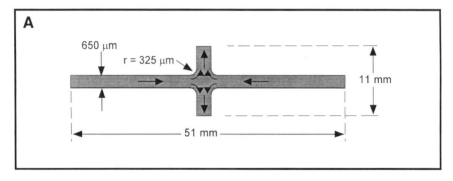

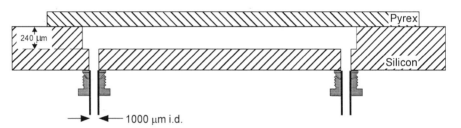

Fig. 10.4 A, B. Flow cell, drawing not to scale

10.3.5
Flow Cell Manufacture

Following the example of Volkmuth et al. [36, 37], the flow cells were etched into silicon wafers and then sealed by anionically bonding [60] Pyrex coverslips to the silicon to seal the top surface of the channels. We etched the crossed channel on the front side and holes for plumbing connections on the back side of the wafers by performing two successive silicon etches in KOH. Typically KOH etching of (100) wafers yields walls that are angled 54° from vertical. But, by rotating the pattern by 45° to the crystal axis [61, 62], we achieved vertical side walls along the inlet and outlet channels.

Nitride is not etched by KOH and serves as the protective layer during the KOH etch. Therefore, an 800 Å nitride layer was grown on 500 μm thick, 4″ dia., L-Prime wafers. To pattern the nitride layer, we used 1 μm thick photoresist as the protective layer and then exposed the sections of the photoresist using a lithographic mask. Subsequent development of photoresist removed the exposed photoresist leaving areas of uncoated nitride and a plasma etch removed this unprotected nitride. After an acid wash in 9:1 H_2SO_4:H_2O_2 to remove the photoresist, the wafers were placed in 30% KOH at 80 °C for 3.5 h, which etched the sections of silicon unprotected by the nitride. This first etching cycle created 200-μm deep features on the back side of the wafer. The wafers were next washed in 5:1:1 H_2O:H_2O_2:HCl to remove any residual KOH. After an alignment between a second lithographic mask and the pattern on the back side of the wafer, we patterned and etched the front surfaces by the same process. The front surface was etched to a depth of 170–250 µm depending on the duration of the etch. During the second etch, the holes on the backside of the wafer were further etched and made connections from the backside to the newly created channels on the front side. The wafers were then taken through a second KOH decontamination and then the front side of the wafers plasma etched to remove the remaining nitride. A final acid wash in 9:1 $H_2SO_4:H_2O_2$ cleaned the wafers. For some unknown reason, some of the wafers were not optically flat but had large (300–500 µm), shallow (1–3 µm) depressions and these wafers were not used.

To seal the flow cell, we anionically bonded Pyrex coverslips to the silicon wafer. First, the Pyrex was cleaned immediate prior to bonding. Next, the Pyrex was placed on the silicon wafer and heated on a hotplate at 400 °C. Upon reaching this temperature, 400 V was applied to the coverslip relative to the wafer, which sealed the Pyrex to the silicon. Pyrex is used instead of another glass because its thermal coefficient of expansion matches that of the silicon. Although it is possible to bond Pyrex to the nitride layer, we found bonding directly to silicon yielded significantly cleaner bonds. The sealed flow cells were cut out of the wafer using a wafer saw to a final dimension of 3″ × 1″ to facilitate mounting.

10.3.6
Pumps and Plumbing

In designing the flow system, the main design consideration was to minimize fluctuations in the flow rate. However, the highest flow rates through the flow cell were 10 µl/min – too slow for even the best syringe pumps to provide truly pulse-free flow. To increase the pump's operating flow rate while keeping the flow rate in the flow cell low, we created a 100:1 bypass shunt for the fluid (Fig. 10.5). The ratio of shunting was constant over the range of pump rates used in the experiment (Fig. 10.6).

The main pump was an Isco Model 100 D syringe pump temperature stabilized at 22.7 °C with a Lauda RM6 circulator. The main pump provided fluid

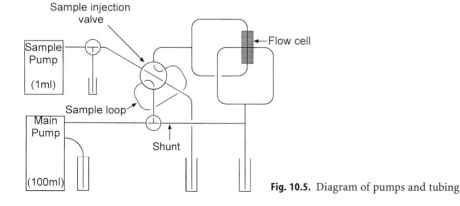

Fig. 10.5. Diagram of pumps and tubing

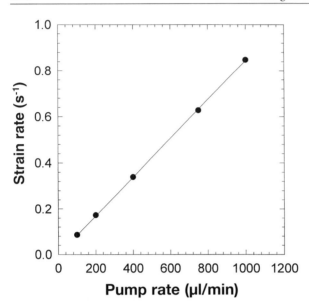

Fig. 10.6. Calibration of elongational flow cell showing the strain rate is linear with pump rate

flows from 100 µl/min to 25,000 µl/min. The maximum flow rate used in this experiment was 1000 µl/min. The sample pump was a Harvard Apparatus 55 syringe pump loaded with a 1-ml syringe (Unimetrics). The tubing (1.0 mm I.D.), connectors, filters, valves, and tees were purchased from Upchurch Scientific. The sample injection valve allows the introduction of the DNA solution without introducing bubbles into the system. To minimize contamination, we loaded the fluid into the main pump through a 2-µm filter.

We attached the tubing to the flow cell using a 125-µm thick, epoxy-fiber glass adhesive film (Ablefilm, Ablestick, Rancho Dominguez, Ca). Donut-shaped pieces were stamped out and placed around the holes on the backside of the wafer. The wafer was then briefly heated to 65 °C. A PEEK ferrule and nut were placed at the end of 1.0 mm I.D. stainless steel tubing which had previously been attached to a stainless steel tee. The tubing was then screwed partway into a 1/2″ thick copper block, which provided the mechanical support to hold the tubing properly aligned to holes in the wafer. The silicon flow cell was then centered over the tubing and was gently clamped down. The nuts were then tightened until the ferrules just made contact with the Ablefilm. The epoxy was cured by baking the whole assembly for 2 h at 120 °C.

Note that silicon wafers are single crystals that are 500 µm thick. They can fracture along a crystal axis under a minimal amount of applied stress. To prevent excess stress, the copper block was smoothed with grinding powder (~10-µm) to generate a bump-free surface. To prevent a metal-silicon-metal contact, the aluminum frame used to clamp the wafer to the copper block was coated with 1/16″ thick Teflon to cushion the applied stress. The affixed tubing is handled gently and not bumped since it is directly coupled to the silicon; otherwise, the flowcell will fracture.

10.3.7
Flow Cell Calibration

The flow cell was calibrated in a three-step process using fluorescent beads. We found that the magnitude of the strain rate is uniform to within 2%. In addition, we determined that the strain rate turns on abruptly ($\dot{\varepsilon}\Delta t \leq 0.02$). Or, in other words, the polymers experienced an approximate step function increase in the strain rate as opposed to a gradual increase when they moved from the uniform-flow, strain-rate free motion in the inlet channel to the elongational flow in the crossed-slot region. This calibrated region is the 22 µm on either side of the center-line of the flow cell and the imaging area (100 × 94 µm).

Although the depth of the flow cell was measured to be ~ 220 µm with a 10 × NA 0.3 air objective, it is important to determine precisely the center of the flow to minimize the shear on the DNA molecules (see Sect. 10.3.4). The index of refraction n_e of the water immersion fluid ($n_e = 1.33$) is not the same as the index of refraction of the glucose/sucrose solution ($n_e \approx 1.44$). Thus, a vertical movement of the objective by 100 µm does not correspond to a change in the focal plane of 100 µm. The vertical center of the flow z_{center} was found by measuring the parabolic velocity profile in the inlet channel as a function of depth. To do this, we injected a high density of beads (0.8 µm dia., Polysciences) into the flow cell such that there were ~ 30 beads in focus on the screen. The flow was turned on and images were digitized every 0.22 s. An autocorrelation between successive images determined the average displacement (Fig. 10.7, inset). This autocorrelation was repeated for a hundred images and the resulting data was averaged to determine the velocity for that depth. Successive sets of data were taken as a function of depth from 20 µm to 160 µm (Fig. 10.7). By fitting this data to a parabolic curve, we determined the center of the flow to

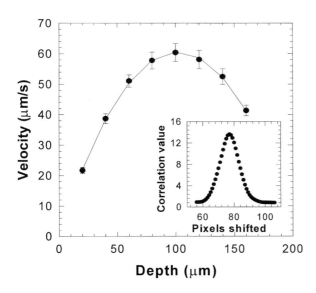

Fig. 10.7. Velocity as a function of depth in the inlet channel

be at 102 µm. We repeated the measurement and, again, determined $z_{center} =$ 102 µm.

The second calibration measured $\dot{\varepsilon}$ over the imaging area (100 µm × 96 µm) around the stagnation point. The stagnation point is centered and images of fluorescent beads are digitized. The y-coordinate of at least 20 beads is then tracked between successive images. When plotted as v_y vs y, this data showed a linear relationship verifying a uniform strain rate $\dot{\varepsilon} \equiv \partial v_y/\partial y$ within the field of view (Fig. 10.8). A linear fit to this data determined the strain-rate for a given pump rate. The process was repeated for different pump rates. The resulting linear relationship between strain-rate and pump rate shows the shunting ratio is constant over these pump rates (Fig. 10.6). For this experiment, a flow rate of 1000 µl/min corresponded to an $\dot{\varepsilon}$ of 0.85 s^{-1}.

The third calibration measured the velocity gradient along the inlet channel. This is important because it located the onset of the elongation flow. Since $t_{res} = 0$ at the onset, this calibration allowed for an accurate determination of the accumulated fluid strain ($\varepsilon = \dot{\varepsilon}t_{res}$) of the polymers. For this measurement, the pump was set to 100 µl/min, which corresponded to a velocity in the inlet (v_{inlet}) of 85 µm/s at the center of the flow. The whole flow cell, which is mounted on a motorized x-y stage, was moved at a constant velocity in relation to the objective. If the stage velocity matched the velocity of the fluorescent beads, the beads are stationary with respect to the imaging optics and do not move on the video image. This measurement can be done along the inlet channel by starting with the image area located 1500 µm upstream from the stagnation point. Next, the velocity of the stage was set to values from 10% to 90% of v_{inlet}. When the fluorescent beads were stationary in the video image, the stage was stopped and the location of the stage as determined by the motor controller was recorded.

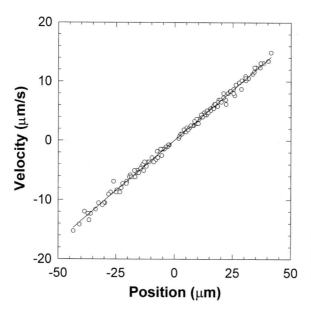

Fig. 10.8. Velocity along the outlet channel as a function of distance from the stagnation point

The resulting measured velocity gradient along the inlet $\partial v_x / \partial x$ was linear (Fig. 10.9). A fit to this data yields a strain rate of $-0.86\ \text{s}^{-1}$ in excellent agreement with calibration of the strain rate $0.86\ \text{s}^{-1}$ as determined by the second method. By extrapolating to v_{inlet}, the onset of the elongational flow x_{onset} was determined to be 960 μm. This value of x_{onset} agrees well with the geometry of the flow cell which would predict 975 μm since the channels are 650 μm wide and there is a 325 μm radius of curvature at the cross.

Taken together, these three calibrations show the elongational flow field is uniform. Since the fluid is incompressible, we have

$$\nabla \cdot v = \frac{\partial v_x}{\partial x} + \frac{\partial v_y}{\partial y} + \frac{\partial v_z}{\partial z} = 0 \tag{10.6}$$

From the geometry of the planar flow cell, we have $\partial v_z / \partial z = 0$. With this information and the calibration of $\dot{\varepsilon}$ along the inlet ($\partial v_x / \partial x = -\dot{\varepsilon}$) and at the stagnation point ($\partial v_y / \partial y = \dot{\varepsilon}$), we know the onset of the elongation flow field is sudden and the polymers experience a constant strain rate. The uniformity of the strain rate is also confirmed by showing the trajectories of the center of mass of individual fluorescent microspheres and individual DNA molecules were hyperbolic. Furthermore, the velocity of the center of mass motion of the individual DNA molecules increased with distance and the strain rate calculated from this motion agreed within 2% of the strain rate calculated from the motion of fluorescent beads (Fig. 10.10).

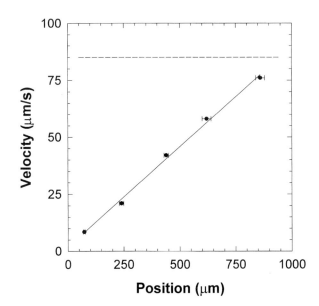

Fig. 10.9. Velocity along the inlet channel as a function of distance from the stagnation point

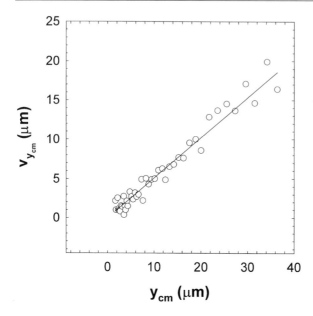

Fig. 10.10. The center of mass motion of an individual DNA molecule. The measured strain-rate from the DNA molecules (*circles*) agrees within 2% of the calibrated strain rate (*line*) determined by tracking fluorescent beads

10.3.8
Solution

To get good images of the DNA, we performed our experiments in a high viscosity aqueous solution. Polymers begin to stretch when $\dot{\varepsilon}\tau_1 \geq 0.5$ or, alternatively, when $\dot{\varepsilon} \geq 0,5/\tau_1$. In water where the $\tau_1 \cong 0.1$ s, this requires a strain rate of at least 5 s^{-1} which necessitates a flow rate of 5 µm/s at 1 µm from the stagnation point and 10 µm/s at 2 µm. At any appreciable distance from the stagnation point, the velocity of the molecule is too great to image with video microscopy ($\frac{1}{30}$ s per video frame). To overcome this limitation, we increased the viscosity η of the solution from 1 cP to 41 cP using a combination of sucrose and glucose. This increase in η led to a corresponding increase in τ_1 and reduction in $\dot{\varepsilon}$ assuming a constant $\dot{\varepsilon}\tau_1$. Thus, an increase in η slows down the bulk and internal motion of the molecules such that rate of motion is amenable to our imaging techniques.

Besides increasing the viscosity, the other main requirement of the solution was to reduce photobleaching of the fluorescent dye. When dye molecules photobleach, two detrimental processes happen. First, the contour length of the DNA shortens (Fig. 10.11) since staining lengthens the DNA. Second, the photobleaching of the dye molecules can fragment the DNA. Without any precautions, YOYO-1 stained DNA will fragment in 1–3 s.

The dominant cause of photobleaching is oxygen radicals (O$^-$). Therefore, we used two standard techniques to eliminate oxygen radicals. First, we added 4 vol.% β-mercaptoethanol, a reducing agent that directly attacks the oxygen radicals. Second, we enzymatically scavenged the oxygen from solution. In this process, two enzymes, glucose oxidase and catalase, are added and the net reac-

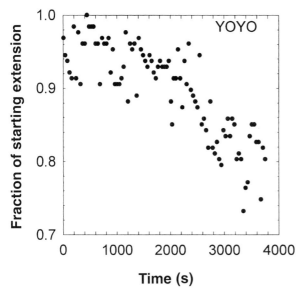

Fig. 10.11. Stability of extension as a function of time at a constant strain-rate to determine if the contour length is a function of time. The extension was measured by trapping a microsphere at the stagnation point in an elongational flow. The flow rate was slightly above the critical strain-rate for the transition from a coiled to the extended state. YOYO-1 stained DNA: the illuminating laser power was at the minimum intensity to observe a molecule for the first 2000 s after which it was increased approximately 10-fold and the measured extension decreases linearly with time

tion is to transfer dissolved oxygen from the solution onto a glucose molecule. The effectiveness of this enzymatic process was much higher than simple physical purging of the oxygen by bubbling dry nitrogen through the solution. Using these techniques, single DNA molecules could be visualized for up to 2 h [26]. These techniques were not as effective in the high viscosity solutions used for this experiment. But, nonetheless, no appreciable fading of the DNA was seen for at least 30 s at the highest illumination power used. Since the DNA was exposed to a 33 ms pulse of light for once every $\dot{\varepsilon}t_{res} \cong 0.1$ (see Sect. 10.3.3) with a maximum value of $\dot{\varepsilon}t_{res} \cong 10$, the total integrated illumination was ~ 3 s. Therefore, the physical properties of the stained DNA remained constant during the course of the experiment.

To achieve a high viscosity ($\eta = 41$ cP) buffer with minimal photobleaching, we used a solution consisting of 10 mmol/l Tris HCl pH 8, 2 mmol/l EDTA, 10 mmol/l NaCl, 4% β-mercaptoethanol, ~ 50 µg/ml glucose oxidase (Behringer-Mannheim) and ~ 10 µg/ml catalase (Behringer-Mannheim), ~ 18 wt% glucose and ~ 40 wt% sucrose. The Tris HCl is a buffering agent. The EDTA binds trace quantities of Mg^{2+} to prevent digestion of the DNA by contaminating enzymes. The NaCl screens the highly charged phosphate on the DNA backbone so the persistence length is not a sensitive function of the ionic concentration. To remove any contamination, solutions were filtered through a 0.4 μm filter.

Note that the viscosity of such a solution is temperature sensitive and we measured the viscosity of each solution in a temperature stabilized viscometer and adjusted the sugar concentration as needed. The flow cell was mounted on copper block and stabilized to 22.7 ± 0.2 °C.

10.3.9
Staining and DNA Preparation

We stained the λ-DNA (New England Biolabs) with 10^{-7} mol/l YOYO-1 (Molecular Probes) at a dye:base pair ratio of 1:4 for >1 h. YOYO-1 was used because it is one of the most sensitive and highest affinity probes available for labeling DNA. The dye exhibits a 100- to 1000-fold increase in its quantum yield upon binding to DNA, giving a strong image signal against a low background [63]. When mixing the DNA into the high viscosity solution, great care was taken to avoid shearing (breaking) the DNA molecules. First, the stained DNA sample was gently pipetted with a wide-bore pipette tip into the high viscosity solution. Next, the solution was slowly mixed with a helix-shaped plastic rod for 10 min at ~ 0.25 revolutions/s. Finally, we visually inspected each solution using fluorescence microscopy to verify that the molecules were homogeneously distributed in the solution and unsheared. The viscosity of each DNA-containing solution was measured and the concentration of solvent was slightly adjusted to produce the desired viscosity.

10.3.10
Measurement of τ_1

To determine τ_1 for our molecules, we needed to establish a relationship between our single molecule results and the classical definition of τ_1 from bulk measurements. Within the dumbbell model, this relation is easily established between the stress relaxation and the relaxation of a single chain via

$$\sigma(t) = n \langle R(t) \cdot F(R) \rangle \tag{10.7}$$

where n is the density of polymers, $R(t)$ is the end-to-end distance and $F(R)$ is the tension between the ends of the polymer. Since for small displacements $F(R)$ is linear in R (Eq. 10.3), we have

$$\sigma(t) \propto n \langle R(t) \cdot R(t) \rangle \tag{10.8}$$

Thus, the classical relaxation time can be directly related to the measurements from single molecules. The relaxation time reported here is from a fit over the region where $x/L < 0.3$ to

$$\langle x(t) \cdot x(t) \rangle = c \exp(-t/\tau_1) - 2R_G \tag{10.9}$$

where $x(t)$ is the maximum visual extension and τ_1, c and R_G were free parameters.

We measured the relaxation of 14 individual molecules from a highly extended ($> 16\,\mu m$) state. The molecules were extended in the elongational flow and relaxed when the flow was stopped. Two sets of data were taken in the first and final preparation of the high viscosity buffer. The statistical nature of the relaxation prevented an accurate calculation of τ_1 from a single relaxation. We therefore averaged the data between relaxations of different molecules. To account for the small differences in initial extension of the molecule, all data sets had their time axis shifted so that $t = 0$ for $x = 15\,\mu m$. We then fit the averaged data over region $x/L < 0.3$ to Eq. (10.9). The relaxation time between the initial and final solution was constant within the statistical error for the measurement. By averaging all data together, we determined $\tau_1 = 3.89 \pm 0.05$ s (Fig. 10.12).

In comparison to other work, the stress relaxation time reported for unstained λ-DNA in water ($\eta = 1\,cP$) is $\tau_1 = 0.067$ s, where we determined τ_1 for λ-DNA by scaling $\tau_1 = 0.046$ s reported by Klotz and Zimm [48] for T7-DNA in a creep recovery experiment and scaled for the slight difference in length between ($L = 13.4\,\mu m$) and λ-DNA ($L = 16.3\,\mu m$) with a scaling exponent of 1.66 [27]. This value of $\tau_1 = 0.067$ s is in agreement, after scaling for length, with $\tau_1 = 0.058 - 0.068$ s from intrinsic viscosity [48], light scattering [49], birefringence [64], and flow dichroism experiments [47].

To compare our measured τ_1 to these previous measurements, we scaled by the change in length $(21.1/16.3)^{1.66}$ and for the change in viscosity $(41/1)$. This calculation yields $\tau_1 \approx 0.061$ s and does not take into account any possible changes in persistence length [54]. The present measurements were done in a

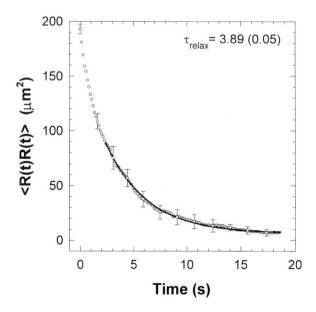

Fig. 10.12. Relaxation of 14 individual DNA molecules averaged together

sucrose-glucose-water solution while the stress relaxation measurements were done in a glycerol-water solution. However, we have measured the relaxation of single DNA molecules in both a glycerol and sucrose enhanced aqueous buffered solution [27] and found that the relaxation time and scaling behavior were the same. Thus, there was no change in solvent quality between the glycerol-water and sucrose-water solutions.

Hence, we see that there is quantitative agreement between τ_1 measured via a single molecule and the classical definition of τ_1 by bulk measurements. Both our single molecule measurements and the stress birefringence experiments determined τ_1 by fitting the last portion of the relaxation to equilibrium to a single exponential. In this limit, the contribution of the higher order modes of the polymer's relaxation are small, and the measured τ_1 is approximately equal to the fundamental or longest relaxation mode of a polymer [48].

It is important to note that our τ_1 reported here is different to the relaxation time determined from $\langle x(t) \rangle = c \exp(t/\tau) - R_G$, which yields $\tau = 6.17 \pm 0.15$ s for free λ-DNA. This is the relaxation time we reported in an earlier experiment where the DNA was tethered to a bead [27]. Also τ measured for a tethered polymer and a free polymer are not the same but, rather, a tethered polymer relaxation approximately corresponds to the relaxation of a free polymer twice as long.

10.4
Experimental Results

10.4.1
Data Reduction

To convert the raw image data to numerical data, a computer program was used to track and to determine the extension of individual molecules as they traversed the field of view. Molecules were required to be in-focus and their center-of-mass was required to start within 22 µm of the center-line of the inlet channel at the edge of the screen (83 µm). This eliminated those molecular paths with fewer than six individual images where $\dot{\varepsilon}\Delta t$ was ~ 0.1 between successive frames. Molecules were tracked until a portion of the molecule went off the screen or the molecules came within ~ 5 µm of another molecule. A schematic sketch of this process is illustrated in Fig. 10.3. For each individual image, nine fields of data were taken that described the extension, time, conformation, center of intensity, total integrated intensity, and positions of the ends of the molecule. The extension of each molecule was checked visually. Further, the conformations were also determined visually. The conformation was specified as one of seven different conformations: kinked, folded, dumbbell, half-dumbbell, uniform coil, or extended. Illustrations of these conformations are shown in Fig. 10.13.

The reduced numerical data was then stored in a database. The raw extension data was smoothed by a weighted average with its nearest neighbors of $x_i = 0.21\, x_{i-1} + 0.58\, x_i + 0.21\, x_{i+1}$.

Fig. 10.13. Illustrations of the seven main conformations observed

To calculate t_{res}, the time a polymer interacts or resides in the elongational flow field, we needed to know the polymer's residency time in the elongational flow field $t_{res}^{initial}$ prior to its appearance in the field of view. This is simple to calculate because the strain rate is constant (see Sect. 10.3.7) and leads to

$$t_{res}^{initial} = \frac{1}{\dot{\varepsilon}} \ln \left(\frac{x_{onset}}{x_{screen}} \right) \tag{10.10}$$

where x_{onset} is the position of the onset of the elongational flow field and x_{screen} is position of the edge of the screen. For our flow cell and imaging system, these values were 960 and 83 μm respectively. Since we do not image the molecules continuously but between illuminated video frames, the molecules are not imaged exactly at the edge of the imaging area. The average distance the molecules travel between illuminated frames is $\langle v\Delta t \rangle$. Thus, our value of x_{screen} takes into account this motion and reduces the true location of x_{screen} by $\langle v\Delta t \rangle/2$. Having calculated $t_{res}^{initial}$, we add $t_{res}^{initial}$ to the time each molecule is visualized in the imaging area to yield t_{res}.

While $t_{res}^{initial}$ is dependent on the flow rate, $\dot{\varepsilon} t_{res}^{initial}$ is dictated by the geometry of the flow cell and the imaging system. Hence, all molecules experience the same amount of accumulated strain before they are visualized. To visualize the earlier time evolution, the position of the objective was moved away from the stagnation point towards the inlet.

10.4.2
Extension vs Residency Time

We performed the experiment at eight different flow rates. In our earlier report on these results [23], we showed the raw data for only the highest strain rate. Here, in Fig. 10.14 we present all of the individual traces of extension vs time. We also plot the average extension $\langle x(t_{res}) \rangle$ as well as highlighting several individual traces. Due to the nature of a stagnation point flow, the number of molecules observable for a given t_{res} decreased exponentially. Our analysis started with ~ 1000 molecules for the five highest strain rates and ~ 400 molecules for the remaining strain rates. For $t_{res} > t_{res}^{initial}$, we calculated averages from at least 40 individual molecules unless stated otherwise.

One immediately striking feature in all of the data sets is the large heterogeneity in dynamics from molecule to molecule. Such heterogeneity is not observable in classical bulk experiments. From an ensemble of individual measurements, we can measure not only $\langle x(t_{res}) \rangle$, we can also measure the time evolution of the full probability distribution for molecular extension (Fig. 10.15).

Clearly, the molecules are not undergoing a simple and simultaneous unwinding as soon as $\dot{\varepsilon} > \dot{\varepsilon}_c$. Since the molecules experienced the same $\dot{\varepsilon}$ and t_{res}, this heterogeneity in dynamics must arise from a combination of the polymer's initial internal configuration and Brownian motion. This heterogeneity underscores the necessity of a monodisperse solution. By using λ-DNA, we know that these diverse dynamics are not caused by a distribution in lengths.

Further, this diversity in dynamics was not caused by pump fluctuations ($\sigma_v/v < 0.028$) but by the dynamics of the molecule. To illustrate this, we plot the extension of two molecules that were imaged at the same time (Fig. 10.16); one molecule is stretching while the other remains a coil and then, while the first molecule is retracting, the second molecule begins to grow.

We now separately discuss the data associated with each strain rate.

For $\dot{\varepsilon}\tau_1 = 0.20$ (Fig. 10.14A), the average extension $\langle x(t_{res}) \rangle$ remains approximately constant but the fluctuations are very large. Since we have $\dot{\varepsilon} = 0.5\,\dot{\varepsilon}_c$, we expect the molecule to fluctuate about coiled configurations. Qualitatively, the magnitude of these fluctuations can be understood as thermal fluctuations within the dumbbell model (Figs. 10.17 and 10.18). The other striking feature of this data set is the apparent shortening of the molecules at larger values of $\dot{\varepsilon}t_{res}$. But we know that the molecules are in a constant $\dot{\varepsilon}$ (see Sect. 10.3.7) and that the molecules are not shortening due to photobleaching (see Sect. 10.3.8). We suggest that this is a misleading visual effect caused by a small number statistics at the larger value of $\dot{\varepsilon}t_{res}$. Since $\dot{\varepsilon}\tau_1 = 0.2$ is below $\dot{\varepsilon}_c$, we approximate the true distribution of molecular extension from a histogram of all of the data (Fig. 10.19). Near $\dot{\varepsilon}t_{res} = 2.5$, there are approximately 400 individual traces while at $\dot{\varepsilon}t_{res} = 3.0$ there are approximately 40.

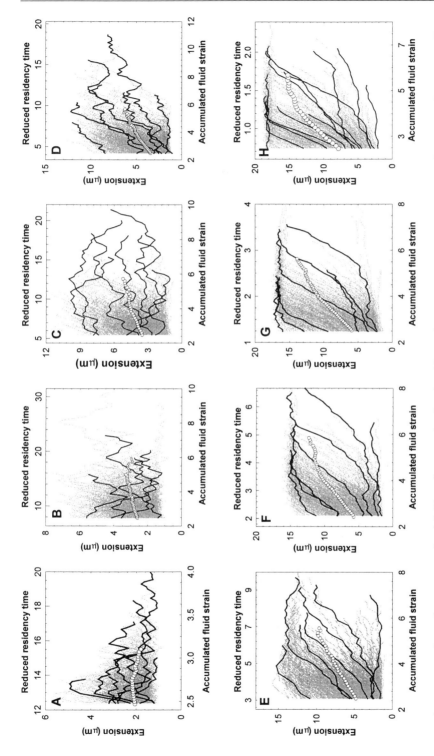

Fig. 10.14A–H. Extension vs accumulated fluid strain $\dot{\varepsilon}t_{res}$ and reduced residency time t_{res}/τ_1; A $\dot{\varepsilon}\tau_1 = 0.2$; B $\dot{\varepsilon}\tau_1 = 0.32$; C $\dot{\varepsilon}\tau_1 = 0.47$; D $\dot{\varepsilon}\tau_1 = 0.58$; E $\dot{\varepsilon}\tau_1 = 0.82$; F $\dot{\varepsilon}\tau_1 = 1.2$; G $\dot{\varepsilon}\tau_1 = 2.0$; H $\dot{\varepsilon}\tau_1 = 3.3$

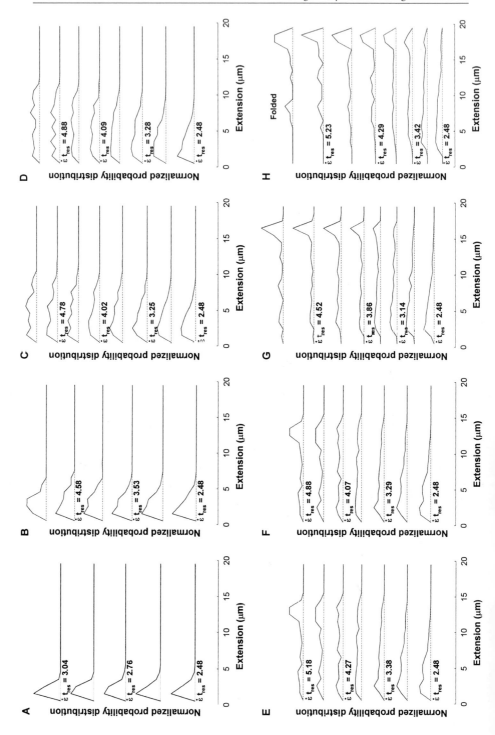

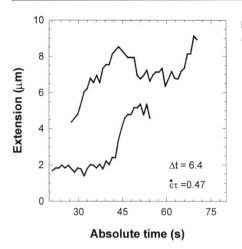

Fig. 10.16. Two molecules stretching at the same time. The difference in their t_{res} is 6.4 s

For $\dot{\varepsilon}\tau_1 = 0.32$, we have $\dot{\varepsilon} = 0.8\,\dot{\varepsilon}_c$ and the fluctuations become more pronounced. Such behavior is often seen at phase transitions. In this case, the fluctuations are increasing because the hydrodynamic force exerted by the elongational flow almost exactly equals the linear elasticity of the chain. Within the dumbbell model, this means the free energy of the chain has a shallow minimum and thermal fluctuations about the mean lead to large changes in x (Fig. 10.18). As shown by the top axis label of Fig. 10.14, these fluctuations are happening on time scales significantly longer than τ_1, the natural time scale for the polymer to relax to equilibrium. $\langle x(t_{res})\rangle$ increases slowly to a value of $\sim 3.2\ \mu m$. To demonstrate that $\langle x(t_{res})\rangle$ does not further increase with t_{res}, we plot the $\langle x(t_{res})\rangle$ averaged for 20–40 molecules in the smaller open circles.

For $\dot{\varepsilon}\tau_1 = 0.47$, we have $\dot{\varepsilon} = 1.1\,\dot{\varepsilon}_c$ and the fluctuations become even more pronounced. Some molecules reach an extension of 9 μm and then relax back towards 6 μm. Other molecules did not stretch appreciably until $t_{res}/\tau_1 > 17$ and then stretched rapidly. The diversity in the dynamics and the magnitude of the fluctuations underscore the difficulty a mean-field theory will have describing this data (see Sect. 10.4.9.6).

For $\dot{\varepsilon}\tau_1 = 0.58$, we have $\dot{\varepsilon} = 1.5\,\dot{\varepsilon}_c$. Once a molecule started to stretch, there was a general trend towards a highly extended, steady-state value with smaller fluctuations about x_{steady}. But there was a large variation in t_{res} at which such stretching begins. To describe this, we define t_{onset} as the t_{res} at which significant stretching began. At this flow rate, t_{onset} can be greater than $10\ t_{res}/\tau_1$ emphasizing how long it takes these polymers to come into equilibrium with the applied velocity gradient.

◄──

Fig. 10.15 A–H. Normalized probability distributions of molecular extension at different accumulated fluid strain $\dot{\varepsilon} t_{res}$: **A** $\dot{\varepsilon}\tau_1 = 0.2$; **B** $\dot{\varepsilon}\tau_1 = 0.32$; **C** $\dot{\varepsilon}\tau_1 = 0.47$; **D** $\dot{\varepsilon}\tau_1 = 0.58$; **E** $\dot{\varepsilon}\tau_1 = 0.82$; **F** $\dot{\varepsilon}\tau_1 = 1.2$; **G** $\dot{\varepsilon}\tau_1 = 2.0$; **H** $\dot{\varepsilon}\tau_1 = 3.3$

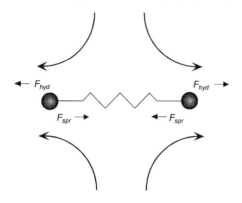

Fig. 10.17. Dumbbell model

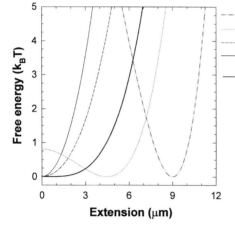

Fig. 10.18. Free energy curves for a simple dumbbell model with no Brownian motion illustrating the large fluctuations near $\dot{\varepsilon}_c$

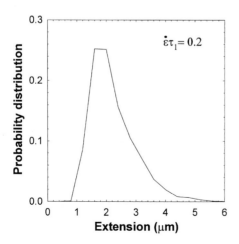

Fig. 10.19. Asymmetric probability distribution

For $\dot{\varepsilon}\tau_1 = 0.82$, we have $\dot{\varepsilon} = 2.1\,\dot{\varepsilon}_c$. The large fluctuations seen near equilibrium were suppressed by the higher tension within the chain associated with larger extensions. However, there was still a large variation in t_{onset}. Interestingly, some molecules overshot x_{steady} and then relaxed back towards x_{steady}.

For $\dot{\varepsilon}\tau_1 = 1.2$, we have $\dot{\varepsilon} = 3\,\dot{\varepsilon}_c$. The large variation in t_{onset} was still very pronounced while the thermal fluctuations during the stretching were further suppressed.

For $\dot{\varepsilon}\tau_1 = 2.0$, we have $\dot{\varepsilon} = 5\,\dot{\varepsilon}_c$. The differences in dynamics between the individual traces and $\langle x(t_{res})\rangle$ illustrates how the large variation t_{onset} alters $\langle x(t_{res})\rangle$. One of the highlighted molecules had significantly slower dynamics. This diversity of dynamics became more pronounced as $\dot{\varepsilon}$ was increased.

For $\dot{\varepsilon}\tau_1 = 3.3$, we have $\dot{\varepsilon} = 8.3\,\dot{\varepsilon}_c$. While the variation in t_{onset} is still present, there is now a large variation in the rate of stretching as well. The extension of some molecules, independent of their t_{onset}, increased exponentially with time. Other, slow-stretching molecules grew linearly in time.

10.4.3
Average and Steady-State Properties; Direct Observation of the Coil-Stretch Transition

The simplest analysis of the x vs t_{res} data is to calculate the average extension $\langle x(t_{res})\rangle$ as a function of $\dot{\varepsilon}$ and t_{res}. To compare the data on an equal basis, we plot the data as a function of the accumulated fluid strain or Henky strain ($\varepsilon = \dot{\varepsilon}t_{res}$) in Fig. 10.20A. This analysis plots the average extension of the polymer in comparison to the deformation of the underlying fluid element where the

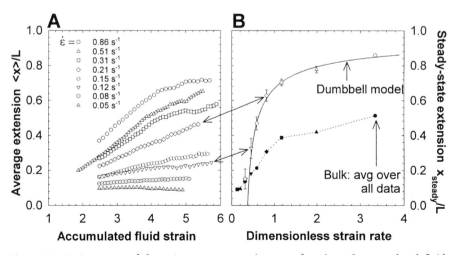

Fig. 10.20A, B. Summary of data: A average extension as a function of accumulated fluid strain $\varepsilon = \dot{\varepsilon}t_{res}$; B steady-state extension (*open symbols*) and a bulk spatio-temporal average (*closed symbols*) determined as a function of the dimensionless strain rate $\dot{\varepsilon}\tau_1$. A fit of the dumbbell model (*solid line*) to the steady-state data

deformation of the fluid element is given by $\exp(\varepsilon)$. $\langle x(\varepsilon) \rangle$ increased monotonically with ε indicating that the average extension had not reached steady-state within the measurement time except for the lowest values of $\dot{\varepsilon}$ where there was no deformation.

We can also determine the steady-state extension x_{steady} by identifying the subset of molecules that reach a steady-state elongation (Fig. 10.21). We plot x_{steady} vs $\dot{\varepsilon}\tau_1$ (Fig. 10.20 B) where $\dot{\varepsilon}\tau_1$ is the appropriate dimensionless strain rate or "Deborah number" which characterizes the rate of deformation of the fluid relative to the relaxation time of the polymer. Note that x_{steady} rises sharply at a critical strain rate of $\dot{\varepsilon}_c \tau_{relax} \cong 0.4$.

Thus, we report the first direct observation of the coil-stretch transition [1] as evidenced by the rapid nonlinear increase in the steady-state extension at a critical strain rate (Fig. 10.20). At $\dot{\varepsilon} \cong 0.9 \, \dot{\varepsilon}_c$, there are large fluctuations ($\sigma_x/x = 0.4$). Similar behavior is seen near phase transitions. The nature of this experiment does not allow us the show that this transition is a first-order phase transition or the presence of hysteresis as proposed by de Gennes [1]. However, future single DNA molecule experiments will be able to resolve this question (see Sect. 10.6).

A comparison between $\langle x(\varepsilon) \rangle$ and x_{steady} shows that $\langle x(\varepsilon) \rangle$ had not yet reached x_{steady} for up to $\varepsilon \cong 5.7$. This corresponds to a deformation of the underlying fluid element by a factor of $\exp(5.7)$ or ~ 300.

In classical bulk measurements by light scattering and birefringence, the response of individual chains is averaged over the diameter of the probe laser beam. Since the measurements are done at the stagnation point, the spatial average due to the diameter of the laser beam leads to an average over a broad distribution in t_{res} as well. To compare our single molecule results to classical bulk measurements, we plot a spatio-temporal average x_{bulk} of all or our data (Fig. 10.20 B). Our imaging area is similar in size to that probed by lasers in previous birefringence experiments [13, 16]. This analysis shows that x_{bulk} is significantly smaller than x_{steady}. This is not unexpected since $\langle x(\varepsilon) \rangle$ is less than

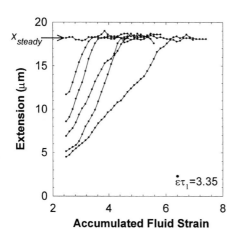

Fig. 10.21. Molecules stretching to steady-state

x_{steady} up to the largest values of ε measured. Thus, our data show that classical bulk measurements in this finite region around the stagnation point do not measure the steady-state properties.

In part, this mismatch between x_{bulk} and x_{steady} arises from the limited t_{res} associated with measurements in the vicinity of the stagnation point. However, if all the molecules started to stretch immediately upon entering the elongational flow field, x_{bulk} would be significantly larger. Thus, visualizing single molecules reveals how the large and previously unobservable variation in t_{onset} reduced x_{bulk} in comparison to x_{steady}.

10.4.4
Steady-State Measurements Fit by the Dumbbell Model

Theoretical descriptions of dilute polymer rheology are often based on constitutive equations. Many of these equations are derived from the simplest of polymer models: a dumbbell model [2]. In this case, the complexity of the complete dynamics is drastically simplified. The hydrodynamic coupling between the fluid and the polymer is represented by two beads while the entropic elasticity is represented by a spring. For flows that generate significant deformation, finitely extensible chains with nonlinear elasticity are used. Fortunately, for modeling DNA with a dumbbell model, the steady-state elasticity of DNA has been measured [33] and found to agree with the elasticity of a worm-like chain [53]. Therefore, our dumbbell model is based on two beads connected by a Marko-Siggia spring (Eq. 10.4). Previously, this simple model has accurately described the steady-state extension of a tethered polymer (DNA) in a uniform flow [28]. In collaboration with R. Larson, we developed a molecular understanding of the origin of this agreement based on simulations [51].

This model does not include Brownian motion, which can lead to a change in x_{steady} even when $\dot{\varepsilon} < \dot{\varepsilon}_c$ (Fig. 10.14). Further, when trying to fit the data to the model, the fit is most sensitive to fluctuations at or near $\dot{\varepsilon}_c$ where x_{steady} is rapidly increasing. Therefore, we fit the model to the data for $\dot{\varepsilon} > 0.15 \text{ s}^{-1}$ to

$$x = L \left(\frac{2b - 1 - \sqrt{4b + 1}}{2b} \right) \tag{10.11A}$$

where b is

$$b = -4 + \frac{12\pi\eta R_{bead}\dot{\varepsilon}Lp}{k_B T} \tag{10.11B}$$

and the two free parameters are R_{bead}, the radius of each bead, and L, the contour length of the polymer. As shown in Fig. 10.20B, our steady-state results for free polymers in an elongational flow is also approximately characterized by this simple model with the parameters ($R_{bead} = 0.15 \text{ μm}$ and $L = 21.1 \text{ μm}$). This value for the stained length agrees closely with previous measurements of the length of stained DNA ($L = 22 \text{ μm}$ [28]). An extrapolation of the model to $x = 0$

gives a critical strain rate of $\dot{\varepsilon}_c \tau_{relax} \cong 0.4$, which is close to the theoretical value of 0.5 calculated from the Zimm model and by the numerical calculation of Larson and Magda [65]. The mismatch between x_{steady} and the model at $\dot{\varepsilon} \cong \dot{\varepsilon}_c$ is caused by the increase in x_{steady} due to Brownian motion which was neglected in this simple model.

10.4.5
Conformational Dependent Dynamics at Highest Strain Rates

The varition in t_{onset} is not the only cause of the slower average dynamics. At higher $\dot{\varepsilon}$ (Fig. 10.14 H), we see a second source: heterogeneity in the rate of stretching. To help understand the origin of this heterogeneity, we analyzed the conformation of the molecule. In general, at each instant in time, the molecules could be classified as one of seven different conformations: kinked, folded, dumbbell, half-dumbbell, uniform, extended, and coiled. Well defined states that were robust to Brownian motion only existed when the molecules were subject to an $\dot{\varepsilon}$ significantly greater than the inverse relaxation time ($\tau_{relax}^{-1} = 0.26$ s^{-1}; $\tau_{relax} = 3.89$ s (see Sect. 10.3.10)). Only at the highest two strain rates did we observe folded and kinked conformations. Several examples of each conformation, except extended, are shown in Fig. 10.13.

Presumably, the differences in conformation as well as the variations in t_{onset} arise directly from the multitude of accessible conformations at equilibrium,

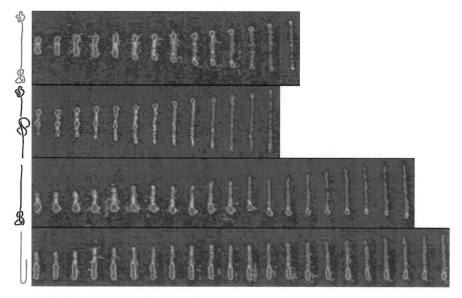

Fig. 10.22. Dynamic unwinding of different conformations. From top to bottom, the images are classified as dumbbell, kinked, half-dumbbell, and folded. Reprinted with permission from Science, 276:2016 (1997). Copyright 1997 American Association for the Advancement of Science

where thermal fluctuations cause instantaneous deviations away from a spherically symmetric distribution. For instance, some of these accessible conformations of an equilibrium coil have both ends on the same side of the center of mass. When subject to an $\dot{\varepsilon} \gg 1/\tau_1$, these ends will not be able to diffuse across the length of the molecule and such types of initial conformations presumably lead to folded configurations.

While this simple classification could describe each of our individual images of a polymer's conformation, we could not uniquely classify the full series of images that showed a polymer stretching from a coiled to an extended state. For example, a molecule at $x = 5$ μm may appear to be in a dumbbell configuration but, by $x = 10$ μm, one of the "balls" of fluorescence disappeared (unraveled) and it is now in a half-dumbbell configuration. Nevertheless, to look for conformation dependent dynamics, we assigned each molecule a conformation to its entire stretching process by creating a hierarchy of conformations. The order for classification was kinked, folded, dumbbell, half-dumbbell, uniform, extended, and coiled. For example, we classified the molecule shown in the second row of Fig. 10.22 as kinked. At the highest strain rate of $\dot{\varepsilon} = 0.86$ s^{-1}, this general classification of all molecules yielded 5.4% kinked, 24% folded, 20% dumbbell, 35% half-dumbbell, 8.3% uniform, 5% extended, and 3% coils (Fig. 10.23). Thus at $\dot{\varepsilon} = 0.86$ s^{-1}, the dominant conformations at the highest strain rate were half-dumbbell, folded, and dumbbell.

The coiled and extended conformations are the starting and ending configurations, necessary to describe those molecules whose evolution in extension are not observed due to the limited range in observation time. A coil is simply a molecule that did not deform from a ball-shaped configuration during its t_{res} and its t_{onset} is larger than its maximum t_{res}. At the other extreme, an extended molecule is a molecule that is already stretched to near its steady-state extension before it was visualized and its conformation during the stretching process was not observed.

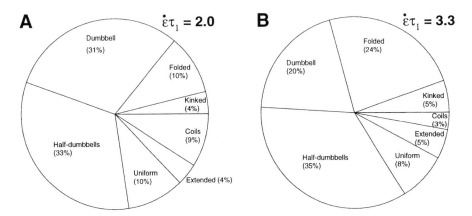

Fig. 10.23 A, B. Pie charts showing the distribution of conformations at two strain-rates

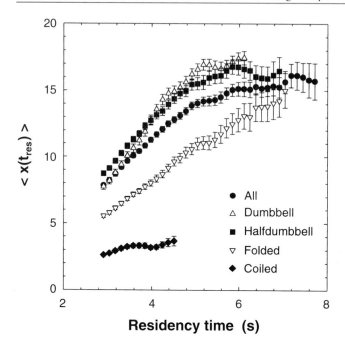

Fig. 10.24. Average extension vs t_{res} for different conformations

A preliminary analysis of the average dynamics based on this first, general classification showed differences between the folded and the dumbbell configuration (Fig. 10.24). But these differences were not as large as observed in some individual traces. However, there was a lot of scatter within each category. The dynamics of a folded molecule whose initial fold was 30% of its extension stretches differently compared to a molecule whose initial fraction of fold was 80%. To highlight the differences between the conformations, we determined those molecules that best typified each conformational class and then re-analyzed these molecules. For classification purposes, the molecules that best typified a dumbbell configuration had approximately symmetric "coils" at each end. For classification as folded, we required the initial percentage of the folded section to be > 75%. This classification yielded 30 dumbbells, 34 folded, and 43 half-dumbbells out of 992 molecules and these molecules were re-analyzed. This data clearly shows a distinct difference in the dynamics of molecules in the folded and dumbbell configuration (Fig. 10.25). Further studies of conformation dependent dynamics were reported by Smith and Chu [D. E. Smith, S. Chu, Science 281, 1335–1340 (1998)].

10.4.6
Master Curves

To compare more accurately the differences in the unwinding dynamics between conformations, we wanted to calculate an average rate of unwinding. However,

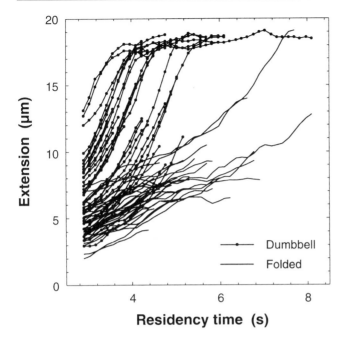

Fig. 10.25.
Comparison between
dynamcis of dumb-
bell and folded confi-
gurations

the large variability in t_{onset} obscures this analysis for the case of a simple time average (Fig. 10.26). Conceptually, we can eliminate t_{onset} by sliding each curve along the time axis until the curves superimpose to form a master curve. An alternative method to accomplish the same thing is to calculate the rate of extension $\dot{x}(x)$ as a function of x and then integrate it to get a "master" curve.

Specifically, we calculate $\dot{x}(x)$ by fitting five successive data points to a line. The average dynamics $\dot{x}(x)$ is calculated by binning the individual $\dot{x}(x)$ every

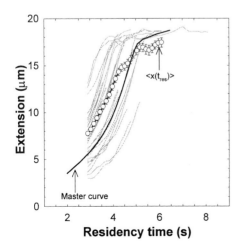

Fig. 10.26. Individual traces of extension vs t_{res} for molecules that best typify the dumbbell configuration. The master curve does a much better job than a simple time average at describing the dynamcis because of the large variability in t_{res}

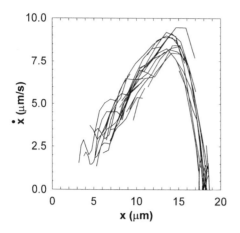

Fig. 10.27. Dynamics of molecules that best typifiy dumbbell molecules

0.5 μm in x and then averaging them to determine $\dot{x}(x)$ for each bin. Note, this calculates the average dynamics by averaging all molecules at a given x, not a given t_{res}, so there is not a distribution about x as there would normally be when calculating such an average.

For those molecules in a dumbbell configuration (Fig. 10.27), this analysis of the rate of stretching $\dot{x}(x)$ as function of x shows that once a molecule starts to stretch, its dynamics follows a specific time evolution. Up to $x/L = 0.6$, we observed a linear increase in $\dot{x}$ with x at $\dot{\varepsilon} = 0.86$ s^{-1}. When integrated, this

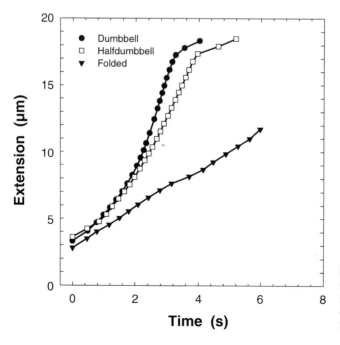

Fig. 10.28. Comparision of master curves for the dominant conformations

yields an initial exponential growth of the master curve. The relationship between the individual curves, the master curve and $\langle x(t_{res})\rangle$ is shown in Fig. 10.26 and shows how the master curve does a much better job capturing the averaging unwinding dynamics than $\langle x(t_{res})\rangle$.

In Fig. 10.28, we show three such master curves generated from the molecules that best typify each of the dominant conformations at $\dot{\varepsilon} = 0.86\ \mathrm{s}^{-1}$. Clearly, there is a strong dependence of the rate of stretching on conformation. For comparison, we show $\langle x(t_{res})\rangle$ in Fig. 10.24 for the full data set as well as for several of the different conformational classes arising from the first, general classification.

In addition to looking for conformational dependent rate of stretching, we calculated $\langle \dot{x}(x)\rangle$ for all $\dot{\varepsilon}$ (Fig. 10.29). This analysis averaged over all conformations. For $\dot{\varepsilon} > 0.21\ \mathrm{s}^{-1}$ and at low extensions, $\langle \dot{x}(x)\rangle$ was a linear function of x. This linear relationship indicates that once the polymers began to deform their deformation increased exponentially with time for small x. This deformation is similar to and caused by the exponential deformation of the underlying fluid element.

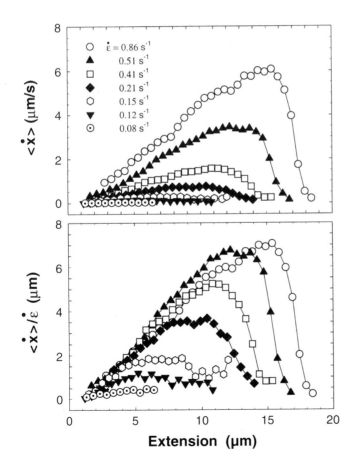

Fig. 10.29. Rates of deformation and normalized rates of deformations for all $\dot{\varepsilon}$

10.4.7
Affine Deformation

A polymer is said to deform "affinely" with the fluid if the molecular deformation equals the deformation of the surrounding fluid element. It has been postulated that when $\dot{\varepsilon} \gg 1/\tau_1$ affine deformation becomes an increasingly valid approximation [66–68]. Our observed exponential growth in the master curves suggests a comparison be made between our data and the approximation of affine deformation.

In the simplest analysis, we note that $\langle x(t_{res}) \rangle$ did not reach x_{steady} even after an accumulated fluid strain of $\varepsilon = \dot{\varepsilon} t_{res} \cong 5.7$, which corresponds to an $e^{5.7} \cong$ 300-fold distortion of the fluid element (Fig. 10.20). For comparison, the required molecular distortion to extend fully stained λ-DNA is $L/R_G \cong 30$ where R_G, the radius of gyration, is 0.73 µm.

In part, this lack of affine deformation in $\langle x(t_{res}) \rangle$ arises from the large variation in t_{onset}. Notwithstanding this variation which is intrinsically non-affine, we wanted to know if molecules deform affinely once they start to stretch. To do so, we analyzed the dynamics of the master curve, which suppresses the variation in t_{onset} by computing $\langle \dot{x}(x) \rangle$ instead of $\langle \dot{x}(t_{onset}) \rangle$. To compare the molecular deformation and the deformation of the underlying fluid element, we define a molecular strain rate $\dot{\varepsilon}_{mol}$ by fitting this linear region of $\langle \dot{x}(x) \rangle$ vs x to $\langle \dot{x}(x) \rangle = \dot{\varepsilon}_{mol} x + h$. The linear portion of $\langle \dot{x}(x) \rangle$ vs x extended up to 5.2, 7.8, 8.8, 10.2, and 10.8 µm in increasing $\dot{\varepsilon}$.

At low strain rates, affine deformation is not expected because there must be some initial slip between the polymer and the fluid to create the hydrodynamic force necessary to overcome the native elasticity of the polymer. This required slip is simply the critical strain rate $\dot{\varepsilon}_c$ identified in the analysis of the steady-state extension below which there is no significant deformation. To account for this threshold, we rescale the rate of deformation of the fluid $\dot{\varepsilon}_{fluid} \equiv \dot{\varepsilon}$ by subtracting off $\dot{\varepsilon}_c$, and plot $\dot{\varepsilon}_{mol}$ vs $\dot{\varepsilon}_{fluid} - \dot{\varepsilon}_c$ (Fig. 10.30 A). At lower $\dot{\varepsilon}$, the molecules are stretching near the theoretically expected limit. At higher $\dot{\varepsilon}$, the rate of elongation shows a marked departure from the theoretical limit at these moderate strain rates. Furthermore, when plotted as $\dot{\varepsilon}_{mol}/(\dot{\varepsilon}_{fluid} - \dot{\varepsilon}_c)$ vs $(\dot{\varepsilon}_{fluid} - \dot{\varepsilon}_c)$, the data decreases at 0.86 s^{-1} (Fig. 10.30 B).

Thus, the average data shows neither an absolute nor a fractional approach toward affine deformation at higher $\dot{\varepsilon}$ even after eliminating the large variation in t_{onset}. Rather, the approximation is becoming increasingly worse at higher $\dot{\varepsilon}$. This failure arises from the introduction of intra-molecular constraints (primarily folds) which dramatically slow down the average dynamics. It must be kept in mind, however, that, in the present experiment, the earliest part of the stretching was not visualized. Later studies showed that the initial ~25% of the stretching was increasingly close to being affine at higher De, though stretching beyond ~25% was increasingly non-affine due to folds. In contrast to the average dynamics, the subset of molecules in a dumbbell conformation stretched almost as fast as can be theoretically expected (Fig. 10.30 B, open symbol). This rapid stretching of the dumbbell configuration makes intuitive sense. The "coils" at each end lead to a large hydrodynamic drag force because they are at the ends of

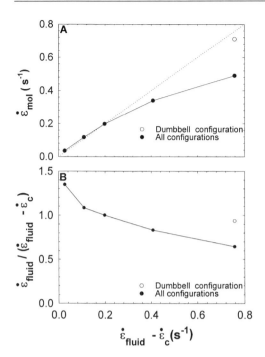

Fig. 10.30 A, B. Comparison of the rate of molecular deformation to rate of deformation of the fluid element minus the necessary fluid slip to overcome the linear elasticity of the polymer

the molecule and therefore experience the full extent of the velocity gradient across the molecule. Furthermore, the force necessary to stretch chain segments from the coil is low because the elasticity force is small at low deformations.

10.4.8
Dynamic Data, Intrinsic Viscosity and the Dumbbell Model

Having seen that the dumbbell model's simplified representation of a polymer described our steady state results, we wanted to see if this model could self-consistently describe the dynamics. The measured dynamics have a large variation in t_{onset} and this model does not include any Brownian motion or a distribution of initial starting configurations. Therefore, we compared the model to the dynamics of the master curve because the master curve suppresses the variation in t_{onset} by calculating $\langle \dot{x}(x) \rangle$ instead of $\langle x(t_{res}) \rangle$ (see Sect. 10.4.6). To self-consistently calculate the predicted dynamics, we use the parameters determined from the steady-state results. Specifically, we calculated the force on each bead as the sum of the entropic elasticity $F_{spring} = F_{MS}(x/L)$ given by Eq. (10.4) and hydrodynamic force F_{hydro} given by

$$F_{hydro} = 6\pi\eta R_{bead}\left(v_{fluid} - \frac{\dot{x}_{predicted}}{2}\right) \qquad (10.12)$$

where the fluid velocity at each bead is given by $v_{fluid} = 0.5\,\dot{\varepsilon}x$. The net force is set to zero

$$F_{total} = F_{spring} + F_{hydro} = 0 \;. \tag{10.13}$$

We can then solve for the predicted rate of stretching of the total chain at each x to be

$$\dot{x}_{predicted} = 2\,(6\,\pi\eta R_{bead} v_{fluid} - F_{spring})/6\,\pi\eta R_{bead} \tag{10.14}$$

Note that $\dot{x}_{predicted}$ is also calculated for a given x, not for a t_{res}, and the comparison to $\langle \dot{x}(x)\rangle$ is valid.

As shown in Fig. 10.31, this simple model overestimates the average rate of deformation. Therefore, we expect difficulty predicting the rheological properties of dilute polymer solutions using this simplified model (see Sect. 10.4.9.4). However, when $\dot{x}_{predicted}$ is compared to the dynamics of molecules in the dumbbell configuration, the disagreement is much smaller. But we must remember that in this analysis the large variation in t_{res} has been suppressed.

Clearly, the average dynamics are decreased from the fastest dynamics by the introduction of folds. In essence, this is analogous to the breakdown in the affine deformation approximation at higher $\dot{\varepsilon}$. But, in that case, the details of the elastic properties of the chain and its hydrodynamic drag were not explicitly written out as a function of R_{bead} and L but incorporated in $\dot{\varepsilon}_c$.

Previously, "internal viscosity" [2, 3, 43] has been added to the dumbbell model to explain the delay in shear thinning [2]. Similarly, one might be tempted to add a term proportional to $-\dot{x}_{predicted}$ because such a term can approximately compensate for the slower average dynamics. Such a term is suggested by the measurements of η_E because it leads a dissipative component of the stress relaxation [20]. In particular, for $\dot{\varepsilon} = 0.86\,s^{-1}$, the average dynamic data is best fit by multiplying $\dot{x}_{predicted}$ in F_{hydro} by $(1 + \alpha)$ where $\alpha = 0.55$ (Fig. 10.31). Note that the trend in Fig. 10.30 away from $\dot{\varepsilon}_{mol} = \dot{\varepsilon}_{fluid} - \dot{\varepsilon}_c$ shows that this coefficient α is dependent on $\dot{\varepsilon}$.

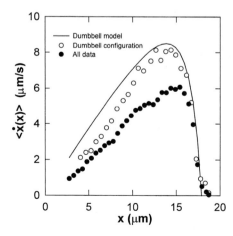

Fig. 10.31. Comparison between the simplified, measured dynamics of the master curve for all molecules and those in the dumbbell configuration at $\dot{\varepsilon} = 0.86\,s^{-1}$ and the prediction of the dumbbell model using the parameters determined from a fit to the measurements of the steady state extension

We stress that the observed slower elongation rates arise from folded configurations and *not* from the monomer-monomer friction typically associated with internal viscosity. We also note that there are addition terms besides $-\dot{x}_{predicted}$ that can lead to dissipative stresses [69–74].

10.4.9
Comparison to Previous Experimental Work

Given our measurements of the dynamic, steady-state, and ensemble-averaged properties of polymers in an elongational flow, we now compare our data to previous experimental and theoretical results.

10.4.9.1
Birefringence

As mentioned in the introduction, birefringence has been the dominant experimental technique for studying polymers in elongational flows. These experiments have shown the birefringence abruptly increases at a critical strain rate followed by saturation at higher strain rates. The classical interpretation has been that this saturation implies the chains are highly extended [5]. However, the implication of this saturation is still debated [16, 75]. To help clarify the relationship between extension and birefringence, we compare our data to the previous measurements of the birefringence of λ-DNA as well as synthetic polymers.

Atkins and Taylor have measured the birefringence of dilute solutions of λ-DNA. This data was also taken in a planar elongational flow. To compare appropriately the birefringence data for unstained DNA and our steady-state extension data for stained DNA, we normalized their birefringence measurements by its saturating value and x_{steady} by L. For both measurements, the strain rate was scaled by the appropriate τ_1 to yield the dimensionless strain rate $\dot{\varepsilon}\tau_1$ (see Sect. 10.3.10 for a complete discussion of τ_1). Clearly, as shown in Fig. 10.32,

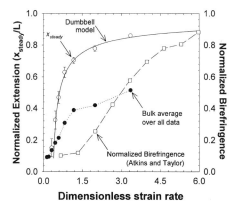

Fig. 10.32. Steady state measurement vs previous birefringence data for λ-DNA by Atkins and Taylor [9]. A spatio-temporal, "bulk" average is plotted for comparision since it averages all molecules within the field of view which is similar to the average of all molecules within the diameter of the laser beam used to measure the birefringence

our ability to select only those molecules that have reached steady-state extensions reveals a much sharper transition occurring at a lower $\dot{\varepsilon}_c \tau_1$. Since birefringence averages over a broad range of positions and t_{res}, we also plot x_{bulk}, a spatio-temporal average of all data at a given $\dot{\varepsilon}$.

The large offset between x_{steady} and the birefringence might lead one to suspect an error in calculating $\dot{\varepsilon}_c \tau_1$ for either our stained λ-DNA or their unstained λ-DNA. However, our measurement of τ_1, after scaling for the $\sim 30\%$ change in length due to staining, is in good agreement with consensus values of τ_1 in the literature (see Sect. 10.3.10). Further, the large offset between x_{steady} and the birefringence is not limited to DNA. For very dilute polystyrene solutions, Nguyen *et al.* determined critical values of $\dot{\varepsilon}_c \tau_1$ ranging from 3 to 8 [16]. Taken together, these results argue that this offset does not arise from a concentration dependent effect or a property of the polymer's local chemical structure.

We therefore conclude that there is no direct correspondence between either x_{steady} or x_{bulk} and the birefringence at the stagnation point. Since birefringence measures the asymmetry in the optical polarizability of the chain segments rather than extension, some disagreement would be expected based on the observed conformational features such as folds. Since folds cause a premature saturation in birefringence with respect to x_{steady}, one might have expected the birefringence to saturate before x_{steady} but our data shows the birefringence increased after x_{steady}.

Our results indicate difficulties in interpreting birefringence and other bulk measurements. Moreover, we suggest that this difficulty may be even greater for synthetic polymers where the larger ratio of L/R_G requires an even larger accumulated fluid strain than is needed to extend λ-DNA.

Other important differences between our data and previous birefringence measurements are the uniformity in the strain rate and the absolute monodispersity of λ-DNA. As discussed in Sect. 10.3.7, our $\dot{\varepsilon}$ is uniform to within 2% over the region of interest. Further, $\dot{\varepsilon}$ turns on abruptly. Both of these reasons may contribute to the sharper, more sudden rise in x_{steady} than is seen in the birefringence.

10.4.9.2
Light Scattering

Previously, light scattering experiments have shown a small increase ($\sim 2X$) in R_G [13]. In particular, Menasveta and Hoagland showed that R_G increased at the same $\dot{\varepsilon}_c$ as determined from birefringence [13]. In contrast to these previous light scattering results on synthetic polymers, our data shows extensions significantly greater than $\sim 2R_G$. We note that a uniformly extended molecule will have an $R_G = x_{steady}/4$. In general, the large difference between R_G and $x_{steady}/4$ is caused by the broad distribution in t_{res} for the population of molecules measured by light scattering. Hence, $x_{bulk}/4$, not $x_{steady}/4$, should be used for comparison. In addition, the highly asymmetric mass distribution of the most common conformations (half-dumbbell) would further reduce R_G.

Since the optical response of our SIT camera is linear, a two dimensional projection of R_G can be directly calculated from an image of a polymer's conformation. For each image, the background is subtracted from the individual intensity values $i_{x',y'}$ where x' and y' are the coordinates of the pixel. R_G is calculated about both the incoming ($\hat{x}$-axis) and the outgoing ($\hat{y}$-axis). Since we are already using x to describe the extension along the outgoing axis, we refer to R_G^+ for the deformation along the outgoing axis and to R_G^- for deformation along the incoming axis. We calculate these values using

$$(R_G^+)^2 = \frac{1}{I_{total}} \sum_{x',y'} i_{x',y'}(y' - y'_{cm})^2 \tag{10.15}$$

$$(R_G^-)^2 = \frac{1}{I_{total}} \sum_{x',y'} i_{x',y'}(x' - x'_{cm})^2 \tag{10.16}$$

where I_{total} is the total intensity, and (x'_{cm}, y'_{cm}) is the center of intensity. $\langle R_G^+ \rangle$ and $\langle R_G^- \rangle$ are the root mean square data calculated from an ensemble average over all molecule in the field of view. In Fig. 10.33, we plot $\langle R_G^+ \rangle$ and $\langle R_G^- \rangle$ as a function of the accumulated fluid strain ($\dot{\varepsilon} t_{res}$) as well as $\langle x \rangle$ at $\dot{\varepsilon} \tau_1 = 1.2\ \text{s}^{-1}$. Clearly, $\langle R_G^+ \rangle$ is much smaller than $\langle x \rangle$ and the ratio between these two measurements is almost constant at 0.29 at each value of $\dot{\varepsilon} \tau_1$ (Fig. 10.33 B). For an object with a uniformly distributed mass, this ratio is 0.25. $\langle R_G^- \rangle$ does not decrease during the measurement because its value is bounded by the resolution limit of the microscope ($\sim 0.4\ \mu\text{m}$).

To compare directly to the light scattering data, we compute a spatio-temporal average over all measurements of R_G^+ at $\dot{\varepsilon} \tau_1 = 1.2\ \text{s}^{-1}$. This yields $R_{G\,bulk}^+ = 2.2\ \mu\text{m}$ which is ~ 7 times smaller than x_{steady} while being ~ 3 times larger than $R_G (= 0.71\ \mu\text{m})$. Thus, our results of single molecule results emphasize that

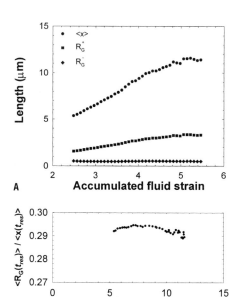

A

B

Fig. 10.33 A, B. **A** The average radius of gyration along the incoming (–) and outgoing axes (+) calculated from the image of the polymer. **B** The normalized average radius of gyration compared to the average extension

$R^+_{G\,bulk}$ is much smaller than x_{steady}. Further, the limited but broad distribution in t_{res} and geometric factors reduce to an $R^+_{G\,bulk}$ that is even smaller and only a small factor larger than the equilibrium value of R_G. We note that "blooming" of the image due to the intensified camera tends to make these calculations overestimates of the true R_G, especially near equilibrium. This agreement between our measurements of $R^+_{G\,bulk}$ and Mensaveta and Hoagland's measurements emphasize the inability of bulk experiments to draw conclusions of steady-state properties of molecules by making measurements at the stagnation point.

10.4.9.3
Stagnation Point Flow Fracture

At $\dot{\varepsilon} \gg 1/\tau_1$, polymers flowing through stagnation point flow fracture at the center of the chain [10–12, 76]. This mid-point chain fracture implies the chains are highly extended. This follows because, if it were not true, the center of the chain would not be a unique point and the chains would fracture into a broad distribution of lengths. These results have been used to argue that all the chains are highly extended. However, our results suggest that mid-point chain fracture does not imply all chains are extended.

The large variability in x shown in Fig. 10.14H indicates a few molecules rapidly reached steady-state. If we extrapolate our results to a 100-times higher $\dot{\varepsilon}$ [12] and if the heterogeneity in the onset of stretching persists, it is these highly-extended, early-stretching molecules that will experience a large enough force to fracture at or near their center. This value of $\dot{\varepsilon}$ is in agreement with the minimum necessary force to rupture DNA when calculated within the dumbbell model assuming a rupture force of 470 pN [77] and assuming the DNA did not have time to undergo a transition to an extended state [78, 79]. Nonetheless the number of such chains that are rapidly stretching and start stretching early is relatively small. Thus, due to the limited t_{res}, only a fraction of the total number of chains fractured in agreement with the results of bulk experiments. The fracture of some chains does not imply that all chains are extended.

10.4.9.4
Rheology

Rheologists often infer molecular deformation from bulk viscoelastic measurements [2, 3]. For instance, within the dumbbell model, one can derive the extensional stress σ_E (Eq. 10.7) and the extensional viscosity $\eta_E = \sigma_E/\dot{\varepsilon}$ using the data in Fig. 10.14, the known elasticity of DNA (Eq. 10.4) and classical results in rheology (Eq. 10.7). However, because the molecules are in highly non-equilibrium configurations (Fig. 10.22), it is inaccurate to used the steady-state elasticity for molecules at $\dot{\varepsilon} \gg \tau_1$. Further, such a simple analysis would fail to predict the stress associated with molecules in the folded conformation. From this and the lack of a physically significant mean described in Sect. 10.4.9.6, our results suggest difficulty in inferring an average conformation from bulk rheological measurements.

10.4.9.5
Filament Stretching

In filament stretching experiments, dilute quantities of high molecular weight polymers are suspended in a fluid of low molecular weight polymers [18, 20, 21, 80]. This composite fluid is then placed between two disks. The distance between the disks is increased exponentially in time while monitoring the force between the polymeric solution and the disk. Such filament stretching experiments are an excellent method for determining the extensional stress within the fluid. While they average over a large number of molecules, most molecules have a well defined t_{res}, unlike stagnation point flows. Therefore, the average properties are well determined. Filament stretching devices have measured extensional viscosities several thousand times greater than the shear viscosity. Also, the stress relaxation from such experiments contained both a strain-rate dependent, "elastic" component as well as a strain-rate independent, "dissipative" component [20, 21].

We do not directly compare our single molecule results to these filament stretching experiments due to our inability to calculate accurately the transient elasticity of the chain in a variety of non-equilibrium configurations (Fig. 10.22). Nonetheless, we can draw several qualitative conclusions. The heterogeneity in $x(t_{res})$ implies that a large value of $\dot{\varepsilon}\tau_{res}$ (> 6) will be needed for the all of the molecules to reach steady-state. Further, this value of $\dot{\varepsilon}\tau_{res}$ is likely to be larger for synthetic polymers, which have a larger ratio of L/R_G than for λ-DNA where the ratio is 30. However, the presence of folded configurations suggests that the stress may plateau before the steady-state extension is achieved.

As stated in the introduction, these filament stretching experiments measure a strain-rate dependent and a strain-rate independent relaxation after the applied deformations is stopped. The presence of the different conformations offers a qualitative explanation to the origin of dissipative or strain-rate dependent stress (see Sect. 10.4.9.8). For example, a molecule folded in half generates stress as the fluid slips past it. A molecule in a dumbbell configuration also leads to a dissipative-like stress. For a given extension, a dumbbell configuration has more tension within its extended portion than a molecule under a uniform tension since a smaller fraction of the molecule accomplishes the same extension. When the applied flow field is stopped, only a small amount of excess chain segment density within the coils rapidly needs to diffuse inward from each end because the elasticity is highly nonlinear at large extensions. This relaxation is akin to the higher order modes in the Zimm [81] and Rouse [82] models. These relaxations are too fast to be measured in the filament stretching device and therefore lead to an apparent dissipative stress [69]. So the measured dissipative stress relaxation most likely arises from both the slip past an extended object and from the higher order modes of molecular relaxation.

10.4.9.6
Mean Field Theories

Mean field theories, such as the Zimm model for dilute solutions as well as the reptation model for concentrated solution, have been very successful in describing polymer dynamics [81,83]. Mean fields are based on the assumption of a well defined mean behavior. However, our data show that the probability distribution for molecular extension is not a narrow distribution about a mean but rather a broad, oddly shaped distribution (Fig. 10.15). This oddly shaped distribution arises from the presence of several distinctly different dynamical processes at high $\dot{\varepsilon}$ as well as the broad distribution in t_{res} at all $\dot{\varepsilon}$ (Fig. 10.14). Further, the differences in $\dot{x}$ and t_{onset} imply a sensitive dependence on the polymer's initial conformation when it enters the velocity gradient.

Specifically, for elongational flows, the Peterlin approximation is often used [84, 85]. In this approximation, $x^2(t_{res})$ is replaced by $\langle x^2(t_{res})\rangle$ in calculating the elasticity of the chain in conjunction with the dumbbell model. Such an approximation yields mathematical closure and enables derivation of constitutive equations from kinetic theory [2,3]. These constitutive equations are then used to predict the stress in the fluid or, alternatively, to determine the molecular configuration from bulk rheological properties.

Our results reveal previously know problems with the Peterlin approximation. The broad distribution in t_{onset} and the conformational dependent dynamics show the evolution of $\langle x(t_{res})\rangle$ is inherently different from the dynamics of individual molecules (Fig. 10.14). Further, even within one conformational class, the dynamics of $\langle x(t_{res})\rangle$ is different from the dynamics of the individual molecules (Fig. 10.26). Thus, our data show that the approximation of $x^2(t_{res})$ by $\langle x^2(t_{res})\rangle$ is a poor one.

The heterogeneity in our data that leads to the breakdown of the Peterlin approximation is also seen in Keunings' stochastic simulations of the finitely extensible dumbbell model [85]. While this simplified model of polymer dynamics based on kinetic theory yields histograms that are in semi-quantitative agreement with our data (Fig. 10.15), simulations using the Peterlin approximation in conjunction with kinetic theory lead to qualitatively different results.

Thus, any closed form analytical solution describing the observed dynamics seems doubtful given the crucial role of fluctuations in the initial conditions. The best opportunity for capturing the diversity in dynamics lies in stochastic simulation of multi-element chains. Such chains are necessary to generate the internal conformations of the molecule. Simulations also offer the best opportunity to investigate the relationship between a polymer's initial conformation when it enters the flow and the resulting dynamics. In particular, there will be regions of a polymer's conformational phase space at equilibrium that directly relate to the observed conformation. For instance, in Fig. 10.34, the molecule will become folded because, at higher $\dot{\varepsilon}$, the ends do not have sufficient time to diffuse during the stretching process. Detailed results of brownian simulations will be described elsewhere [R.G. Larson et al., to be published].

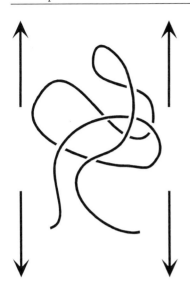

Fig. 10.34. Schematic of an equilibrium configuration likely to become folded at high strain rates

10.4.9.7
Comments on Dumbbell Model

The dumbbell model is the simplest of models to describe a polymer. It can extend and orient with the flow. This model, when using a Marko-Siggia spring, has been remarkably successful in describing our current as well as previous steady-state results [28]. Such steady-state results are in equilibrium and justify the use of the steady-state elasticity to describe the spring.

However, the dumbbell model cannot model the dynamics over a broad range of $\dot{\varepsilon}$. In part, the dumbbell model fails due to its inherent simplicity; it cannot model folds. Yet folds are becoming increasingly prevalent as $\dot{\varepsilon}$ is increased. Specifically, our data show that, at lower values of $\dot{\varepsilon}$ where there is not conformational dependent rate of stretching, the dumbbell model approximately described the average unwinding dynamics of the master curve, where the large variation in t_{onset} was suppressed. But at the highest $\dot{\varepsilon}$ investigated, the dumbbell model overestimated even these simplifed, average dynamics. This overestimation occurred, in part, because of the appearance of the folded configuration, which greatly slowed down the average dynamics. The fraction of molecules in the folded configuration dramatically increased from 9% to 24% as $\dot{\varepsilon}\tau_1$ was increased from 2 to 3.3. Therefore, our data suggests that the range of applicability of the dumbbell model to describe dynamics is limited to small values of $\dot{\varepsilon}\tau_1$. Simple modifications to the dumbbell model, such as the introduction of a dissipative term – $\dot{x}(1 + \alpha)$, are most likely to be of limited use since α would have to be a function of $\dot{\varepsilon}$ (Fig. 10.30). More importantly, the origin of such a dissipative term from internal viscosity is not physically justified in light of our data.

Another limitation of this model at higher $\dot{\varepsilon}$ is the assumption of adiabaticity. For $\dot{\varepsilon} \gg 1/\tau_1$, molecules do not have time to sample the full, accessible configurational space at each x. Therefore, it is unrealistic to expect the steady-state elasticity, which is derived from the change in the number of accessible states with x, to describe the elasticity of a rapidly deforming chain. For processes that occur on time scales much faster than τ_1, the fundamental time scale of the polymer, the polymer is not in equilibrium with the flow. Figure 10.22 emphasizes the highly non-equilibrium configurations observed at $\dot{\varepsilon}\tau_1 = 3.3$ and Fig. 10.14 emphasizes how long it can take the polymers to come into equilibrium with the flow. Yet the dumbbell model assumes the deformation is adiabatic by using the steady-state elasticity to describe the spring. Therefore, in general terms, we expect difficulty in predicting the dynamics of polymers at higher $\dot{\varepsilon}$ using the dumbbell model.

While the dumbbell model cannot robustly model the average dynamics, it describes approximately the dynamics of the dumbbell configuration. In this case, there are no internal constraints to slow down the dynamics. But the overall usefulness of this approximate agreement may also be limited since the fractional number of molecules in the dumbbell configuration decreased from 31 % to 20 % as $\dot{\varepsilon}\tau_1$ was increased from 2 to 3.3.

10.4.9.8
Proposed Conformations

The dynamics of polymers in elongational flows has been a challenging theoretical problem for several decades [1–3]. To simplify the full complexity of the dynamics, several different theoretical models have been investigated, the most popular of which is the finitely extensible dumbbell model [1–3]. But, as noted above, this model cannot describe a polymer's internal conformation, which can dramatically effet the dynamics. Therefore, over the last 15 years, several additional models have been developed [22, 71–74]. These models make assumptions about the conformation of the polymer and then derive the consequence of these assumptions. These models were developed, in part, to account for the excess stress measured by James and Saringer [17].

Ryskin developed the "yoyo" model which describes a molecule as two coils connected by a highly taught region. This model is very similar to the observed dumbbell configuration while the half dumbbell model lacks a second coil at one end [71, 72]. Larson and Hinch developed "kink dynamics" models by assuming the polymer is compressed into a one-dimensional object and the dynamics of this object are dominated by its kinked or multiply folded structure [22, 73]. This conformation is similar to our folded configuration, but these simulations in which there were multiple folds were done at much higher values of $\dot{\varepsilon}\tau_1$ (~ 40). King and James proposed a conformation based on internal entanglements [74]. This configuration may be similar to our kinked conformation. But, within the resolution of our microscope, it is not possible to assert whether kinked conformations arise from internal constraints or, rather, they are arise from variation of a half dumbbell conformation in which a filament of taut DNA comes from each side of a coil.

No one of these theories describes the complete range of observed conformations. Rather, the individual conformations assumed in these theories represent one of the several observed conformations. Notwithstanding, the presence of these observed conformations provides a qualitative explanation for the dissipative component of stress found in measurements of η_E.

10.4.10
Limitation of Applicability

The results presented here should not be generalized to polymers in a mixed elongational and shearing flow or to polymers in an elongational flow that were pre-sheared. Our data indicate that the processes involved in the diverse dynamics arise from the variation in t_{onset} and from internal configurations (i.e., folds). In mixed flows, a large fraction of the molecules are partially extended due to shearing and this may eliminate some of the internal constraints that led to the observed dynamics.

However, the dynamics molecules in mixed flows can easily be studied. Preventing preshearing of the molecules along the inlet was the main design criterion for the flow cell (see Sect. 10.3.4). To study pre-sheared molecules, the microscope simply needs to be focused to a different depth. Molecules in a mixed flow can be studied in a similar manner.

10.5
Summary

We have presented the first results of imaging single DNA molecules in an elongational flow where the velocity gradient is uniform and the polymers are isolated. The results reveal complex, heterogeneous dynamics. Such results were previously unobservable in bulk measurement and current theoretical models only capture a portion of the observed complexity. Through these methods, we have illustrated the use of DNA as a model polymer for investigating a long-standing problem in polymer dynamics. In using DNA, one gains a detailed knowledge of the conformation and internal dynamics of individual polymers. This knowledge eliminates many of the ambiguities associated with classical bulk experiments that average over a macroscopic large number of molecules.

Unique to this method we were able to identify those individual molecules that reach steady-state elongations. This analysis has led to the first direct observation of the coil-stretch transition in the steady-state extension as evidenced by the sudden, nonlinear increase in the steady-state extension of polymers at a critical strain rate. This steady-state extension is well characterized by a simple finitely extensible dumbbell model using a Marko-Siggia spring [53] and the critical strain rate agrees with theoretical and numerical calculations [65].

Our data indicate that the concept of a discrete and abrupt coil-stretch transition is limited to the steady-state. One cannot think of polymers undergoing a simple, collective and simultaneous unwinding as soon as $\dot{\varepsilon} > \dot{\varepsilon}_c$. The mismatch between $\langle x\left(t_{res}\right)\rangle$ and x_{steady} implies that the non-Newtonian properties of dilute polymer solutions in most practical elongational flows (where $\dot{\varepsilon}t_{res} < 5.5$) are dominated by the dynamic and not the steady-state properties.

However, even for molecules of identical length and strain history, the dynamics are complex. By visualizing single molecules, we identifed distinct conformational classes with differing dynamics. These conformations provide a qualitative explanation for the high stress observed by James and Saringer [17] and the dissipative stress observed in the measurements of η_e [18, 20, 21]. Furthermore, the variety of conformations and the large variation in t_{onset} imply difficulties with any mean field description of polymers in an elongational flow.

The data presented here should serve as a guide in developing improved microscopic theories for the polymer dynamics and the bulk rheological properties of such solutions.

10.6
Future Prospects

Within the context of elongational flows, the appearance of conformational dependent rate of stretching at $\dot{\varepsilon}\tau_1 > 2$ suggests that these experiments be extended to higher strain rates and longer chains. At higher strain rates, the fraction of chains in each conformation can be measured, while longer chains allow for multiple folds. Furthermore, by positioning a molecule at the stagnation point [86] and by pulsing the pump, the evolution of the conformation from equilibrium can be imaged.

We have confirmed the presence of a sharp coil-stretch transition but de Gennes postulated that polymers in an elongational flow will undergo a hysteresis due to an increase in the drag of an extended state relative to the coiled state. Our present experiment is unable to probe this question. However, the presence of such hysteresis is being experimentally investigated by holding one end of a long DNA molecule at the stagnation point. This geometry mimics co-moving with the fluid and allows for an indefinite residency time.

Imaging single DNA molecules is not limited to a simple system such as an isolated polymer in an elongational flow. By coupling fluorescence microscopy with optical tweezers, one can go beyond passive observation of molecules undergoing Brownian motion by driving the dynamics with forces and stresses applied at the level of single chains and then measure the resulting response. Quantitative measurements of extension [23, 27, 28], internal dynamics [54], and forces [52] are now possible.

For instance, DNA offers the opportunity to measure the non-equilibrium force generated in a polymer. DNA can be attached between two microspheres and held in tension between two optical traps. By monitoring the position of the first trap while the second trap is rapidly moved, the non-equilibrium tension within the chain can be measured [87]. Alternatively, the force exerted on the bead can be measured during the relaxation of a chain from a highly extended state. Such non-equilibrium measurements will be a critical step towards developing an understanding of dynamics of polymers at the very high strain rates.

Moreover, the application of this technique is not limited to single, isolated molecules. For example, we have observed the tube-like motion of individual molecules in concentrated solutions by staining one DNA molecule in a background of unstained DNA [26]. Future prospects include studying concentrated polymers under flow, DNA-based polymer brushes, and polymer dynamics in reduced dimensions [36].

Note added in proof: Additional results at higher strain rates were recently reported in [D.E. Smith, S. Chu, Science 281, 1335–1340 (1998)]. Additionally, these techniques were extended by Smith et al. to study single polymers in a shear flow [D.E. Smith, H.P. Babcock, S. Chu, Science 283, 1724–1727 (1999)].

Acknowledgements. We acknowledge helpful discussions with G. Fuller, D. Hoagland, R. Larson, R. Pecora, and B. Zimm. We thank J. Shott for his generous aid in lithographic design. This work was supported in part by grants from the U.S. AFOSR, the NSF, and the Human Frontiers Foundation. We acknowledge the generous assistance of J. Spudich, including support through the NIH grant GM33289 to J.S. D.E.S was supported by a Fellowship from the Program in Mathematics and Molecular Biology at the Univ. of California Berkeley through NSF grant DMS 9406348. S.C. was funded, in part, by the Guggenheim-foundation.

List of Symbols and Abbreviations

v_y	Velocity in $\hat{y}$
$\dot{\varepsilon}$	Elongational velocity gradient, $\dot{\varepsilon} \equiv \bar{\partial}v_y/\bar{\partial}y$
$\dot{\varepsilon}_c$	Critical strain rate
τ_1	Longest relaxation time
$\dot{\varepsilon}\tau_1$	Dimensionless strain rate or Deborah number
R_{bead}	Radius of bead in the dumbbell model
L	Polymer's contour length
F_{hydro}	Hydrodynamic force exerted on the polymer/dumbbell
F_{spring}	Entropic elasticity of polymer/dumbbell
k_{spring}	Spring constant characterizing a polymer's linear elasticity
x	Visual extension of the polymer
η	Viscosity
t_{res}	Residency time in elongational flow
ε	Accumulated fluid or Henky strain, $\varepsilon = \dot{\varepsilon}t_{res}$
R_G	Radius of gyration
η_E	Extensional viscosity
p	Persistence length
F_{MS}	Marko-Siggia force law for the elasticity of a worm-like chain (DNA)

λ-DNA   DNA from lambda bacteriophage
$\dot{\gamma}$ Shear rate, $\dot{\gamma} \equiv \bar{\sigma}v_x/\sigma z$
Wi Weissenberg numer, $Wi \equiv \dot{\gamma}\tau_1$
v_{inlet} Velocity at the center of flow in the $\hat{x}$ direction
z_{center} Distance from the top surface to the middle of the flow
Tris HCl Trizma hydrochloride, tris[Hydroxymethyl]aminomethane hydrochloride
EDTA   Ethylenediamine tetraacetic acid
σ Stress in the fluid
n Density of polymers
R End-to-end distance
$t_{res}^{initial}$ Polymer's t_{res} when it is first visualized at the edge of the imaging area
x_{onset} Position of the onset of the elongational flow field
x_{screen} Positon of the edge of the imaging area
x_{steady} Steady-state extension at a particular strain rate
x_{bulk} Spatial-temporal average of all extension measurements

References

1. de Gennes PG (1974) J Chem Phys 60:5030
2. Larson RG (1988) Constitutive equations for polymer melts and solution. Buttersworths, New York
3. Bird RB, Curtiss CF, Armstrong RC, Hassager O (1987) Dynamics of polymeric liquids, 2nd edn, vol 2. Wiley, New York
4. Larson RG, Magda JJ (1989) Macromolecules 22:3004
5. Keller A, Odell JA (1985) Coll & Polym Sci 263:181
6. Fuller GG, Leal LG (1980) Rheol Acta 19:580–600
7. Cathey CA, Fuller GG (1990) J Non-Newtonian Fluid Mech 34:63–88
8. Dunlap PN, Leal LG (1987) J Non-Newtonian Fluid Mech 23:5–48
9. Atkins EDT, Taylor MA (1992) Biopolymers 32:911–23
10. Odell JA, Keller A, Rabin Y (1988) J Chem Phys 88:4022–4028
11. Odell JA, Keller A, Muller AJ (1992) Coll & Polym Sci 270:307–324
12. Odell JA, Taylor MA (1994) Biopolymers 34:1483–1493
13. Menasveta MJ, Hoagland D (1991) Macromolecules 24:3427–3433
14. Smith KA, Merrill EW, Peebles LH, Banijamali SH (1975) Colloq Int C.N.R.S. 233:341
15. Lumley JL (1977) Phys Fluids 20:64
16. Nguyen TQ, Yu G, Kausch H-H (1995) Macromolecules 28:4851–4860
17. James DF, Saringer JH (1980) J Fluid Mech 97:655–671
18. Tirtaatmadja V, Sridhar T (1993) J Rheol 37:1081–1102
19. James DF, Sridhar T (1995) J Rheol 39:713–724
20. Spiegelberg SH, McKinely GH (1996) J Non-Newtonian Fluid Mech 67:49–76
21. Orr NV, Sridhar T (1996) J Non-Newtonian Fluid Mech 67:77–103
22. Hinch EJ (1994) J Non-Newtonian Fluid Mech 54:209–230
23. Perkins TT, Smith DE, Chu S (1997) Science 276:2016–2021
24. Ashkin A, Dziedzic JM, Bjorkholm JE, Chu S (1986) Optics Lett 11:288–290
25. Chu S (1991) Science 253:861–866
26. Perkins TT, Smith DE, Chu S (1994) Science 264:819–822
27. Perkins TT, Quake SR, Smith DE, Chu S (1994) Science 264:822–826
28. Perkins TT, Smith DE, Larson RG, Chu S (1995) Science 168:83–87
29. Crocker JC, Grier DG (1994) Phys Rev Lett 73:352–355

30. Finer JT, Simmons RM, Spudich JA (1994) Nature 368:1134–1139
31. Svoboda K, Schmidt CF, Schnapp BJ, Block SM (1993) Nature 721–727
32. Yin H et al. (1995) Science 270:1653–1657
33. Smith SB, Finzi L, Bustamante C (1992) Science 258:1122–1126
34. Smith SB, Aldridge PK, Callis JB (1989) Science 243:203–206
35. Schwartz DC, Koval M (1989) Nature 338:520–522
36. Volkmuth WD, Austin RH (1992) Nature 358:600–602
37. Volkmuth WD, Duke T, Wu MC, Austin RH (1994) Phys Rev Lett 72:2117–2120
38. Mitnik L, Heller C, Prost J, Viovy JL (1995) Science 267:219–222
39. Dinsmore AD, Yodh AG, Pine DJ (1996) Nature 383:239–242
40. Hagerman PJ (1988) Annual review of biophysics and biophysical chemistry 17:265–286
41. Eimer W, Pecora R (1991) J Chem Phys 94:2324–2329
42. Pecora R (1991) Science 251:893–898
43. de Gennes PG (1979) Scaling concepts in polymer physics (Cornell University Press, Ithica)
44. Doi M, Edwards SE (1986) The theory of polymer dynamics (Oxford Press, New York)
45. Smith DE, Perkins TT, Chu S (1995) Macromolecules 29:1372–1373
46. Crothers DM, Zimm BH (1965) J Mol Biol 12:525
47. Callis PR, Davidson N (1969) Biopolymers 8:379–390
48. Klotz LC, Zimm BH (1972) J Mol Biol 72:779–800
49. Schmitz KS, Pecora R (1975) Biopolymers 14:521–542
50. Musti R, Sikorav J-L, Lairex D, Jannink G, Adam M (1995) Comptes Rendus De L'Academie des Sciences Series II 352:599–605
51. Larson RG, Perkins TT, Smith DE, Chu S (1997) Phys Rev E55:1794–1797
52. Bustamante C, Marko JF, Siggia ED, Smith SB (1994) Science 265:1599–1600
53. Marko JF, Siggia ED (1995) Macromolecules 28:8759–8770
54. Quake SR, Babcock HP, Chu S (1997) Nature 388:151–154
55. Zimm BH (1998) Macromolecules 31:6089–6098
56. Vologodskii A (1994) Macromolecules 27:5623–5625
57. Kas J et al. (1996) Biophys J 70:609–625
58. Matsumoto S, Morikawa K, Yanigida M (1981) J Mol Biol 110:501
59. Morikawa K, Yanigida M (1981) J Biochem 89:693–696
60. Wallis G, Pomerantz D (1969) J Appl Phys 40:3946–3949
61. Hu C, Kim S (1976) Appl Phys Lett 29:582–585
62. Petersen KE (1982) Proceedings of the IEEE 70:420–457
63. Rye HS, Dabora JM, Quesada MA, Mathies RA, Glazser AN (1993) Anal Biochem 208:144–150
64. Thompson DS, Gill SJ (1967) J Chem Phys 47:5008–5017
65. Magda JJ, Larson RG, Mackay ME (1988) J Chem Phys 89:2504–2513
66. Daoudi S, Brochard F (1978) Macromolecules 11:751–758
67. de Gennes PG (1986) Physica 140A:9–25
68. Hinch EJ (1977) Phys Fluids 20:522
69. Rallison JM (1997) J Non-Newtonian Fluid Mech 68:61–83
70. Doyle PS, Shaqfeh ESG, Gast AP (1997) J Fluid Mech 334:251–291
71. Ryskin G (1987) Phys Rev Lett 59:2059–2062
72. Ryskin G (1987) J Fluid Mech 178:423–449
73. Larson RG (1990) Rheol Acta 29:371–384
74. King DH, James DF (1983) J Chem Phys 78:4749–4754
75. Carrington SP, Odell JA (1996) J Non-Newtonian Fluid Mech 67:269–283
76. Rabin Y (1988) J Non-Newtonian Fluid Mech 30:119–123
77. Bensimon D, Simon AJ, Croquette V, Bensimon A (1995) Phys Rev Lett 74:4754–4757
78. Cluzel P et al. (1996) Science 271:792–794
79. Smith SB, Cui Y, Bustamante C (1996) Science 271:795–799
80. Tirtaatmadja V, Sridhar T (1995) J Rheol 39:1133–1160

81. Zimm BH (1956) J Chem Phys 24:269–278
82. Rouse PE (1953) J Chem Phys 21:1272
83. de Gennes PG (1971) J Chem Phys 55:572–579
84. Peterlin A (1961) Makro Chem 44
85. Keunings R (1997) J Non-Newtonian Fluid Mech 68:85–100
86. Smith DE, Perkins TT, Chu S (1996) Macromolecules 29:1372–1373
87. Seifert U, Wintz W, Nelson P (1996) Phys Rev Lett 77:5289–5292

CHAPTER 11

The Rheology of Polymer Solutions in Porous Media [1]

A. J. Müller and A. E. Sáez

11.1
Introduction

The flow of polymer solutions through porous media is relevant to a wide variety of practical applications in processes such as enhanced oil recovery from underground reservoirs, gel permeation chromatography, and filtration of polymer solutions. At this point in time, there exist many aspects of the non-Newtonian effects in the flow of polymer solutions through porous media that are not understood well enough for a quantification for modeling purposes to be effected. This is basically due to the intrinsic difficulties of characterizing non-Newtonian effects in the flow of polymer solutions and to the complexities of the local geometry of the porous medium, which give rise to a flow field that is far from being ideal shear or ideal extension, for which the understanding of polymer solution rheology has greatly advanced in the last three decades. It is evident that a substantial improvement in the actual level of knowledge will result in important technological implications for the applications mentioned.

One of the most interesting and studied aspects of the flow of polymer solutions through porous media is the increase in flow resistance obtained beyond a critical flow rate for solutions of flexible polymers. This increase has been attributed to the extensional nature of the flow field in the pores caused by the successive expansions and contractions that a fluid element experiences as it traverses the pore space, and by the presence of multiple stagnation points. Even though the flow field at the pore level is not an ideal extensional flow due to the relatively large lateral velocity gradients that occur close to the walls of the pores, which impose an important degree of shear and rotation, the increase in flow resistance can be referred to as an extension thickening effect. The extension thickening nature of the flow of polymer solutions through porous media will be a central point in this chapter.

Flexible polymers are used in many applications that involve the flow of polymer solutions through porous media. A particularly interesting application is the use of high molecular weight polymers in enhanced oil recovery [1-3], where both flexible polymers (typically polyacrylamides in brine) and semi-rigid polymers (typically polysaccharides) are employed. In petroleum production, polymers are used mainly for two specific purposes. The first use is as

---

1 The authors wish to dedicate this chapter to the memory of Prof. Andrew Keller.

part of a strategy for near-well treatments, in which the performance of water flooding operations is improved by using a polymer solution to block high permeability zones adjacent to the wells. In some cases this blocking is a consequence of the relatively high apparent viscosity of the polymer solution employed, whereas in other cases cross-linking agents are used in situ to cause gelation of the polymer and thus a complete blockage. The second use is as agents that lower the mobility of the aqueous phase in displacement operations. It is widely accepted that the reduction of the mobility is caused by the increase in the apparent viscosity of the aqueous phase brought about by the presence of the polymers.

The second application mentioned is termed polymer flooding, a process in which a polymer solution is injected in the reservoir to displace movable oil in the formation. Polymers are also used in what is known as micellar-polymer flooding, in which the polymer solution is used to drive a surfactant solution into the reservoir whose purpose is to lower the interfacial tension between the aqueous phase and the residual oil with the object of enhancing recovery [2].

It has been argued that the extension thickening nature of the flow of flexible polymer solutions in porous media has no practical impact on oil recovery operations [1]. In fact, specialized monographs that deal with enhanced oil recovery [2, 3] do not establish a link between extension thickening and oil recovery efficiency with polymer solutions. The reason for this is the relatively low strain rates encountered in reservoir flows, which are thought to be much lower than those required for the onset of extension thickening effects. On the other hand, a simple analysis might lead one to think that extension thickening would not be advantageous in oil recovery operations since in this process typically the highest strain rates are obtained near the injecting wells, and the lowest strain rates are present deep into the reservoir where the viscosifying effect of the displacing phase is more important. However, some authors have remarked on the advantages of the extension thickening. For example, Slater and Farouq Ali [4] argued that extension thickening might improve oil production by retarding the flow of the displacing phase in regions of high permeability (which typically exhibit large strain rates) thus avoiding the by-passing of low permeability, oil rich regions by the displacing phase. A more thorough exposition of arguments along these lines was presented by Savins [5]. Recently, Gleasure and Phillips [6] have studied injectivity and oil displacement with poly(ethylene oxide) solutions under conditions for which extension thickening occurs.

This chapter will follow the development of the works in the literature that have studied the non-Newtonian flow of polymer solutions through porous media with special emphasis on those works that have attempted to explain the observed phenomena in terms of molecular behavior. An extensive discussion on the current state of the several molecular hypotheses that have been advanced will be the main thrust of the central section of the review. We believe that an understanding of macromolecular dynamics at the pore level is essential to obtain a comprehensive view of the physics of the process.

This review is organized as follows. We start in Sect. 11.2 with a brief description of the characterization of the fluid dynamics of porous media flows, placing emphasis on Newtonian flows. Section 11.3 presents an extensive review of the non-Newtonian behavior of the flow of polymer solutions through porous media. Since the main emphasis of this work is on the extension thickening behavior, we start by establishing the mechanisms that have been proposed in the literature as the underlying causes of the observed phenomena, and continue with an assessment of the effects that the main independent parameters have on this behavior, focusing on a molecular view point. To end Sect. 11.3, we analyze the behavior of semi-rigid polymers, which exhibit apparent shear thinning in porous media flows. Section 11.4 is devoted to studies of flow-induced degradation of the macromolecules in the porous medium. In Sect. 11.5 we analyze less conventional aspects of the flow of polymer solutions through porous media, including the flow of solutions of polymer blends and the effect of the presence of cross-linking agents on the fluid dynamic behavior of the polymer solutions.

11.2
Fluid Dynamics Characterization of Porous Media Flows

The traditional standpoint in the study of fluid dynamics through porous media has been to treat the medium as a pseudo-continuum in which macroscopic properties are assumed to exhibit smooth spatial variations over length scales which are much greater than the size of the pores. If we denote by L a characteristic length of the macroscopic scale and by l a characteristic length associated with pore size, then it is assumed that $l \ll L$. This approach avoids the tremendous practical difficulties involved in the description of point variables, owing to the complexity of the solid phase geometry at the local or pore level. The macroscopic variables used in the characterization of the pseudo-continuum are usually volume averages of point variables over averaging volumes whose size (characterized by a length r_0) is typically much greater than the pore size but much smaller than the scale of observation ($l \ll r_0 \ll L$). In the hydrodynamics of one-phase flow through porous media, the main macroscopic variables are the superficial velocity vector (v), defined as the average of the point velocity vector over an averaging volume, and the volume-averaged modified pressure (P), which includes gravitational potential contributions. When conservation principles are expressed in terms of macroscopic variables, they contain effects that are a consequence of interactions that occur at the pore level and thus are usually characterized in the pseudo-continuum from a phenomenological standpoint. In this section we will present a brief description of the parameters required for the characterization of the flow of Newtonian fluids through porous media.

The basic theory of the flow of Newtonian fluids through porous media is well established. Incompressible viscous fluids in isothermal flows satisfy the continuity and Navier-Stokes equations at the pore level. Any macroscopic representation of the principles of mass and momentum conservation should therefore be a consequence of these equations.

We can define a Reynolds number based on the pore scale as

$$Re_p = \frac{\varrho v_p l}{\mu} \qquad (11.1)$$

where ϱ and μ are the density and viscosity of the fluid, respectively, v_p is the interstitial velocity of the fluid, which is related to the superficial velocity by $v_p = v/\phi$, where ϕ is the porosity of the medium. When this Reynolds number is sufficiently small (strictly when $Re_p \ll 1$), inertial effects are negligible at the pore level and if the flow is at steady state, the macroscopic representation of the conservation principles is given by the following equations [7]:

– macroscopic continuity equation

$$\nabla \cdot \boldsymbol{v} = 0 \qquad (11.2)$$

– Darcy's law

$$\boldsymbol{v} = -\frac{K}{\mu} \cdot \nabla P \qquad (11.3)$$

where μ is the viscosity of the fluid and K is the permeability tensor. It has been well established that the permeability tensor is only a function of the geometrical structure of the porous medium at the pore level. When the porous medium is isotropic with respect to the flow process, the permeability becomes a scalar tensor and Eq. (11.3) simplifies to

$$\boldsymbol{v} = -\frac{K}{\mu} \nabla P \qquad (11.4)$$

If the pore Reynolds number is of O(1) or larger, inertial effects in the pores become important and Darcy's law is no longer applicable. In this case, empirical evidence supports the use of the vectorial form of Forchheimer's equation for isotropic flow processes, given by

$$-\nabla P = \frac{\mu}{K} \boldsymbol{v} + \frac{\varrho}{M} v\boldsymbol{v} \qquad (11.5)$$

where ϱ is the fluid density and M is the inertial parameter, which only depends on the local geometry.

For packed beds of spherical particles, K and M have been correlated in terms of the porosity and the particle size as follows:

$$K = \frac{d^2 \phi^3}{A(1 - \phi)^2} \qquad (11.6)$$

$$M = \frac{\phi^3 d}{B(1 - \phi)} \tag{11.7}$$

where d is the particle diameter, and A and B are empirical constants.

The particular form of Eq. (11.5) for one-dimensional flow in the x direction is

$$-\frac{dP}{dx} = \frac{\mu}{K} v + \frac{\varrho}{M} v^2 \tag{11.8}$$

The substitution of Eqs. (11.6) and (11.7) into Eq. (11.8) leads to the well-known Ergun equation. In the original empirical relation, the parameters A and B were set to: $A = 150$, $B = 1.75$ for disordered packings [8]. Later, after a study of a large amount of previous data, Macdonald et al. [9] suggested that the best fit was provided by $A = 180$, $B = 1.8$, although both parameters are sensitive to the procedure employed to pack the particles.

The relation between pressure gradient and superficial velocity given by Eq. (11.8) is commonly represented in dimensionless form in terms of the resistance coefficient, which is a dimensionless ratio between pressure drop and superficial velocity, defined by

$$\Lambda = \frac{d^2 \phi^3 (\Delta P/L)}{\mu v (1 - \phi)^2} \tag{11.9}$$

In this equation, ΔP is the pressure drop over a length L of porous medium. If the Reynolds number is defined as

$$Re = \frac{\varrho v d}{\mu (1 - \phi)} \tag{11.10}$$

then the Ergun equation can be written as

$$\Lambda = A + BRe \tag{11.11}$$

According to Eq. (11.11) the resistance coefficient is a linear function of Reynolds number. For low values of this parameter (typically $Re < 1$), the constant term dominates (A). This is the Darcian regime. At higher Reynolds number, when inertial effects at the pore level become important, the second term (BRe) becomes appreciable. An illustration of this behavior is shown in Fig. 11.1 for water in two packed beds with monodisperse spheres of different sizes. The curve obtained is correlated by

$$\Lambda = 134 + 1.7 \, Re \tag{11.12}$$

These values of A and B are within the range commonly encountered in sphere packings [9].

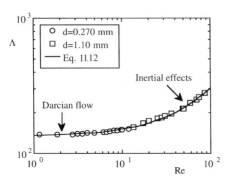

Fig. 11.1. Resistance coefficients for water flowing through disordered monodisperse sphere packings. Data from Sargenti et al. [10]

The results presented in Fig. 11.1 show that the effect of particle diameter on pressure drop for disordered monodisperse sphere packings is well represented by the empirical Ergun equation.

In the macroscopic representation of Newtonian flows, a relation such as Eq. (11.11) is enough for a characterization of the hydrodynamics of the process, since the main interest is the description of the relation between pressure drop (and thus energy consumption) and total flow rate. In non-Newtonian flows, however, features of the local flow field that are not reflected on superficial velocity and average pressure gradients might be important. In the flow of solutions of flexible polymers there occur changes in macromolecular conformation which are induced by the local flow field. As will be discussed below, the macroconformation changes are primarily induced by the elongational nature of the flow field at the pore level, which is a consequence of the existence of periodic contractions and expansions in the flow, and the presence of stagnation points.

11.3
Non-Newtonian Behavior in the Flow of Polymer Solutions Through Porous Media

Polymer solutions exhibit non-Newtonian behavior in flow through porous media. The nature of the non-Newtonian behavior is dramatically affected by parameters such as chain conformation, polymer concentration, and molecular weight. In terms of resistance coefficient vs Reynolds number, four types of behavior have been reported. They are illustrated in Fig. 11.2. In what follows, Λ and Re will represent the resistance coefficient and Reynolds number, respectively, calculated by using the solvent physical properties (density and viscosity). In Fig. 11.2c, the resistance coefficient and Reynolds number are calculated in terms of a zero-shear rate viscosity, which is assumed to be a linear function of polymer concentration for dilute solutions [13]. They are denoted by Re^* and Λ^*, and are defined as follows:

$$\Lambda^* = \frac{d^2 \phi^3 (\Delta P / L)}{\eta_s (1 + C[\eta]) \, \upsilon (1 - \phi)^2} \tag{11.13}$$

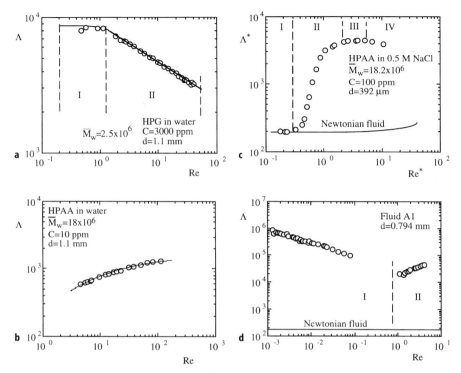

Fig. 11.2 a–d. Non-Newtonian behavior observed when polymer solutions flow through disordered packings of spherical particles: **a** shear thinning data from [11]; **b** non-critical extension thickening data from [12]; **c** critical extension thickening data from [13]; **d** shear thinning and extension thickening data from [14]

$$Re^* = \frac{\varrho v d}{\eta_s (1 + C[\eta]) (1 - \phi)} \tag{11.14}$$

where η_s is the viscosity of the solvent, C is the polymer concentration and $[\eta]$ is the intrinsic viscosity of the polymer.

The characteristics and occurrence of each of the four types of behavior are summarized in what follows.

(a) Shear thinning behavior (Fig. 11.2a). This behavior is typical of semi-rigid polymers that adopt an expanded coil conformation in solution at moderate to high polymer concentrations. It is characterized by a resistance coefficient curve that is divided in two regions. Region I (low Re) is pseudo-Newtonian: the resistance coefficient does not vary with Reynolds number, and region II (moderate to high Re) is a shear thinning region that can be described by power law models. A third region at higher Reynolds numbers in which a second pseudo-Newtonian plateau is reached has also been reported [15]. Behavior such as that exemplified in Fig. 11.2a has been observed in aqueous solutions of hydroxyethyl cellulose [16], carboxymethyl cellulose [17, 18], hydroxypropyl guar (HPG) [11, 19], xanthan gum [15, 18, 20–22], and sclero-glucan [23].

(b) Gradual extension thickening (Fig. 11.2b). This behavior has been observed in hydrolyzed polyacrylamide (HPAA) solutions in deionized water, for polymers with molecular weights greater than 8×10^6 [12, 24]. Presumably this behavior would be characteristic of very high molecular weight polymers that exhibit an expanded coil conformation in solution. This point will be further discussed in Sect. 11.3.2.

(c) Critical extension thickening (Fig. 11.2c). This behavior is typical of high-molecular weight flexible polymers. It contains four different regions. In region I, at low Reynolds number, the solution behaves as a Newtonian fluid (if the polymer concentration is low enough the resistance coefficient in this region is identical to that of the solvent). In region II there is a sudden increase in flow resistance beyond a critical Reynolds number, denoted onset Reynolds number (Re_o). In region III, at even higher Reynolds numbers, the value of the resistance coefficient becomes practically constant, reaching a pseudo-Newtonian plateau. In region IV, the resistance coefficient decreases as the Reynolds number increases. The critical extension thickening has been extensively studied in the literature, although most of the works report results in regions I and II. Some polymer solutions that exhibit this behavior are: aqueous poly(ethylene oxide) (PEO) [1, 25–32], aqueous solutions of polyacrylamide and hydrolyzed polyacrylamide in ionic environments [1, 5, 12, 13, 24, 27, 33–38], polyisobutylene (PIB) in various organic solvents [1, 34], and atactic polystyrene (aPS) in various organic solvents [13, 39].

(d) Shear thinning and extension thickening (Fig. 11.2d). This behavior is a combination of (a) and (c). At low Reynolds numbers, a decrease in resistance coefficient is observed (region I) and, after an onset Reynolds number, extension thickening sets in (region II). This behavior is characteristic of flexible polymers in concentrated solutions. It has been observed for solutions of low molecular weight PIB in decalin [14], solutions of HPAA in ionic environments [40], and solutions of mixtures of a semi-rigid polymer (HPG) and a flexible one (PEO) [41] (see Sect. 11.5).

The four types of behavior described above are a result of the combination of three basic characteristics of the flow of polymer solutions: Newtonian behavior, shear thinning, and extension thickening. In what follows we will concentrate on aspects related to extension thickening.

The first investigations that reported substantial increases in flow resistance beyond critical Reynolds numbers when solutions of high molecular weight flexible polymers were made to flow through porous media are those of Pye [33], Sandiford [42], Jones and Maddock [43], Gogarty [44], Dauben and Menzie [25], and Marshall and Metzner [34]. The work of Savins [5] provides a comprehensive review of the literature prior to 1969. As pointed out by Marshall and Metzner [34] most of the earlier works concentrated on flow through porous media at flow rates low enough that viscous effects dominate the flow.

In this section, we start by discussing the possible mechanisms that lead to extension thickening in porous media flows of polymer solutions. Afterwards we present a summary of the effects of independent parameters such as polymer concentration, polymer molecular weight, solvent quality, and ionic environment on the extension thickening behavior. We then assess the effect of the microstructure of the porous medium on the extension thickening. Finally, we briefly discuss aspects related to the apparent shear thinning behavior of polymer solutions in porous media flows.

11.3.1
Review of Mechanisms Proposed to Explain Extension Thickening

The increase in flow resistance with Reynolds number in porous media flows is a consequence of the extensional nature of the flow field at the pore level. The first attempts to interpret the increase in flow resistance did not recognize this fact explicitly, but approached the analysis from a macroscopic standpoint. Marshall and Metzner [34] argued that these viscoelastic effects were produced by increased elasticity as the Deborah number increases. The Deborah number is defined by

$$De = \frac{\lambda}{t_c} \tag{11.15}$$

where t_c is a characteristic time of the flow and λ is the relaxation time of the fluid. Marshall and Metzner defined t_c as

$$t_c = \frac{d\phi}{v} \tag{11.16}$$

where the ratio v/ϕ is the interstitial velocity of the liquid in the porous medium. The relaxation time of the liquid was determined from measurements of the first normal stress difference $(\tau_{11} - \tau_{22})$ and the shear stress (τ_{12}) as a function of shear rate $(\dot{\gamma})$ in simple shear experiments by means of the use of a convected Maxwell constitutive model [45]

$$\lambda = \frac{1}{\dot{\gamma}} \frac{\tau_{11} - \tau_{22}}{\tau_{12}} \tag{11.17}$$

Marshall and Metzner argued that, as the Deborah number reaches and exceeds values of $O(1)$, elastic effects set in and they lead to a large increase in the pressure drop. A similar interpretation was followed by Michele [36] and Franzen [46] to analyze experimental results on the flow of hydrolyzed polyacrylamide solutions through disordered packings of spheres and structural models of periodic packings of spheres, respectively.

Dauben and Menzie [25] are the first investigators, to our knowledge, to try to explain the extension thickening effect in terms of macromolecular conformation changes. They argued that the macromolecules passed from a spherical coil conformation to an ellipsoid due to the application of stresses in the flow. This generates normal stresses that, according to them, lead to higher pressure drops. This interpretation was later proved to be inadequate by James and McLaren [26] who, based on a detailed analysis of the flow field at the pore level, concluded that normal stresses due to shear would result in negligible changes in the pressure drop. Furthermore, the effect of normal stresses on the solid particles would lead, according to James and McLaren, to a decrease in pressure drops, if pure extensional flow is considered to occur in the pore space. In fact, James and McLaren attribute the decrease in resistance coefficient with Reynolds number in region IV (Fig. 11.2c) partly to this effect. They concluded that the extensional nature of the flow field at the pore level was crucial in explaining the increase in flow resistance.

It is interesting to point out that Dauben and Menzie [25] acknowledged that the extensional nature of the flow at the pore level (based on the existence of successive expansions and contractions) might be partly responsible for the increase in flow resistance, although they attribute only a slight impact to this fact.

Jones and Maddock [47] associated extension thickening in porous media flows to a destabilization of the flow attributed to the existence of stretched chains in the flow field. They argue that partial stretching of the macromolecules is enough to produce the onset of extension thickening. Even though this explanation has not been pursued further, their work is the first to invoke chain stretching and kinetic theory arguments to explain extension thickening.

Maerker [48] recognized the role of macromolecular deformation induced by the elongational nature of the flow in the pores. Maerker pictures the randomly coiled flexible molecules in solution as being entangled with one another or with themselves; when these molecules are made to flow through a constriction Maerker postulates that they are stretched. Furthermore, Maerker considers that very large stresses will only build up at sufficiently large strain rates when the time scale of deformation is short and the molecules do not have time to disentangle. These arguments only seem appropriate when the solution is in the semi-dilute regime at equilibrium but they bear a similarity to more recent hypotheses used to explain the extension thickening behavior, as will be established later in this section.

Theoretical considerations by de Gennes [49] and Hinch [50] predict that high molecular weight flexible molecules should undergo a coil-stretch transition in elongational flow fields, when the applied strain rate exceeds a critical value which is of the order of the inverse of the longest relaxation time of the coil. According to de Gennes a sudden transition is expected, which is consistent with the hysteresis of molecular relaxation time with chain extension that arises due to the change of the draining characteristics of the stretching molecules. In other words, sudden extension implies a sudden conformational change with the random coils greatly aligned in the elongational flow direction, changing their draining characteristics and causing the frictional contact between inner chain segments and the solvent to increase dramatically (i.e., changing from a non-free draining coil to a free draining coil). It is therefore expected that a coil-stretch transition will lead to an increase in the elongational viscosity of the solution, a fact that has been experimentally verified in ideal elongational flow situations, such as in the flow through opposed jets [51].

Elata et al. [28] invoked the coil-stretch transition theory to explain the sudden increase in flow resistance observed for PEO solutions in porous media. The authors acknowledged, however, that the coil-stretch theory could not explain the observed dependence of onset strain rates for extension thickening on polymer concentration. If the solutions were truly dilute, the critical strain rate for coil-stretch transition would be independent of polymer concentration. The experimental data obtained by Elata et al. showed that the onset strain rate decreased with an increase in polymer concentration.

Naudascher and Killen [29] performed porous media flow experiments with PEO solutions which they interpreted by means of an analysis that closely followed a suggestion presented previously by Batchelor [52]. Batchelor argued that the stress levels that had been commonly observed for polymer solutions in elongational flows were too large to be ascribed to the stretching of isolated macromolecules. He proposed that hydrodynamic interactions between stretched macromolecules could account for the observations. In the same publication, Batchelor [52] developed a model to calculate stresses in fiber suspensions subjected to ideal extensional flow by including fiber-fiber interactions. The model shows that substantial increases in elongational viscosities can be obtained at very low fiber concentrations, provided that the fibers are aligned in the stretching direction, and that their length is much larger than the mean lateral separation distance between fibers. Naudascher and Killen [29] proposed that the extension thickening in porous media flows of solutions of flexible polymers was a consequence of the hydrodynamic interactions among stretched and aligned macromolecules in the manner described by Batchelor's model for fiber suspension. Even though this theory could predict the scaling between onset strain rates and polymer concentration (see Sect. 11.3.2), it predicts an extension thickening behavior that is more gradual with respect to changes in strain rates than what is observed experimentally.

After the first attempts to characterize the mechanisms behind extension thickening described above, a series of work invoked the coil-stretch theory as the preferred explanation for this behavior. This explanation was adopted by making the analogy that the flow at the pore level could be treated in the same manner as idealized elongational flow. Following this line of thought, Durst and Haas [53] proposed that the coil-stretch theory could be applied for quantification purposes by assuming that the macromolecules behaved as dumbbells. Using the finitely extensible nonlinear elastic dumbbell (FENE) model [54–56], they employ the following definition of relaxation time:

$$\lambda = \frac{[\eta]\,\eta_s M}{AkT} \tag{11.18}$$

where k is Boltzmann's constant, A is Avogadro's number, M is the molecular weight of the polymer, and T is the temperature.

The characteristic time of the flow through the porous medium is the inverse of an average strain rate, defined as

$$\dot{\varepsilon} = \frac{1}{t_c} = k_1 \frac{\upsilon}{d} \tag{11.19}$$

where k_1 is a geometric parameter that depends on the microstructure of the porous medium. These parameters are used to define the Deborah number, according to Eq. (11.15).

Durst and Haas [53] performed experiments with polyacrylamides in flow through porous media varying the polymer concentration, polymer molecular weight, particle diameter, and solvent quality. They reported that a representation of the results in terms of an effective extensional viscosity (proportional to resistance coefficient) as a function of the Deborah number, as defined above, led to curves with an onset point which they claimed to be approximately the same. They used k_1 as an adjustable parameter so that the onset of extension thickening occurred at $De = 0.5$, which corresponds to the expectation of the dumbbell theory [56]. A value of $k_1 = 4$ was found. In later works [37, 57] they found that the experimental data gathered by Durst and Haas [53] and Durst et al. [24] led to $De = 0.5$ at onset if a value $k_1 = 8$ was used. However, it is necessary to point out that the spread of the experimental data around this onset point spanned almost an order of magnitude, if all the independent variables studied (concentration, molecular weight, particle size, temperature, and solvent quality) are taken into account.

Hoagland and Prud'homme [58] observed that the apparent macromolecular size in hydrodynamic chromatography of polymer samples in porous columns depends strongly on flow rate. Following Kulicke and Haas [37], they used the coil-stretch transition theory to interpret these results.

The use of a single Deborah number as a correlation parameter for extension thickening has been challenged for various reasons. Deiber and Schowalter [59] recognized that a single Deborah number cannot account for all the effects that the local geometry might have on the elongational nature of the flow field. In calculations involving the simulation of the flow of a Maxwell fluid through a tube with sinusoidal axial variations in diameter, they found that the critical Deborah number for the onset of elastic effects depended strongly on the amplitude of the diameter oscillations. A single value of the critical Deborah number can account for neither the molecular weight distribution of the polymer nor the spatial distribution of local strain rates encountered in a porous medium. Most of the experiments reported in the literature have dealt with polymers of polydispersity greater than 2.5. On the other hand, wide distribution of elongation rates exists even for model porous media with the simplest structures such as regular arrays of particles and periodically constricted tubes. The distribution is more complex in porous media of practical relevance [60]. It has been widely recognized that a necessary condition for the achievement of a coil-stretch transition is the fact that the macromolecules require a sufficient amount of molecular strain. Some investigators have argued that the use of a Deborah number does not take into account the residence time of the macromolecules under strain rates beyond that required for the coil-stretch transition [61–63]. Skartsis et al. [64] have proposed the use of an alternative Deborah number that considers only the extensional component of the velocity gradient tensor, based on an analysis performed by Dunlap and Leal [65] in extensional flows. However, their approach does not lead to a substantial improvement in the correlation of extension thickening in flow through beds of fibers.

Odell et al. [66] proposed an alternative explanation for the extension thickening behavior of flexible polymers in porous media flows. In experimental results obtained in opposed-jets flow with hydrolyzed polyacrylamide at concentrations within the typical ranges used in porous media flows, they determined that after the macromolecules underwent a coil-stretch transition, the solution exhibited only small departures in flow resistance with respect to the solvent. On the other hand, beyond a critical strain rate, strong extension thickening effects were observed. These effects only occurred when transient networks of polymer molecules were formed in the flow, arising from entanglements becoming mechanically effective at time scales shorter than the network disentanglement time. Odell et al. proposed that the degree of extension thickening observed in porous media flows was more consistent with the formation of transient entanglement networks than with the occurrence of only a coil-stretch transition.

Rodríguez et al. [32] performed experiments in porous media flows with PEO solutions in a wide concentration range. They interpreted the extension thickening on the basis of the transient network hypothesis. They provide quantitative information that supports the transient network hypothesis and invalidates the coil-stretch theory as the only explanation of the extension thickening observed in porous media flows (see Sect. 11.3.2). The argument is based on two facts. First of all, they show that the strong dependence of onset strain rate on polymer concentration is consistent with that observed in opposed-jets flow when transient entanglement networks are formed. The magnitude of the change of the onset strain rate with concentration is much larger when networks are present than when the macromolecules are simply stretched in isolation in opposed jets flow. Second, a quantification of the criticality of the increase in resistance coefficient with strain rate after the onset of extension thickening is too abrupt to be explained by the spectrum of relaxation times of the polymer resulting from its molecular weight distribution.

It is worthwhile to point out that most of the work previously mentioned has used polydisperse polymers. Sáez et al. [39] employed closely monodisperse samples of polystyrene with various molecular weights, dissolved in organic solvents. The use of monodisperse polymers provides a more meaningful comparison between the results obtained in ideal elongational flow experiments and porous media flow. Sáez et al. compared the extension thickening obtained in porous media flows with pressure drop measurements performed in opposed jets. They found that sizable increases in resistance coefficient in porous media flow could only be observed at concentrations greater than the overlap concentration required for the formation of transient entanglement networks (C^+), as detected from opposed jets rheometry [51]. This concentration has been found to be much lower than the static coil overlap concentration (C^*) so that a solution perceived to be dilute from a C^* point of view could still contain weak molecular entanglements that are active only on extremely short time scales, such as those that arise in strong elongational flows [67]. These results give strong support to the transient network hypothesis. Further details can be found in Sect. 11.3.2.

The onset of extension thickening for polystyrene solutions in porous media was detected by Sáez et al. [39] by performing simultaneous measurements of resistance coefficients and flow-induced birefringence, averaged over several pore sizes. The increases of both parameters at the onset were practically perfect steps, a fact that lends support to the transient network theory, since the polydispersity of the polymer molecular weight (which is small but finite) and the strain rate distribution in the porous medium would lead to a measurable spread in the criticality of the effect if it were a result of isolated stretching of the macromolecules.

Evans et al. [68] performed experiments by passing polyisobutylene (PIB) solutions through a bed of fibers with random orientations on the plane perpendicular to the flow direction, and measuring simultaneously pressure drops and flow induced birefringence. The PIB was dissolved in a viscous solvent consisting mainly of low molecular weight polybutene, so that the solution resembled the M1 fluid [69]. The volume fraction of fibers in the bed was 2.47%. They found that extension thickening and birefringence occurred simultaneously, and they interpret their observations in terms of the stretching of the macromolecules in the direction of the mean flow. As evidence of the fact that the macromolecules undergo substantial stretching the authors cite the occurrence of degradation, the relatively large pressure drops obtained, and the observation that the maximum level of birefringence obtained is of the same order of magnitude as that reported in previous work [70] for the same polymer in elongational flow in a two-roll mill. The experiments performed by Evans et al. were at polymer concentrations spanning dilute and semi-dilute regimes according to their observations of variations of specific birefringence with concentrations.

11.3.2
The Nature of Extension Thickening in Porous Media Flows

In this section we will describe the effect that various independent variables have on the extension thickening in porous media flows. The effects described below include those of polymer concentration, molecular weight, solvent quality, and temperature.

The effect of polymer concentration on the resistance coefficient vs Reynolds number curves is illustrated in Fig. 11.3 for aqueous solutions of PEO flowing through a disordered sphere packing in a cylindrical bed with diameter 1.9 cm and 30 cm length [32]. The deviation from the Newtonian behavior starts to be appreciable at concentrations as low as 5 ppm (Fig. 11.3b). At low concentrations and low Reynolds numbers, the polymer solution behaves

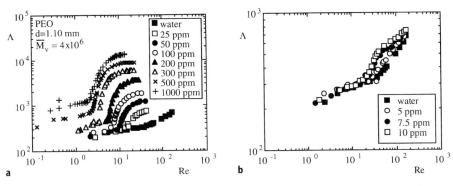

Fig. 11.3 a, b. Resistance coefficients for aqueous solutions of PEO [32]; a high concentrations; b low concentrations

exactly as water. As the Reynolds number is increased, an increase in flow resistance appears. At higher concentrations (50 ppm and higher, Fig. 11.3 a) an increase in flow resistance occurs abruptly until a plateau value is reached. Notice that increases in resistance coefficient of an order of magnitude and more are present at concentrations over 100 ppm. At the largest concentrations ($C \geq 200$ ppm) the curves corresponding to the solutions do not coincide with the water curves at low Reynolds number. However, the resistance coefficient approaches a constant value as Re decreases, indicating a Newtonian behavior. The higher values of Λ at low Re for high concentrations reflect an increase in the shear viscosity of the solution. Notice that we have used the viscosity of the solvent to calculate the resistance coefficient. If the shear viscosity of the solution were used then the values of Λ would be the same for all the solutions as Re becomes small.

The results presented here for one-dimensional flow through porous media are consistent with those reported in previous works for PEO solutions and for other flexible polymers, such as HPAA in excess salt environments, PIB and aPS [1, 12, 13, 24–26, 28, 29, 31, 34, 37–39, 53, 60]. An additional observation, how-ever, in the work of Rodríguez et al. [32] is the occurrence of the increased flow resistance effect in a range of Reynolds numbers that is within the inertial flow regime (Fig. 11.3 b), for which the microscopic velocity profiles in the porous medium become drastically different than those observed in the viscous flow regime, with presence of vortices and steeper velocity gradients [71]. These results indicate that the increase in the elongational viscosity of the solution can occur under different microscopic flow patterns. Note that, for low concentrations, the curves eventually almost coincide again with the water curves at high Reynolds numbers. This behavior might be attributed to the degradation of the polymer, the degree of which increases as the local strain rates increase (see Sect. 11.4).

It is interesting to point out that similar extension thickening behavior has been reported for other types of flow that include significant extensional components of the velocity gradient tensor, such as flow through periodic contractions and expansions, capillary tube entrance flow, orifice entrance flow, Pitot tube entrance flows, and several geometric arrays of different types of obstacles [62, 64, 68, 72–81].

It should be noted that in the data presented in Fig. 11.3 the onset Re at which the extension thickening suddenly appears (Re_o) decreases as concentration increases. Such an effect cannot be accounted for by the increase in shear viscosity with solution concentration [39, 82]. The decrease of Re_o with concentration for PEO solutions was noticed by James and McLaren [26] who did not expect this result given the low level of concentrations employed which were thought by the authors to be in the dilute regime. James and McLaren proposed that this result could be due to the wide molecular weight distribution of the polymer. Similar results led Elata et al. [28] to propose an empirical correlation to fit all concentration curves by plotting Λ vs a dimensionless factor of the form: $De\,C^{1/2}[\eta]^{1/2}$. Using such plots their data on PEO solutions flowing through porous media were less dependent on concentration. However, their replotting of the data of James and McLaren using the same factor did not yield a very good correlation. Naudascher and Killen [29] found a theoretical justification for the use of the $C^{1/2}$ scaling of the extension thickening onset based on the theory of Batchelor [52] of fiber interaction (assuming the macromolecules to be completely extended) but their data were somewhat less successfully fitted to the above-mentioned representation than those of Elata et al. [28].

As mentioned in Sect. 11.3.1, Haas and coworkers [24, 37, 53, 57] proposed that only the Deborah number was required to make the representation of the data independent of polymer concentration, molecular weight, and temperature, as long as the adequate value of k_1 was chosen (Eq. 11.19); their data, however, show considerable spread around the chosen critical value of Deborah number equal to 0.5. Nevertheless, they claimed that they could identify a concentration below which the Re_o does not vary with concentration, and attribute any variation of Re_o with concentration to molecular interactions. The results of their study of the effect of concentration on porous media flow for a high molecular weight HPAA in 0.5 mol/l NaCl is shown in Fig. 11.4, where the values of Λ^* are plotted as a function of Re^* for a wide concentration range.

As pointed out in Fig. 11.4, Haas and Kulicke [13] indicate that molecular interactions are present above 25 ppm for the HPAA solutions, and that below this concentration their Re_o^* does not depend on concentration. They calculate the Re_o^* by extrapolating the region II of their curves up to the point where it meets region I (see Sect. 11.3.1). Using their definition of Re_o^*, we have extracted the onset values of Fig. 11.4, converted them to Re_o (by multiplying them by the value of $1 + C[\eta]$, see Eq. 11.14) and plotted them as a function of concentration in Fig. 11.5, along with the data of Müller et al. [82] for aqueous solutions of PEO with different molecular weights. The straight lines drawn in these log-log plots have a –0.5 slope. It is evident that the Re_o obtained from the data of Haas and Kulicke are not independent of concentration in all the range of concentrations reported by these authors. In fact, all the experimental data compiled in Fig. 11.5 seem to be well fitted by the following empirical relation:

$$Re_o = J C^{-c} \tag{11.20}$$

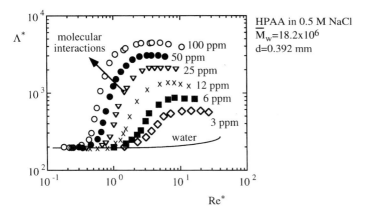

Fig. 11.4. Resistance coefficients for solutions of HPAA in 0.5 mol/l NaCl [13]

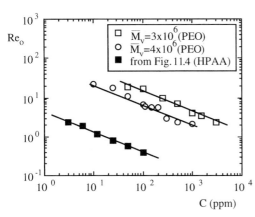

Fig. 11.5. Dependence of onset Reynolds number on polymer concentration in flow through porous media (PEO data taken from [82]; HPAA data obtained from Fig. 11.4)

where J is a constant, and c has a value close to 0.5 (Müller et al. [82] reported that a value of 0.6 would better fit their experimental data). This is an interesting observation since it reflects a behavior similar to that reported by Elata et al. [28] and Naudascher and Killen [29] for PEO (see above), and also because this trend seems to be independent of molecular weight. However, this result would have to be validated with the use of better characterized polymer samples.

The concentration dependence of onset Reynolds number discussed above is not consistent with the trends observed for the coil-stretch transition in ideal elongational flows. Figure 11.6 shows the results obtained by Chow et al. [83] for flow between opposed jets of a polydisperse PEO solution. The flow between opposed jets can be considered close to an ideal elongational flow in the sense that there are no appreciable rotational components in the velocity gradient, since the flow approximates a uniaxial extension field. The results presented in

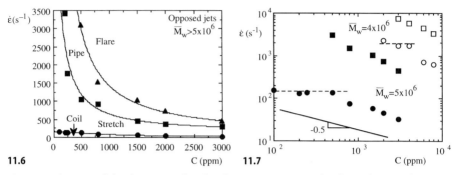

Fig. 11.6. Diagram of development of molecular connectivity in the flow of PEO solutions through opposed jets [83]

Fig. 11.7. Effect of concentration on development of molecular connectivity in opposed jets flow for PEO solutions; *circles – coil-stretch transition; squares – pipe-flare transition* [67, 83]

Fig. 11.6 indicate that the strain rate at which molecules undergo a coil-stretch transition is a weak function of polymer concentration. The reason for this is that the coil-stretch transition is a process that pertains to isolated, non-interacting molecules, when the concentration of polymer is within the dilute limit. At higher strain rates, the molecules go from the stretched state to a regime called "pipe" in which, essentially, significant flow modification occurs between the jets due to the presence of polymer molecules [84]. At still higher strain rates a different regime is attained (termed "flare" because of the particular form of birefringence observations) at which, according to Chow et al., the formation of transient networks of molecules is evident. The transition to the flare regime is characterized by a strong interaction among polymer molecules that have been stretched in the flow field. For this reason, these transitions are very sensitive to polymer concentration: a larger concentration implies an increased probability of molecular interaction and, therefore, it leads to greater interaction at lower strain rates. In the flare regime, the polymer exerts strong flow modification, leading to unstable velocity profiles and the loss of the stagnation point [51].

The data of Fig. 11.6 are plotted in log-log scale in Fig. 11.7, along with the opposed jets data of Keller et al. [67] which correspond to another sample of PEO of lower molecular weight. This representation is analogous to that of Fig. 11.5. In Fig. 11.7 the onset strain rate for coil-stretch transition is practically constant with concentration, at low polymer concentrations, for which the solution viscosity is nearly identical to that to the solvent (dashed lines). At higher concentrations, the critical strain rate for the coil-stretch transition varies with concentration due to changes in the viscosity of the solution. The transition to the flare state decreases more rapidly as polymer concentration increases. A qualitative comparison between Fig. 11.5 and 11.7 indicates that the Reynolds number at which the onset of non-Newtonian behavior in porous media flow occurs is more consistent with molecular interaction effects than with the coil-stretch transition. In Fig. 11.7, the solid line represents the variations of Reynolds onset in porous media flows. Notice that, even though the – 0.5 slope seems to be lower than the variations observed for the pipe-flare transitions, the flow in the porous medium is more complex than in the opposed jets, and facts such as strain rate distribution and presence of shear might smooth changes in the onset of transient entanglements, as compared with purely elongational flows. Furthermore, even though the polymers used by Chow et al. [83] and Keller et al. [67] have nominal average molecular weights close to the ones used by Rodríguez et al. [32] and Müller et al. [82], we have no detailed information on the molecular weight distributions, which strongly affects the elongational flow behavior of the solutions.

Sáez et al. [39] have explored the origin of the increase in flow resistance observed in the flow of polymer solutions through porous media, by studying the behavior of closely monodisperse atactic polystyrene. They used a disordered packing of 5-mm glass spheres inside a parallelepiped of 2.5 cm × 2.5 cm × 10 cm with a porosity of 0.4. The effect of polymer concentration on the resistance coefficient can be seen in Fig. 11.8. The values of Λ are very high, even at low Re (in the Newtonian limit). This is a consequence of the fact that the shear viscosity of the solution is higher than that of the solvent, and that the flow distribution is not uniform. The flow is not uniform due to the limited size of the cell used. The fact that entrances and exits have a cross section that is smaller than the 2.5 cm × 2.5 cm section of the cell implies that the average velocity field in the porous medium is not completely one dimensional and this has been proven to lead to higher resistance coefficients and lower onset Reynolds numbers as compared with uniform flow [32, 82]. It should be noted that the values of Λ and Re were calculated using a superficial velocity based on the total available cross-sectional area of the porous medium. In view of previous results, we can treat the data presented in Fig. 11.8 as qualitatively similar to uniform flow data (for instance, the solvent behavior is accurately described by Eq. (11.11) but with different values of the constants A and B).

The general trend of changes in Λ with concentration observed in Fig. 11.8 is similar to that reported for other flexible polymer solutions (see, for example, Fig. 11.2c). However, the curves in Fig. 11.8 exhibit a more critical extension thickening behavior than any other curve previously reported using poly-

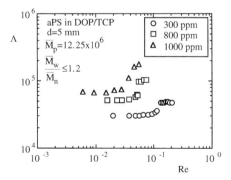

Fig. 11.8. Resistance coefficients for solutions of a PS in a viscous solvent (80.5% dioctyl phthalate/19.5% tricresyl phosphate v/v) at various concentrations [39]

disperse polymers, a fact that is clearly related to the narrow molecular weight distributions of the samples of aPS used by Sáez et al. [39]. It should be pointed out that the behavior of solutions of 300 and 800 ppm was reported to be perfectly critical within the sensitivity range of the measuring devices employed; that is, there was a sudden jump in pressure drop at the onset Reynolds number and that is the reason why there is a discrete step in the Λ values at that point, with no intermediate values recorded. When the concentration of 1000 ppm was used, Sáez et al. were able to record a data point in the middle of the range where the pressure drop increases, thus indicating a somewhat less critical behavior, which they attributed to flow induced degradation (see Sect. 11.4).

The fact that the curves in Fig. 11.8 do not converge to the same values of Λ at low Re is a consequence of the differences in solution shear viscosity, as pointed out before in the case of PEO solutions. Another important fact gathered from Fig. 11.8 is that the absolute change in Λ is not as large for aPS solutions as it is for PEO solutions of roughly comparable molecular weight and concentration. This is to be expected on the basis of the much greater contour length of the PEO molecules (about 3.5 times longer) as compared to aPS of the same molecular weight [11].

As already pointed out, it can be seen in Fig. 11.8 that the onset Reynolds number decreases with increasing concentration. Such an effect is not merely due to an increase in shear viscosity. It can be seen that the Re_o decreases by a factor of two upon changing the concentration from 300 ppm to 800 ppm. The shear viscosity, however, only changed by a factor of 1.3 as indicated by the change in Λ values at low Re. The formation of transient entanglement networks, as pointed out above, would explain this trend since it is expected that the flow rate for the onset of such an effect change very rapidly with varying concentration [67, 83].

The effect of polymer molecular weight was recognized early in the literature as one of the most important aspects that affect extension thickening, since it is well known that the higher the molecular weight the lower the onset Reynolds number and the higher the resulting resistance coefficient for the same polymer concentration [25, 26, 37, 39, 57, 60, 82].

The effects of molecular weight on the resistance coefficient of polymer solutions in uniform flow were investigated by Müller et al. [82] using two different PEOs ($\bar{M}_v = 3 \times 10^6$ and 4×10^6). Figure 11.9 shows that the porous media flow behavior of these polymers is substantially different even at low concentrations. The onset Reynolds number is higher for the lower molecular weight polymer. This trend is clearly seen in Fig. 11.5, where the onset Reynolds number is plotted as a function of concentration for the solutions of both polymers. It should be noted that the two 100 ppm solutions used in Fig. 11.9a had identical shear viscosities [82]; this is also indicated by the fact that the two values of resistance coefficients at very low Reynolds numbers are the same. When higher concentrations are used, the curves corresponding to the two polymers differ from the lowest Re numbers (see Fig. 11.9b). The difference in the plateau value of resistance coefficient (Λ_p) is evident even for the lowest concentrations: the lower molecular weight polymer causes a lower increase in flow resistance. This plateau value should be influenced by two opposing ef-

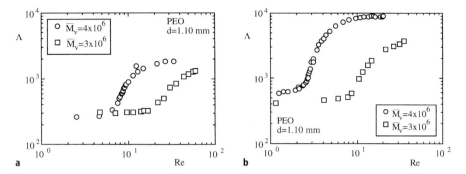

Fig. 11.9a, b. Resistance coefficients for aqueous solutions of PEO of the indicated molecular weights: **a** 100 ppm; **b** 500 ppm [82]

fects: the shear viscosity of the solution might decrease with increasing shear rate (PEO/water solutions are known to be shear thinning, and this effect is more noticeable at moderate concentrations, and it also depends on the molecular weight of the polymer) and the elongational viscosity increases with increasing strain rate. It is clear by the shape of the curves in Fig. 11.9b that the increase in shear viscosity upon increasing solution concentration is responsible for the differences observed in the Λ values at low Reynolds number (Newtonian regime), while the elongational viscosity is dominating the behavior of the resistance coefficient at Reynolds numbers larger than the onset value.

Other important aspects of comparing Λ vs Re curves of the two different molecular weight polymers used by Müller et al. [82] can be appreciated in Fig. 11.10. To obtain the results in Fig. 11.10a two different concentrations have been deliberately chosen for which the two polymer solutions exhibit the same resistance coefficient plateau (which should be proportional to the elongational viscosity), approximately the same onset Reynolds number and different low Reynolds number resistance coefficients (which should be proportional to the

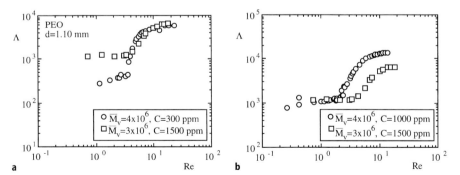

Fig. 11.10a, b. Resistance coefficients for aqueous solutions of PEO of the indicated molecular weights and concentrations: **a** solutions with similar saturation limit of resistance coefficient; **b** solutions with similar Newtonian limit of resistance coefficient [82]

shear viscosity). This case illustrates that two solutions exhibiting very different shear viscosities might produce the same extension thickening action in porous media flow. Figure 11.10b, on the other hand, depicts another possible case in which the shear viscosities of both polymer solutions are identical but their elongational response varies according to the molecular weight of the polymer. These results lead to conclude that the importance of the elongational viscosity should not be overlooked if polymers are to be selected for applications that involve complex flow situations like porous media flow. Figure 11.10 clearly shows that the selection should not be based exclusively on the shear flow behavior. This result is in line with previous observations in the literature where degradation studies have shown that screen factors are more sensitive to mechanical degradation than shear viscosity measurements [1, 48] (see also Sect. 11.4).

If the increase in flow resistance were due solely to the coil-stretch transition of the polymer molecules, then the strain rates at which this happens (which are proportional to Re_o) should be inversely proportional to the longest relaxation time of the coil. For flexible molecules, the longest relaxation time is found to depend on the viscosity of the solvent and on the molecular weight according to [85]:

$$\lambda \alpha \frac{\eta_s M^a}{kT} \qquad (11.21)$$

where a has value that ranges from 1.5 to 1.8 depending on the polymer-solvent interactions [86, 87]. This relaxation time is inversely proportional to the critical strain rate for the coil-stretch transition and Eq. (11.21) has been verified experimentally using purely elongational flow technique such as opposed jets and cross slots [86].

Sáez et al. [39] studied the molecular weight dependence of the onset Reynolds number for extension thickening in porous media flows using three different closely monodisperse aPS samples of molecular weights 9.8×10^6, 11.25×10^6, and 20×10^6 ($\bar{M}_w/\bar{M}_n \leq 1.2$ for all samples). If the coil-stretch transition is the mechanism responsible for the non-Newtonian behavior, then Eq. (11.21) should be obeyed and the Re_o should scale with M as:

$$Re_o \alpha M^{-b} \qquad (11.22)$$

where b is 1.5 for aPS in a θ solvent. The available data of Sáez et al. followed a much greater exponent (of the order of 3.0 to 4.0), even allowing for a shear viscosity correction. An exponent greater than 2 is not compatible with the coil-stretch transition hypothesis and would probably indicate the molecular interactions are taking place and this would then give support to the transient network hypothesis. Müller et al. [82] worked with polydisperse PEO samples (see Fig. 11.5). The onset Reynolds number data on these two polymers yields b = 3.5, which is consistent with the aPS results of Sáez et al. [39].

Haas and Durst [57] and Kulicke and Haas [37] studied the molecular weight dependence of the Λ vs Re curves in porous media flow of PAA solutions. They claimed that the Re_o scales with polymer molecular weight according to the expectations of the coil-stretch transition theory (i.e., with $b = 1.5$ in Eq. 11.22). Figure 11.11a shows the data of Haas and Durst [57] plotted as modified resistance coefficient (Λ^*) vs modified Reynolds number (Re^*) (Eqs. 11.13 and 11.14), and Fig. 11.11b shows the same data in terms of a reduced effective extensional viscosity (η_e^*) as a function of an effective Deborah number, defined as follows:

$$\eta_e^* = \frac{\Lambda - \Lambda_s}{\Lambda_0 - \Lambda_s} \tag{11.23}$$

$$De_e = \frac{k_1 \eta_s^2 (1 - \phi)(1 + [\eta]C)[\eta]M}{\varrho d^2 A k T} Re^* \tag{11.24}$$

where Λ_s is the resistance coefficient of the solvent, Λ_0 is the resistance coefficient of the polymer solution at $Re \to 0$ (notice that, for dilute solutions, $\Lambda_0 = (1 + [\eta]C)\Lambda_s$). For the data in Fig. 11.11, a value of $k_1 = 8$ was used.

It should be noted that even if the representation suggested by Haas and Durst [57] and plotted in Fig. 11.11b makes the data appear more uniform, the value of the onset Deborah number is not unique for the four samples shown and will also depend on how the onset of the extension thickening is calculated. Furthermore, the data correspond to different concentrations, and even if the authors also claimed that samples of different concentrations for the same molecular weight should also scale so that $De_e = 0.5$ (see Fig. 11.5 and its discussion), their scaling plots show a variation of the onset Deborah number of an order of magnitude when all the variables are taken into account. It is also worth mentioning that they used specially prepared PAA with polydispersity of 2.5, except for the PAA with $\bar{M}_w = 18.2 \times 10^6$ which is a polydisperse commercial sample [37] that should have an appreciably larger polydispersity. The difference in polydispersity is not taken into account by the authors in their scal-

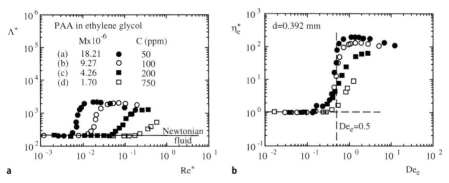

Fig. 11.11a, b. Effect of molecular weight on resistance coefficients: **a** porous media flow behaviour of PAA solutions in ethylene glycol with different concentrations and molecular weights; **b** Deborah number scaling of effective extensional viscosity [13]

ing arguments. It is interesting to point out that the sample with the highest molecular weight and polydispersity exhibits the most critical behavior in Fig. 11.11 b.

It has been argued [66] that the observed criticality of the extension thickening effect in porous media flows would not be expected in the context of coil-stretch transition since the polymers typically used are polydisperse PEO and HPAA. A polymer with a wide distribution of molecular weight will have a wide spectrum of relaxation times and therefore should have a gradual extension thickening behavior. This fact was recognized by early investigators who could not offer an explanation for the observed criticality [28, 37]. Recently, Rodríguez et al. [32] have studied the flow of polydisperse PEO ($\bar{M}_w/\bar{M}_n > 5$) solutions through non-consolidated porous media. They calculated that a polydispersity of less than 1.04 would have to be used in order to explain the observed criticality of the thickening effect in terms of only coil-stretch transitions of macromolecules. This result also supports the transient network hypothesis.

Further evidence that supports the hypothesis of the formation of transient networks is provided by Fig. 11.12 [39]. Simultaneous results on the measured transmitted intensity due to flow induced birefringence through crossed polars, and resistance coefficient of a closely monodisperse aPS solution flowing through porous media are reported. The intensity in Fig. 11.12 is an average value of the optical retardation through several pores. The refractive index of the glass beads that constituted the porous medium was matched to that of the solvent. Perhaps the most striking feature of Fig. 11.12 is the almost perfect criticality of both the increase in Λ and the simultaneous increase in the intensity. Figure 11.12 demonstrates that when a major increase in flow resistance occurs, a major change in the molecular orientation is also occurring. However, the criticality of the effect is such that it resembles a step function; a perfectly monodisperse polymer would be required to rationalize these results in terms of the stretching of isolated molecules. Even though the polymer standard used

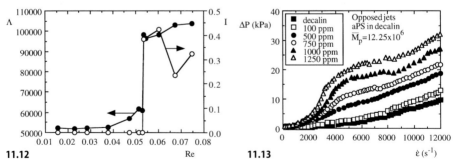

11.12   Re   **11.13** $\dot{\varepsilon}\,(s^{-1})$

Fig. 11.12. Resistance coefficient and intensity of optical retardation as a function of Reynolds number for an 800 ppm aPS solution in a viscous solvent (80.5% DOP/19.5% TCP v/v) [39]

Fig. 11.13. Pressure drop in opposed jets as a function of strain rate for decalin and aPS solutions in decalin [51]

here is closely monodisperse, its residual polydispersity ($\bar{M}_w/\bar{M}_n = 1.2$) is enough to invalidate this explanation [32]. Again it is found that the formation of transient entanglement networks could explain better the criticality of the effect.

Müller et al. [51] have studied the behavior of closely monodisperse aPS solutions in stagnation point elongational flows produced by opposed jets. In these experiments, the authors measure the pressure drop across the jets while simultaneous birefringence visualizations of induced molecular strains are performed. The same samples were used in porous media flow experiments in the work of Sáez et al. [39] mentioned previously. Figure 11.13 shows pressure drop vs strain rate curves for opposed jets flow of $\bar{M}_p = 12.25 \times 10^6$ aPS in decalin solutions. When the solution is dilute (i.e., 100 ppm), there is a small increase in the pressure drop as compared to pure decalin, only after the isolated molecules become stretched. This small increase in pressure drop is consistent with a higher degree of stretching of isolated molecules over a small area surrounding the stagnation point. Optical detection of the coil-stretch transition for this solution occurs at $\dot{\varepsilon} = 1500$ s^{-1}. Large increases in pressure drop, however, are found only beyond a critical concentration and strain rate that are associated with the stretching of transient entangled molecules [51, 88]. Such large increases in pressure drop cause significant perturbation of the flow field, leading ultimately to the loss of stagnation point flow [83, 89].

Müller et al. [51] have developed a method for determining the overlap concentration necessary to produce the transient network behavior in the opposed jets device. It consists of subtracting the solvent pressure drop to all the polymer solution curves to obtain a corrected pressure drop ($\Delta P'$) and then differentiating with respect to strain rate in order to obtain an effective excess elongational viscosity:

$$\eta'_e = \frac{d(\Delta P')}{d\dot{\varepsilon}} \tag{11.25}$$

All the curves of effective excess elongational viscosity vs strain rate above a critical concentration exhibit a pronounced maximum that diminishes in size as the concentration is lowered. If the value of that maximum is plotted as a function of concentration, a representation such as Fig. 11.14 is generated. Extrapolating to zero effective excess elongational viscosity allows one to calculate a value of concentration below which the extension thickening is not observed. This concentration corresponds to the minimum overlap concentration for the formation of transient entanglement networks (C^+). The value obtained in flow through opposed jets for aPs with $\bar{M}_p = 12.25 \times 10^6$ in decalin (a θ solvent), is approximately 335 ppm (Fig. 11.14). This concentration is much lower than the conventional C^*, values of which would range from 50- to 70-times higher than 300 ppm, depending on the method used to assess it. The observation of molecular interactions at concentrations much lower than C^* has been attributed to weak molecular entanglements which are active only on extremely short time scales, such as those that arise in strong elongational flows [51, 67, 83, 86, 89].

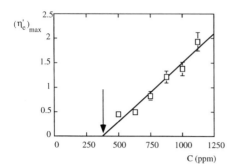

Fig. 11.14. Maximum value of excess effective elongational viscosity (in arbitrary units) as determined in opposed jets flow for aPS solutions in decalin (obtained from the data in Fig. 11.13). The extrapolated value of C^+ is indicated by the *arrow*

Evans et al. [68] performed simultaneous measurements of pressure drop and birefringence in the flow of a PIB solution ($\bar{M} \approx 4-6 \times 10^6$) in a viscous solvent (mixture of trichloroethylene and polybutene) through a fixed bed of solid cylindrical fibers arranged in a disorderly fashion with their axis perpendicular to the direction of flow. The volume fraction of fibers was 2.47%. The birefringence was measured after the solution left the fiber bed. Figure 11.15 and 11.16 show pressure drop and birefringence data. In these figures, De_p is the pore Deborah number, defined by Evans et al. as

$$De_p = \frac{v\lambda}{K^{1/2}} \tag{11.26}$$

In this equation, $K^{1/2}$ is used as a characteristic length of the pore scale. The quantity $vt/K^{1/2}$ in Figs. 11.15 and 11.16 is a dimensionless time representing the number of pore lengths an average polymer molecule (in the hydrodynamic sense) emerging from the bed has traversed.

Figure 11.15 shows the evolution of macromolecular orientation as the polymer advances through the pore space. The fact that the birefringence increases monotonically with number of pores traversed is an indication of the accumulation of molecular strain as the solution experiences successive exposure

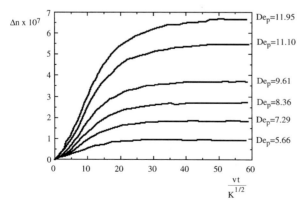

Fig. 11.15. Transient birefringence for a 333 ppm PIB ($\bar{M} \approx 4-6 \times 10^6$) in a mixed solvent (low molecular weight PB/trichloroethylene) solution at various values of the pore-size Deborah number in a disordered fiber bed [68]

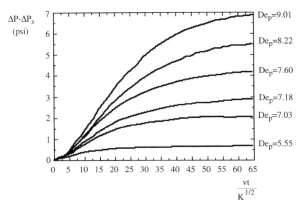

Fig. 11.16. Transient pressure drop for a 1000 ppm PIB ($\bar{M} \approx 4 - 6 \times 10^6$) in a mixed solvent (low molecular weight PB/ trichloroethylene) solution at various values of the pore-size Deborah number in a disordered fiber bed [68]

to local elongational flow fields. This is consistent with observations performed in other works in spatially periodic stretching flows of polymer solutions [90, 91]. Notice that a saturation in birefringence is obtained at large times, and that its value increases as the Deborah number increases. The pressure drop results (Fig. 11.16) exhibit similar behavior. Evans et al. [68] interpreted these results as the consequence of coil-stretch transition of the macromolecules.

Figure 11.17 shows a plot of specific saturation (steady-state) birefringence as a function of pore Deborah number for various concentrations. At low concentrations (50 and 100 ppm) the specific birefringence is the same function of Deborah number. At larger concentrations, the curves depart from the low concentration limit at values that decrease as the polymer concentration increases. This change in behavior was explained in terms of a change from dilute to semi-dilute regimes, since the loss of linearity of the relation between birefringence and concentration implies intermolecular interactions which, according to Evans et al., produce a larger stretching of the polymer molecules. However, the results in Fig. 11.17 closely resemble the behavior of birefringence and pressure drop curves in opposed jet flow of solutions of flexible polymers [51, 67, 89], for which dilute solutions exhibit only modest increases in pressure drop and a

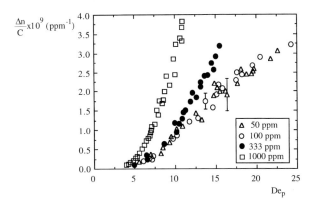

Fig. 11.17. Specific steady-state birefringence for PIB ($\bar{M} \approx 4 - 6 \times 10^6$) in a mixed solvent (low molecular weight PB/trichloroethylene) solutions as a function of pore-size Deborah number in a disordered fiber bed [68]

specific birefringence that is independent of concentration, and semi-dilute solutions $(C > C^+)$ exhibit large increases in both parameters with concentration (including flow instabilities and delocalization of birefringence) which are a consequence of the formation of transient entanglement networks, and not further polymer stretching. The specific birefringence for 333 and 1000 ppm in Fig. 11.17 seem to undergo a transition from the stretching of isolated molecules to the transient network formation.

The results presented previously encourage a comparison between flow in porous media and flow in ideal elongational flows. For instance, the Λ vs Re curve in Fig. 11.12 shows a small continuous rise in Λ at low Re (between 0.04 and 0.05), then at $Re = 0.053$ a sharp increase in Λ occurs. It is tempting to interpret the first small rise in Λ as the contribution of the coil-stretch transition, and the big rise as caused by the formation of transient networks. Such an explanation would parallel the experimental results found measuring pressure drops through opposed jets [51, 66] (see Fig. 11.13).

A direct comparison between opposed jets flow and porous media flow is presented in Fig. 11.18, which shows results obtained when PEO solutions of the same concentrations are passed through a disordered sphere packing and an opposed jets device. Kauser et al. [12] and Rando et al. [92] used a porous medium and opposed jets device with identical dimensions to those employed in previous work [11]. Notice that $\Delta P/\dot\varepsilon$ is a parameter similar to Λ, since, for a given set of results, both represent the ratio between pressure drop and average velocity. The similarity between the porous media flow results and the opposed jets case is remarkable. Both sets of curves have the same overall behavior, except at low strain rates where the opposed jets results do not achieve a plateau, a fact that is caused by inertial effects in the region between the jets. The sudden increase in $\Delta P/\dot\varepsilon$ in the opposed jets is a result of transient network formation. This represents more evidence linking the extension thickening in porous media flows to transient network formation due to the elongational flow at the pore level.

An interesting observation from the results in Fig. 11.18 is the difference in the criticality of the extension thickening between opposed jets and porous media flows. Although the extension thickening in the porous medium is more

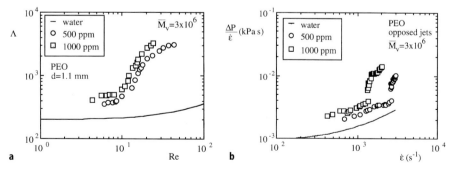

Fig. 11.18a, b. Comparison between porous media and opposed jets flow for PEO solutions: a porous media flow [12]; b opposed jets flow [91]

critical than would be expected from coil-stretch transition theory (as argued above), the behavior in opposed jets flow is even more critical. The smoothing of the resistance coefficient changes in porous media flow is a consequence of the existence of a wide range of strain rates and local geometries.

The formation of transient networks in opposed jet flow is characterized by a strong degree of flow modification, leading to flow instabilities and the loss of the stagnation point [51]. Observations of the flow field in the pores in flow through porous media reported in the literature are contradictory. James and McLaren [26] reported that the streamlines patterns for flow through two layers of spheres placed between flat plates did not experience any changes even after extension thickening occurred. On the other hand, Chauveteau and Kohler [93] report substantial flow modification in flow through sand packs with extension thickening, including vortex formation, and point out that the streamline pattern obtained differs from that observed with Newtonian fluids at high enough Reynolds numbers so that inertial effects are relevant. It is interesting that Chauveteau and Kohler argue that flow modification is a consequence of the same elastic effects that give rise to extension thickening. Deiber and Schowalter [59] performed flow visualization experiments in flow through periodically constricted tubes with solutions of flexible polymers. At Reynolds numbers at and slightly above the onset for extension thickening effects, they do not observe a change in the streamline patterns. A point is reached at higher Reynolds numbers at which secondary flow sets in, and vortices are formed in the expanding zones of the tubes. This occurs as the resistance coefficient vs Reynolds number curves start to exhibit significant extension thickening behavior. Even though the evidence available is still not conclusive, and more flow visualization experiments are required, one might speculate that there should be a degree of flow modification in the streamline patterns in porous media flow, especially in the regions of strong elongational flow (stagnation points and sudden contractions). This would be consistent with the transient network hypothesis.

An interesting observation regarding macroscopic flow stability was reported by James and McLaren [26]. Working with PEO solutions in porous media flows they found that, in region IV of the extension thickening behavior (at high Reynolds number, see Fig. 11.2c) a point was reached at which the pressure drop became practically constant with flow rate (which would imply a resistance coefficient inversely proportional to Re). At this point, the flow rate in the experiments suddenly jumped to a higher value and a velocity gap appeared in which no steady state measurements of pressure drop could be performed. This velocity gap was later observed under similar conditions by Durst et al. [30]. Presumably, the reduction in resistance coefficient in region IV is due to an increase in the degree of polymer degradation with an increase in Re (see Sect. 11.4). Assuming this to be the case, this implies that the rate of degradation reaches a value so high that an increase in the flow rate does not lead to an appreciable change in pressure drop. This is an inherently unstable situation if the flow is controlled by externally imposing a pressure drop (as is commonly the case) since when the point in question is reached, a slight increase in pressure drop leads to a jump in flow rate up to a value in which the pressure drop starts to increase monotonically with flow rate again.

Kulicke and Haas [37] have reported that the criticality of extension thickening also depends on the solvent quality; they find more critical resistance coefficient vs Reynolds number curves as the solvent approaches the θ state, and they rationalize their results in terms of the coil-stretch transition theory. In one particular example (Fig. 11.19), they find that only a slight change in the quality of the solvent changes the criticality of the extension thickening. The Mark-Houwink exponent is equal to 0.54 for ethylene glycol (nearly θ condition) and equal to 0.5 in the methanol-water-NaCl mixture (perfect θ condition). Such extreme sensitivity on the solvent quality has not been observed in the stretching of isolated macromolecules in idealized extensional flow experiments, where even the detection of a change in the power law of the molecular weight dependence on the longest relaxation time of the coil upon changing the solvent quality has been very difficult to obtain and the subject of some controversy in the literature [86, 87]. Another interesting point is that the Λ^* vs Re^* curve for the PAA in the θ solvent is nearly a step function, a fact not compatible with the polydispersity of 2.5 present in the sample they employed. Kulicke and Haas also reported that the maximum value of Λ^* increases as the solvent quality increases from a nearly θ solvent to a good one, a fact difficult to explain if the coil-stretch transition is invoked since once the molecules have been stretched, the value of the final elongational viscosity will be primarily determined by the solvent viscosity and the molecular length. This result could be explained in terms of the formation of transient networks since there will be more chain interpenetration in a good solvent compared to a poor one and therefore a stronger network (i.e., a network with a greater number of entanglements per unit volume) will be produced beyond the onset of extension thickening.

Few data exist on the effect of temperature on extension thickening. Haas and Durst [57] have reported that changing the temperature of the experiment from 25°C to 5°C for solutions of PAA ($\bar{M}_v = 18.2 \times 10^6$) flowing through a porous medium, shifts the onset Reynolds number to lower values, while the maximum resistance coefficient remains nearly constant. It is interesting to note that the authors explain these results in terms of a change in solvent vis-

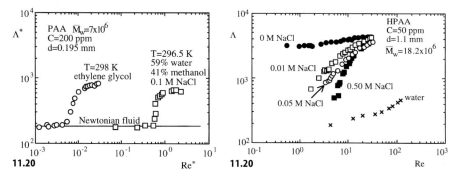

Fig. 11.19. Effect of solvent quality on resistance coefficient for PAA solutions [37]

Fig. 11.20. Effect of ionic environment on resistance coefficient for HPAA solutions [10]

cosity and solvent quality but neither the criticality nor the maximum value of resistance coefficient are affected. This perhaps could be due to the moderate change in temperature.

The effect of the ionic environment on porous media flow of polyelectrolyte solutions like hydrolyzed polyacrylamides in water-NaCl mixtures has been widely studied in the literature [1, 12, 24, 40, 57, 94]. The contrasting behavior of flexible and semi-flexible polymers flowing through porous media (see Sect. 11.3.1, Figs. 11.2b and 11.2c) can be controlled in polyelectrolytic chains by changing the ionic environment of the solvent. This behavior is followed by HPAA, as shown in Fig. 11.20. For these experiments, the HPAA was dissolved in the NaCl solution with gentle stirring for 24 h (see below). As the NaCl concentration is increased, thereby increasing the ionic strength of the solution, the behavior changes from that of a semi-flexible polymer to that of a flexible polymer, since the HPAA molecule tends to adopt a coiled conformation as a result of the complete screening of its anionic groups by the sodium cations [37, 66].

Solutions of hydrolyzed polyacrylamide have shown anomalous effects that depend on the storage time of the solution [95]. Haas and Kulicke [96] determined that storage time of an HPAA solution in brine led to a decrease in the intrinsic viscosity over a period of several days. At the same time, the effective excess elongational viscosity in porous media flows at the pseudo-Newtonian plateau (region III, Fig. 11.2c) increased until it reached a maximum and then decreased with storage time. These results were interpreted in terms of changes in macromolecular conformation with storage time, superimposed with thermal degradation.

Kauser et al. [12] and Rando et al. [92] have found that the procedure for dissolving HPAA in NaCl solutions affects the results obtained when the solution flows through opposed jets and porous media. Figure 11.21 shows results for HPAA solutions, in which it is evident that the moment at which the NaCl is added affects the final result: when the HPAA is dissolved in the 0.5 mol/l NaCl solution the flow resistance is appreciably larger and the onset of extension thickening occurs at lower strain rates than when the NaCl is added just before the flow experiments, even though the solution was optically homogeneous in both cases (the solution was left to stand for at least 2 h after the NaCl was added). These results might be due to conformational changes of the macromolecules during storage, although no precise explanation is available at present. Further work should be performed on this.

It is interesting to notice that both in the porous medium (Fig. 11.21a) and the opposed jets (Fig. 11.21b), the flow resistance obtained when the polymer is dissolved in the NaCl solution seems to converge at high strain rates to the values obtained in the absence of salt. This would seem to indicate that in both the solution without salt and the solution originally prepared in 0.5 mol/l NaCl, the polymer achieves a similar macromolecular conformation at high enough strain rates (see also Fig. 11.20). Notice that, as it happened with PEO solutions (Fig. 11.18), the opposed jets and porous media flow results are qualitatively similar for HPAA (Fig. 11.21), including the case in which the extension thickening occurs gradually (in the absence of NaCl).

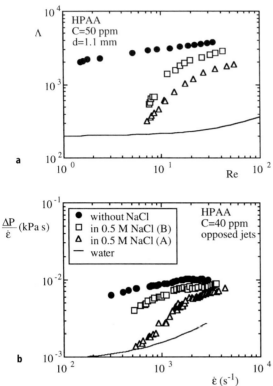

Fig. 11.21 a, b. Effect of solution preparation procedure on the flow resistance of HPAA in electrolyte solutions; (A) NaCl added 24 h after polymer dissolution in water, (B) polymer dissolved in NaCl solution: **a** porous media flow [12]; **b** opposed jets flow [92]

11.3.3
Effect of Porous Media Microstructure

The study of how the microstructure of the porous medium affects the extension thickening behavior has important practical implications. For unconsolidated porous media, the microstructure is affected by the size and shape of particles constituting the porous matrix, and the procedure that led to the spatial distribution of particles in the porous medium. Most work dealing with laboratory studies of the flow of polymer solutions through porous media has employed disordered packings of spherical particles. If the packing procedure is always the same, one would expect the average particle size to be the appropriate scaling parameter for such porous media; for instance, packings attained with monodisperse spheres of different sizes should be, on average, similar to each other but at different scales, controlled by the particle size. It might be argued that, since the extension thickening behavior of flexible polymer solutions is a product of the extensional flow at the pore level, its onset should occur at a critical strain rate that is independent of particle size. This is not so simple, since there is a complex distribution of strain rates at the pore level. However, one would expect the strain rate field to obey a self-similar mapping controlled by the value of a representative strain rate, such as that defined in Eq. (11.19); i.e., for a given value of $\dot{\varepsilon}$ but different sphere diameters, one would expect to have in the porous medium the same spatial distribution of strain rates. It is then to be expected that the value of the critical strain rate for extension thickening, $\dot{\varepsilon}_o$, calculated from Eq. (11.19) for a given constant value of k_1, should be independent of particle diameter.

James and McLaren [26] worked with PEO solutions flowing through disordered packings of monodisperse spheres. They concluded that the onset strain rate for extension thickening was roughly constant for a given solution and various particle sizes. Even though calculated values of the critical strain rates differed sometimes by a factor of two between experiments with different particle size, James and McLaren argued that the differences observed in critical strain rates were always within the experimental error incurred in their determination.

Durst and Haas [53, 57] also observed that the critical strain rate is independent of particle size for solutions of a hydrolyzed polyacrylamide in different solvents. They also concluded that the plateau value of resistance coefficient is independent of particle size, a fact that is consistent with the pseudo-Newtonian character of the plateau, where presumably the maximum extensional viscosity of the solution is attained.

The behavior described above is also followed by the results presented in Fig. 11.22, obtained by Sargenti et al. [10] for PEO solutions. Figure 11.22a shows the curves of resistance coefficient vs Reynolds number for three different particle sizes in a disordered packing of spheres inside a cylindrical bed of 1.9 cm diameter and a length of 30 cm. The same data are represented in Fig. 11.22b using as independent variable a strain rate evaluated from Eq. (11.19) taking $k_1 = \phi^{-1}$, so that $\dot{\varepsilon}$ represents the ratio between interstitial velocity and particle diameter. When represented in terms of strain rates, the three curves collapse to one curve, except for a slight deviation at low strain rates of the curve corresponding to $d = 1.1$ mm.

Even though the evidence presented above indicates that the critical strain rate is independent of particle diameter for disordered packings of monodisperse spheres, there are some results in the literature that do not follow this

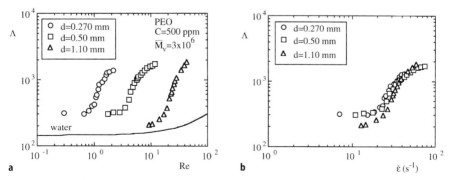

Fig. 11.22 a, b. Effect of particle size on resistance coefficient for PEO solutions: **a** as a function of Reynolds number; **b** as a function of average strain rate [10]

trend. Flew and Sellin [40] performed experiments with a high molecular weight hydrolyzed polyacrylamide (ALCOFLOOD 1145A provided by Allied Colloids, Inc.). They found that the onset of extension thickening did not occur at the same strain rate when the particle size was varied. In fact, smaller particles had lower values of the onset strain rate. They attributed this behavior to the fact that there was more polymer retention in the beds with smaller particles. The retention of the polymer, caused presumably by mechanical blocking of the pores, led to essentially pore spaces with different conformations for different particle sizes. However, Sargenti et al. [10], working with the same polymer, also found that the critical strain rate varied with particle size (Fig. 11.23) in such a way that smaller particles exhibited smaller onset strain rates. Sargenti et al. discarded polymer retention as an explanation for this result since flowing pure water through the porous medium after completing the experiment with the polymer solution led to the results obtained with "clean" packings, i.e., packings through which polymer solution had not been passed. The cause for the lack of scaling observed is not clear, and should be investigated further. It should be pointed out that the HPAA solutions in Fig. 11.23 were prepared by adding the NaCl after the polymer was dissolved (see Sect. 11.3.2). This might have a bearing on the particle size scaling.

The fact that the three curves in Fig. 11.23a start from the same value of resistance coefficient and reach the same plateau indicates that Λ contains the appropriate scaling with respect to particle size. The correlations for K and M given by Eqs. (11.6) and (11.7), which lead to the definition of Λ, are based upon the use of a characteristic length for the pore scale that is equal to the hydraulic diameter of the packing [10]. This leads to the presence of the porosity in those equations. In the results presented above, the particle diameter changed but the porosity remained constant ($\phi = 0.37$). One way of changing the porosity of disordered sphere packings is to use different particle size distributions. Sargenti et al. [10] explored the effect that changes in the porosity have on the extension thickening behavior of HPAA solutions. With this purpose, they used disordered packings of mixtures of spheres with two different sizes, in which

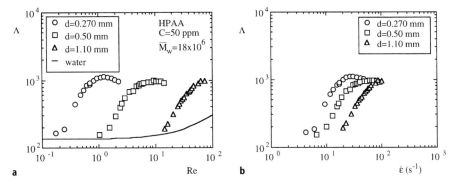

Fig. 11.23a, b. Effect of particle size on resistance coefficient for HPAA solutions in 0.5 mol/l NaCl: **a** as a function of Reynolds number; **b** as a function of average strain rate [10]

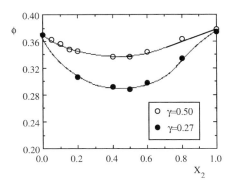

Fig. 11.24. Porosities for disordered bi-disperse sphere packings [10]

the porosity was varied by changing the relative proportion of each sphere size in the mixture. The porosities obtained are shown in Fig. 11.24, for two different mixtures, characterized by two different particle size ratios $\gamma = d_2/d_1$, where $d_1 > d_2$. In this figure, X_2 represents the volume fraction of small particles in the mixture. The porosity has a minimum value at volume fractions close to 0.5, and it decreases as the particle size ratio decreases. Sargenti et al. measured pressure drops as a function of flow rate for the bidisperse packings characterized in Fig. 11.24, for water and HPAA solutions. The resistance coefficients and Reynolds numbers were calculated by means of Eqs. (11.9) and (11.10), using as particle diameter an equivalent volume-to-surface ratio for the bidisperse mixture, defined as

$$d = \frac{n_1 d_1^3 + n_2 d_2^3}{n_1 d_1^2 + n_2 d_2^2} \tag{11.27}$$

where n_i is the number of particles of size d_i in the mixture.

When the approach stated above was employed, the resistance coefficient vs Reynolds number relations for the bidisperse mixtures with water as working fluid were approximately equal to those of the monodisperse packings. This indicates that the definitions of Λ and Re approximately provide the appropriate scaling for porosity variations with Newtonian fluids. The results corresponding to HPAA solutions in 0.5 mol/l NaCl at a polymer concentration $C = 50$ ppm are presented in Fig. 11.25. The onset strain rate changes with the composition of the mixture: as the composition of any of the two particle size ranges increases, the onset strain rates decreases, goes through a minimum, and then increases. This trend is consistent with the changes in the porosity: the porosity is lower in the intermediate composition range (see Fig. 11.24) and this reduces the effective pore length scale over which extensional flow occurs, thus reducing the strain rate for the onset of extension thickening. Notice that the resistance coefficient values and onset strain rates of the curves corresponding to intermediate composition of the mixture are closer to the small sphere curve ($X_2 = 100\%$). When a sizable amount of small spheres are present in the mixture, the relatively large strain rates associated with the small pores are part of the strain rate distribution, and the results indicate that this portion of the

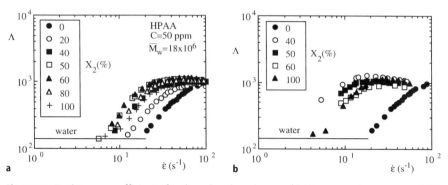

Fig. 11.25. Resistance coefficients for disordered packings of bidisperse spheres: **a** $\gamma = 0.50$; **b** $\gamma = 0.27$ [10]

distribution dominates the non-Newtonian behavior. A similar result was obtained by Tsiklauri [97] regarding the onset of extension thickening for a mixture of monodisperse spheres of two different sizes.

For a lower particle size ratio (Fig. 11.25b) the changes of the onset strain rate with composition are relatively larger than those obtained for the larger particle size ratio (Fig. 11.25a) since the porosity decreases more in the former case (Fig. 11.24). An interesting observation from Fig. 11.25 is the fact that the low-Reynolds number and plateau values of the resistance coefficients seem to be independent of particle mixture composition. This means that the resistance coefficient contains the appropriate scaling with particle diameter and porosity in the Newtonian and pseudo-Newtonian limits.

The microstructure of monodisperse sphere packings can be altered by changing the packing procedure or the spatial arrangement of the particles. Haas and Durst [76] compared the resistance coefficients obtained when passing a PAA solution through three different sphere arrangements: a disordered packing, and two spatially periodic packings: cubic and orthorhombic arrays. The results are shown in Fig. 11.26. It should be pointed out that the periodic arrays used by Haas and Durst were not tridimensional arrays but an axial repetition of a unit cell consisting of a square arrangement of four particles in the cubic packing and a triangular arrangement of three particles in the orthorhombic packing. Haas and Durst used the results presented in Fig. 11.26 to evaluate empirically the value of the structural coefficient k_1 to be used in Eq. (11.19) so that the onset strain rate led to an onset Deborah number equal to 0.5.

The onset of the extension thickening behavior for the data in Fig. 11.26 can be expressed in terms of a global strain rate evaluated from Eq. (11.19) taking $k_1 = \phi^{-1}$, so that once again $\dot{\varepsilon}$ represents the ratio between interstitial velocity and particle diameter. The product between onset Reynolds number and the factor $(1 - \phi)/\phi$ yields a parameter proportional to this $\dot{\varepsilon}$. The results of these calculations are shown in Table 11.1. The onset strain rate is not the same for the three arrays. The definition employed for strain rate considers a velocity change between the interstitial value and zero over a distance equal to the

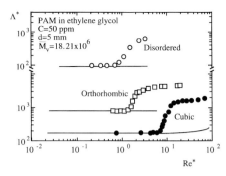

Fig. 11.26. Effect of packing structure on resistance coefficients [76]

Table 11.1. Onset of extension thickening in various sphere packings [76]

Type of Packing	Re_o	ϕ	$Re_o(1 - \phi)/\phi$
Disordered	1	0.37	1.703
Orthorhombic	1.5	0.395	2.297
Cubic	6.5	0.476	7.15

particle diameter. However, in the orthorhombic and cubic arrays, there are no stagnation points. It would be more convenient in these cases to define a strain rate considering the velocity change between the maximum cross sectional value (obtained at the sections with minimum area) and the minimum value (at sections with maximum area). When this is performed, the corrected values of $Re_o(1 - \phi)/\phi$ are 11.5 for the cubic array and 8.8 for the orthorhombic arrays. These values are still different from each other, and different from the value corresponding to disordered packing (Table 11.1). The fact that larger strain rates are needed for the onset of extension thickening in the periodic arrays than in the disordered packing might be a consequence that the latter has stagnation points that are absent in the periodic arrays.

Other works in the literature have dealt with the extension thickening in the flow of polymer solutions through spatially periodic structures. These include periodically constricted tubes and channels [18, 46, 59, 90], and periodic arrays of cylinders [80, 81, 91].

11.3.4
Shear Thinning

The characteristic behavior of solutions of semi-rigid polymers like HPG, xanthan gum, or even HPAA in non-ionic environment in flow through porous media is that of apparent shear thinning (Fig. 11.2a).

Tatham et al. [11] have studied the flow of a semi-rigid polymer, HPG, through a non-consolidated porous medium consisting of a disordered packing of 1 mm diameter glass spheres inside a cylinder (1.9 cm internal diameter and 30 cm length). The porosity of the medium was 0.37 and the HPG had a nominal molecular weight of 2.5×10^6. Figure 11.27 shows the resistance coefficients for HPG solutions. At low to moderate HPG concentrations, the solution behaves as a Newtonian fluid with increased shear viscosity (which results in higher Λ values). At high concentrations (≥ 1500 ppm) there is a definite shear-thinning effect.

Opposed jets results on HPG solutions presented by Tatham et al. [11] have demonstrated that the HPG molecule has a certain degree of flexibility. This has been ascertained from birefringence measurements and from the fact that HPG solutions exhibit extension thickening in opposed jets flow. One would then expect to observe extension thickening effects in porous media flows of HPG solutions. The fact that this is not apparent from the results shown in Fig. 11.27 does not mean that this effect is absent but, as will be shown below, that it could be masked by the change in apparent viscosity due to shear thinning. The expected degree of extension thickening is not as noticeable as that obtained for more flexible molecules in view of the smaller number of equivalent flexible units in the HPG molecule as compared to, for example, PEO or aPS of the same molecular weight [11].

The results presented in Fig. 11.27 are similar to those obtained by Chakrabarti et al. [19] for HPG solutions. They corrected their resistance coefficient by taking into account the effect of variable viscosity due to shear and showed that there is extension thickening. To do this, Chakrabarti et al. work with an apparent shear viscosity calculated from a cylindrical pore model of the porous medium. However, the use of capillary models to represent shear rates in a porous medium has been questioned on the basis of the wide distribution of shear rates present in a real porous medium [18].

In order to find out whether there is extension thickening in the porous medium, Tatham et al. [11] performed a comparison between shear viscosities and apparent viscosities in the porous medium. First of all, they fitted shear viscosity data obtained by simple shear flow to a Cross model,

$$\eta = \eta_\infty + \frac{\eta_0 - \eta_\infty}{1 + G\dot{\gamma}^p} \tag{11.28}$$

where η is the shear viscosity at the shear rate $\dot{\gamma}$, η_0 is the viscosity at zero shear

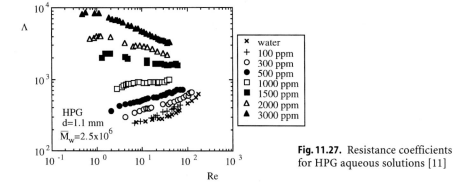

Fig. 11.27. Resistance coefficients for HPG aqueous solutions [11]

rate, η_∞ is the viscosity at infinite shear rate, and G and p are adjustable parameters that depend upon polymer concentration. Assuming that the pressure drop through the porous medium is directly proportional to the shear viscosity (neglecting inertial effects) yields

$$\Lambda = H\eta \tag{11.29}$$

where H is a constant. By considering that the Reynolds number is directly proportional to an effective shear rate, we have

$$Re = J\dot{\gamma} \tag{11.30}$$

Combining Eqs. (11.28–11.30) leads to

$$\frac{1 - \dfrac{\Lambda}{\Lambda_0}}{\dfrac{\Lambda}{\Lambda_0} - \dfrac{\Lambda_\infty}{\Lambda_0}} = DRe^p \tag{11.31}$$

where D is a constant. If the assumptions made were correct, then, by plotting $\log W$ vs $\log Re$, where

$$W = \frac{1 - \dfrac{\Lambda}{\Lambda_0}}{\dfrac{\Lambda}{\Lambda_0} - \dfrac{\Lambda_\infty}{\Lambda_0}} \tag{11.32}$$

one should obtain a straight line with slope p. Figure 11.28 shows the plot obtained in this way for a 3000 ppm HPG solution. The value $p = 0.62$, which represents the data at low Reynolds numbers, is the one obtained from the shear viscosity data of Tatham et al. [11]. At higher Reynolds numbers, the porous media flow results seem to undergo a transition to a lower exponent,

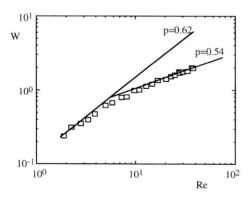

Fig. 11.28. Determination of the Cross model exponent (p) from porous media data (Fig. 11.27) of 3000 ppm HPG solution in water

$p = 0.54$. These results show that the flow at low Reynolds numbers is dominated by shear. However, a Reynolds number is reached beyond which the degree of shear thinning decreases as indicated by the lower exponent. This behavior could be a result of an extension thickening effect.

Sorbie and Huang [22] used the power law model to fit the shear thinning behavior of xanthan gum solutions in flow through porous media and found that the power law exponent was lower than that found in shear rheometry. The lower degree of shear thinning was explained by Sorbie and Huang in terms of the existence of a size-exclusion layer close to the solid surface of the packing, a concept previously employed by Chauveteau [20]. According to Sorbie and Huang, since the xanthan molecules behave as rigid rods, they experience steric hindrances near the wall due to the reduced number of probable conformations that they can assume in that region. This results in a layer adjacent to the solid walls in which the polymer concentration is lower than the bulk value. This, in turn, causes a reduced apparent viscosity. Sorbie and Huang did not observe, however, different exponents at different flow rates as Tatham et al. [11] found for HPG. This might be a consequence of the fact that xanthan is a more rigid molecule than HPG, although there is experimental evidence that indicates that xanthan possesses dynamic properties between those of flexible and fully rod-like molecules [90]. We can speculate that at least part of the effect observed by Sorbie and Huang could be a consequence of a hidden extension thickening in porous media flow.

One important aspect in the analysis of the degree of shear thinning in porous media flows is the fact that a single value of shear rate is used to represent the data when making the analogy with shear viscometry. Due to the geometric nature of the pore scale in real porous media, a distribution of shear rates exists for a given flow experiment, so that the use of a shear rate defined by an equation analogous to Eq. (11.19) might not be adequate. This point has been argued before in the literature [18]. A detailed knowledge of the flow field at the pore level could provide an adequate average shear rate for this kind of analysis. This point has not been satisfactorily addressed in the literature.

11.4
Flow-Induced Degradation

One of the most important limiting factors in the application of polymer additives is their susceptibility to mechanical degradation. In the field of porous media flow this is a subject on which many investigations have been performed [1, 15, 26, 31, 38, 40, 41, 48, 60, 78, 82, 93, 98–105]. It has been widely recognized in the literature that semi-rigid polysaccharides such as xanthan gum or HPG are more stable to flow-induced mechanical degradation than flexible polymers like PEO or HPAA. This is a consequence of the usually smaller contour length of semi-rigid polymers as compared to flexible ones of the same molecular weight, since the molecular length is the key parameter when comparing resistance to degradation of different macromolecules [11, 106]. In this section we will discuss relevant results on mechanical degradation in porous media flows.

In one of the first studies to acknowledge the relevance of flow-induced degradation in porous media flows, Jennings et al. [1] designed a viscometer in which they measured the flow time of a given volume of solution through a pack of five 100-mesh screens. This viscometer has been widely used in the petroleum industry. The results are reported as a "screen factor", which is defined as the ratio of the flow time of the polymer solution to the flow time of the solvent. Jennings et al. found, using HPAA solutions in brine, that the screen factor was more sensitive to molecular degradation than shear viscosity measurements.

A key study on the degradation of partially hydrolyzed polyacrylamide solutions when they flow through sand cores is that of Maerker [48]. Maerker recognized that the degradation of viscoelastic polymer solutions as they flow through porous media is caused by elongational stresses as opposed to shear stresses. Maerker was the first, to our knowledge, to invoke entanglements in order to rationalize the degradation experienced by flexible molecules during porous media flow. Maerker characterized degradation by measuring the screen factor of the HPAA/salt solutions employed before and after passing them through the sand cores and calculating the percent loss of screen factor. The results obtained indicate that the degradation only occurs beyond a certain strain rate and increases gradually with strain rate after this onset. It was found that degradation has a greater effect on reducing the extension thickening effects as characterized by screen factor losses than on reducing the shear viscosity of the solution, a fact attributed by Maerker to the importance of the high molecular weight tail of the molecular weight distribution in inducing elastic effects.

Maerker [48] found no effect of concentration on flow-induced degradation during porous media flow for a limited concentration range (300–600 ppm) of HPAA in brine. This result is similar to that of Chauveteau [78] who found no concentration dependence of degradation in the limited range between 300 and 800 ppm for solutions of PAA with $\bar{M}_w = 4.5 \times 10^6$ in water. Martin [104] reported experiments in which solutions of high molecular weight HPAA ($\bar{M}_w$ ranged from 2.2×10^6 to 6.2×10^6) in brine were made to flow through sand packs and their degradation behavior was deduced in a similar way to Maerker by the use of the percentage loss in screen factor. Martin found, employing a wide concentration range (100–3000 ppm), some evidence suggesting that screen factor loss increases with polymer concentration up to a certain value after which it decreases slightly with further increases in concentration.

Farinato and Yen [38] performed a study of polymer degradation in porous media flow. Their porous medium was a disordered packing of 590-μm glass beads with a porosity of 0.34. They used high molecular weight HPAA with $\bar{M}$ of 6.1 and 8.5×10^6 in a 1 mol/l NaCl brine. Their degradation experiments consisted of measuring the resistance coefficient vs Reynolds number curve of a particular solution concentration in order to determine its onset Reynolds number; then the solution was passed through the medium at a constant Reynolds number and finally its new Re_o determined by once again recording its Λ vs Re curve. Any shift in Re_o was interpreted as an evidence of flow-induced degradation. Their results suggested that at low concentrations, in the dilute concentration regime, the degradation was greater at specific strain rates than at higher concentrations in the semi-dilute concentration regime.

It is interesting to point out that in idealized extensional flow experiments, the molecular degradation rate also depends on the concentration. For truly dilute solutions, degradation in opposed jets flow is characterized by central scission and no apparent concentration dependence on degradation rate, while for semi-dilute solutions the degradation rate is lower as the concentration increases (a fact connected to flow induced modifications in the flow field) and it is also increasingly non-central [107, 108].

Maerker [48] found that screen factor loss increased as the ionic strength of the solvent increased upon addition of NaCl to HPAA solutions. When CaCl$_2$ is used instead of NaCl, the increase in screen factor loss was stronger than that expected on the basis of an increase in ionic strength. Smith [98] had also indicated that HPAA suffered more severe degradation in calcium-containing brines than in sodium-containing brines, a fact later corroborated by Sandiford [42]. Maerker explained the sensitivity of degradation on ionic strength by considering that a more efficient neutralization of the negative charges in the HPAA molecules would lead to the formation of tighter coils. If such coils form entanglement networks, these entanglements will not be as easily removed by thermal Brownian motion, and mechanical rupture of the polymer chains during elongational flow will be more probable. Maerker also proposed an empirical correlation for screen factor loss by plotting this parameter as a function of the product of the stretch rate and the one-third power of a dimensionless flow distance. Further extensions of Maerker's original work to improve empirical correlations of degradation data from unconsolidated porous media, as well as different consolidated sand cores, with several experimental variables were later performed by Maerker [99], Morris and Jackson [100], Seright [101], Seright et al. [109] and Ghoniem [62].

The influence of particle size on flow-induced degradation was studied by James and McLaren [26], who found the surprising result that degradation effects for PEO solutions ($\bar{M}_w$ of $0.96 - 9.7 \times 10^6$) increased as the particle size in the porous medium increased from 110 µm to 450 µm. They quantified degradation by comparing the values of pressure drop across different lengths of the porous bed. The pressure gradient progressively decreased along the length of the medium as a result of degradation. Later, Käser and Keller [31] reported in a very similar study that the observation of James and McLaren could be erroneous because their measurements did not include the extensive degradation that was occurring upstream of their first measuring point, which is located 25 mm after the entrance to the porous medium. The experiments of Käser and Keller indicated that degradation extent increased as the particle size decreased, not only by measurements of resistance coefficients along the length of the medium but also by the change in shear viscosity of the solutions after they were circulated through the porous media. Maerker [48] found that screen factor loss increased as the permeability of a sand core decreased, which is consistent with the particle size effect reported by Käser and Keller. Chauveteau [78] reported similar results to those of Käser and Keller with a PAA of $\bar{M}_w = 4.5 \times 10^6$ in water flowing through disordered packings of glass spheres of

150 and 270 µm diameter. However, Moreno et al. [105] have recently reported an increase in the extent of degradation with increases in pore size, as demonstrated by measurements of the elongational flow properties of HPAA solutions in brine degraded in porous media.

Haas and Kulicke [13] have reported data on porous media flow-induced degradation of solutions of HPAA and aPS. They have argued that the mechanical degradation is induced by the drastic decoiling of the macromolecules (onset of coil-stretch transition) enforced by the elongational nature of the flow. Figure 11.29a presents their data for HPAA ($\bar{M}_w = 27 \times 10^6$) in 0.5 mol/l NaCl and Fig. 11.29b presents data for aPS ($\bar{M}_w = 23.6 \times 10^6$) in toluene. Figure 11.29a shows how the extension thickening behavior of the original HPAA solution (run A) changes after passing it once through the porous medium at a constant $Re^* = 15$ (run B). Run A in Fig. 11.29a exhibits the four zones described in Sect. 11.3.1 (Fig. 11.2c), the fourth zone indicates a decrease in Λ^* upon increasing Re^* that is probably due to mechanical degradation. After run B, the solution was passed once through the medium at $Re^* = 20$ and the porous media flow behavior of the resulting solution is presented in curve C. It is clear that flowing the solution at high Re^*, a substantial decrease in the mean molecular weight occurred as indicated by the shift in the onset Re^* to higher values and the decrease in the magnitude of the

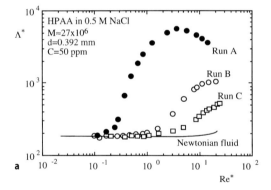

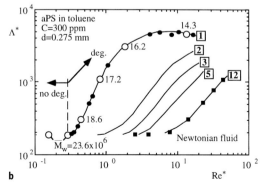

Fig. 11.29a, b. Mechanical degradation in porous media flow: a HPAA in 0.5 mol/l NaCl solutions; b aPS in toluene solutions. In b, numbers enclosed by squares represent number of passes of the solution through the porous medium [13]

maximum resistance coefficient. A quantitative demonstration of molecular weight reduction was provided for aPS by Haas and Kulicke (Fig. 11.29b) by determining the weight-average molecular weight of solutions of closely monodisperse aPS ($\bar{M}_w/\bar{M}_n \leq 1.2$) through light scattering measurements of samples after flowing them through the porous medium up to a given Reynolds numbers. Figure 11.29b shows that they did not find appreciable degradation at Re^* below the onset of extension thickening, but above it the degradation increased upon increasing Re^* even during the first run of the sample (run 1). Furthermore, Haas and Kulicke stated that molecular degradation is present in all polymer solutions that exhibit extension thickening (at $De \geq 0.5$, according to their interpretation based on the coil-stretch transition theory).

In order to investigate the mechanical degradation induced by the flow, two PEO ($\bar{M}_v = 3 \times 10^6$) solutions ($C = 100$ and 1000 ppm) were repeatedly passed by Müller et al. [82] through a porous medium at a constant flow rate while the pressure drop was being recorded. Figure 11.30 shows the variation of the resistance coefficient with the number of passes (N) at constant Reynolds number. It can be seen that, at the lowest Reynolds number, there is no appreciable variation in the value of Λ with the number of passes (Fig. 11.30a). At higher Reynolds numbers the value of Λ rapidly decreases with N, achieving a nearly constant value after three to four passes. The decrease in the value of Λ at constant Reynolds number indicates that the solution is being degraded at a particular average strain rate, which is proportional to Reynolds number. Such strain rate for fracture is expected to be dependent on the polymer molecular weight [107]. Hence when most of the molecules that can be broken at that particular strain rate have been degraded, the resistance coefficient will become constant.

The flow-induced degradation can also be assessed by the relative displacement of the Λ vs Re curves measured after the solution was circulated at constant Reynolds number until no further degradation occurred (i.e., when the resistance coefficient approached a constant value with N, Fig. 11.30). Figure 11.31

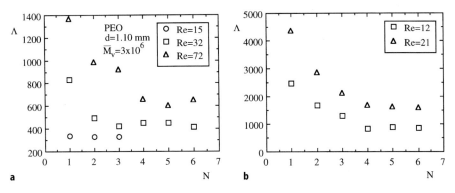

Fig. 11.30a, b. Decrease in resistance coefficient during mechanical degradation for PEO solutions: **a** $C = 100$ ppm; **b** $C = 1000$ ppm [82]

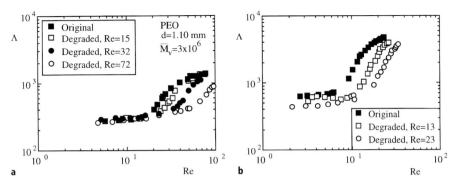

Fig. 11.31a, b. Resistance coefficients of original and degraded PEO solutions: **a** $C = 100$ ppm; **b** $C = 1000$ ppm [82]

shows how the degradation of the polymer affects the Λ vs Re curves in the whole range of Reynolds number in which the polymer induces a non-Newtonian effect. The onset of non-Newtonian behavior of the solution is shifted to higher values of Reynolds number after degradation occurs. The increase in onset Reynolds number (Re_o) can be interpreted as an increase in the critical strain rate for the onset of transient entanglement network formation [51]. Any increase in onset Reynolds number will therefore indicate a decrease in the average molecular weight of the polymer if the same porous medium is used and the solution concentration remains constant.

Figure 11.31 also shows that the solution is significantly degraded only if the Reynolds number at which it is circulated through the porous medium is higher that the original Re_o of the undegraded solution. In fact, the value of onset Reynolds number after degradation is close to the Reynolds number at which the solution was degraded. This result implies that most of the molecules break once they are in the stretched state as part of a transient entanglement network. Parallel results were obtained by Sáez et al. [39] for the flow of closely monodisperse atactic polystyrene dissolved in an organic solvent mixture, and by Moreno et al. [105] for the flow of HPAA in NaCl solutions. It should be pointed out that a similar conclusion has been reported in idealized elongational flow experiments, where degradation is always encountered when transient networks cause extension thickening effects [51, 106]. The observation that for flexible polymers large viscosification effects in porous media flow are always accompanied by degradation is in itself not consistent with the coil-stretch hypothesis [39, 66]. For isolated molecules the fracture strain rate scales with molecular weight with a higher exponent than the coil-stretch strain rate, so that at least at lower molecular weights fracture should only occur at much higher Reynolds number than those required for stretching, especially when a series of molecular weights are compared [86, 107]. This trend is not observed in porous media flow [13].

11.5
Porous Media Flows of Polymer Blends and Cross-Linked Polymers in Solution

In this section we present experimental results on non-conventional aspects of the flow of polymer solutions through porous media. We start by analyzing the effect of cross-linking agents on flow resistance. Afterwards we present results regarding the flow of solutions of polymer-polymer blends.

11.5.1
Flow of Cross-Linked Polymer Solutions Through Porous Media

Certain polysaccharides, such as guar gum and HPG, can be cross-linked in solution in the presence of various transition metal ions, and also in the presence of borate ions. The nature of the cross-linking process strongly depends on polymer concentration. Three different regimes with respect to the nature of the cross-links can be distinguished.

1. For dilute solutions, it is expected that intermolecular cross-links cannot be formed, but the macromolecule may exhibit intramolecular cross-links. There is evidence that intramolecular cross-links might decrease the relaxation time of the macromolecule since its radius of gyration can be reduced [11].
2. At intermediate concentrations, intermolecular cross-links are formed but, if the concentration is small enough, there might not be sufficient intermolecular connectivity to form a three dimensional network. In this case, the cross-links lead to a dissolved structure that has a macromolecular conformation with an apparent molecular weight higher than the original polymer and possibly a different degree of flexibility.
3. At sufficiently high concentrations so that at least two intermolecular cross-links per chain can be formed, a point is reached in which the macromolecules form a three dimensional associated network (gel state).

Cross-linking is used in the oil recovery industry as a means of achieving gel formation for fracturing operations or the blocking of high permeability paths near well bores. In these cases, the gels do not flow through the reservoir but the goal is to form a static network. In this section we will deal with the intermediate concentration range. This has two important practical implications. First, it might be of interest in the characterization of the flow of a solution during the in situ cross-linking process before the gel state is achieved. Second, it is worthwhile to explore the possibility of an intentional modification of the macromolecular conformation in solution as a means of modifying the potential applications of the polymer solution in oil recovery or fracturing operations.

In what follows, we will present results regarding the flow of aqueous hydroxypropyl guar solutions in the presence of excess borate ions as cross-linking agent (provided by dissolved borax in a $pH = 11$ environment). The results presented here are a summary of the study performed by Tatham et al. [11],

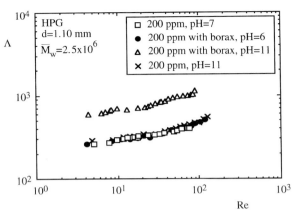

Fig. 11.32. Effect of pH on resistance coefficient for 200 ppm HPG solutions in water with and without cross-linking agent [11]

where more details about the experiments can be found. The porous medium used was a disordered packing of 1-mm spheres in a cylinder with diameter of 1.9 cm and a length of 30 cm.

Figure 11.32 shows results corresponding to 200 ppm HPG solutions. First of all, it can be seen that the value of pH, which is a parameter that can be used for controlling the cross-linking process, does not affect the resistance coefficients of the HPG solutions, since the curves for pH = 7 and 11 without borax co-incide. The presence of cross-linking agent at pH = 11 leads to substantially higher resistance coefficients as a consequence of the increase of the apparent viscosity of the solution due to the presence of intermolecular cross-links. This is confirmed by the fact that, when the cross-links are "switched off" by lowering to 6 the pH of the solution with borax, the curve goes back to that of the original, uncross-linked solution.

Tatham et al. also explored the transition to the dilute regime. Evidence of intramolecular cross-links was sought in flow through porous media since they were detected in elongational flow through opposed jets. However, decreasing the polymer concentration below 200 ppm led to an absence of change in the resistance coefficient when cross-linking agent was added since the solution without cross-linking agent had a resistance coefficient very similar to that of water. The sought evidence of intramolecular cross-links was obtained after filtering the 200 ppm solution through a 20-μm mesh. The results are presented in Fig. 11.33. Filtering the solution in the absence of cross-linking agent effected no change on the resistance coefficient. However, when cross-linking agent is added, the filtered solution exhibits lower resistance coefficients which are very similar to those of water. The fact that cross-linking the solution leads to lower pressure drops is consistent with the formation of intramolecular cross-links. The results indicate that the filtration process removes the high molecular weight end of the molecular weight distribution which is apparently responsible for the formation of intermolecular cross-links.

At higher concentrations (Fig. 11.34) the addition of cross-linking agent does not simply lead to a higher apparent viscosity. Notice that, as the HPG concentration increases, the addition of cross-linking agent leads to resistance

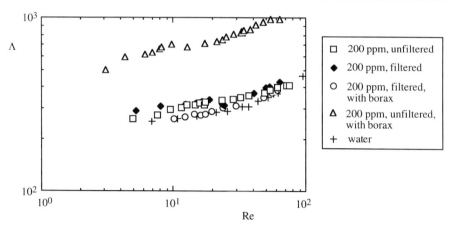

Fig. 11.33. Effect of filtering on the addition of cross-linking agent to 200 ppm HPG solutions in water [11]

coefficients that increase more rapidly with Reynolds number. This indicates that the macromolecular structure of the polymer in solution is changing drastically in the presence of intermolecular cross-links. The curve for 3000 ppm without cross-linking agent in Fig. 11.34 is presented for comparison purposes. This curve exhibits a noticeable degree of shear thinning whereas the 1000 ppm cross-linked solution, which achieves values of the resistance coefficient of the same order of magnitude as the 3000 ppm solution, exhibits a large degree of extension thickening. The modification of the molecular structure induced by the cross-linking process leads to an increase in the apparent molecular weight of the molecule and possibly to a branched macromolecular structure. The increase in the apparent molecular weight leads to a greater number of equivalent flexible units that results in a more pronounced extension thickening effect, which becomes dominant over shear thinning.

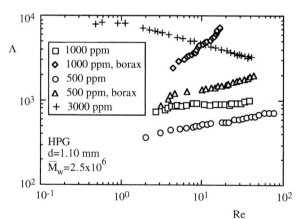

Fig. 11.34. Effect of the addition of cross-linking agent to solutions with high HPG concentrations on the resistance coefficient, pH = 11 [11]

The results presented above show how the intermolecular cross-links can be used as a means of changing the rheological behavior of polymer solutions in flow through porous media. This has interesting practical implications, especially if one considers that the cross-linking process can be controlled by changing the pH of the solution, or even the concentration of cross-linking agent.

11.5.2
Flow of Solutions of Polymer Blends Through Porous Media

Solutions of polymer blends have been studied in shear and elongational flows mainly to ascertain the effect of the flow on the miscibility of the system. For perfectly miscible mixtures, drag reduction experiments in turbulent flows through pipes [110, 111] have shown that the mixture has drag reducing capabilities that exceed that of solutions of each of the individual polymers, especially when at least one of the polymers has a semi-rigid structure. This has motivated studies of the flow of solutions of polymer blends through porous media by Gamboa et al. [41] and Kauser et al. [12]. In both works, porous media made up of disordered packings of 1-mm spheres in a cylinder with a diameter of 1.9 cm and a length of 30 cm have been employed. Gamboa et al. [41] explored the behavior of aqueous solutions of blends of PEO (as the flexible polymer) and HPG (as the semi-rigid polymer). Kauser et al. [12] studied aqueous solutions of PEO and HPAA. The molecular equilibrium conformation of the latter was controlled by changing the ionic strength of the solvent with the addition of NaCl.

Figures 11.35 and 11.36 show the resistance coefficients obtained by Gamboa et al. [41] for solutions of mixtures of PEO and HPG. In Fig. 11.35 the results corresponding to a constant PEO concentration of 100 ppm are presented. The extension thickening induced by the PEO becomes less critical with respect to Reynolds number as the HPG concentration is increased. At 3000 ppm HPG, the

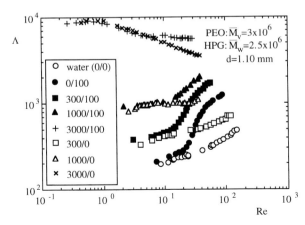

Fig. 11.35. Resistance coefficients for aqueous solutions of HPG/PEO mixtures for low PEO concentrations. Legend shows HPG/PEO concentrations in ppm [41]

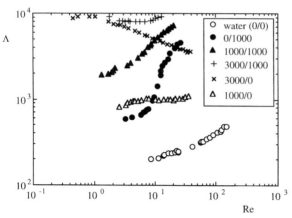

PEO merely inhibits to some extent the shear thinning behavior of the HPG solution. Notice that the resistance coefficients of the solutions of blends are equal to those of the HPG solution of the same concentration at low Reynolds numbers. This is consistent with the fact that the PEO does not contribute to the shear viscosity of the blend solutions. The results corresponding to a higher concentration of PEO (1000 ppm) are shown in Fig. 11.36. At low HPG concentrations and Reynolds numbers, the PEO induces a noticeable change in the apparent viscosity of the solution. As the concentration of HPG is increased in the blend, the criticality of the extension thickening decreases. Note that, at 3000 ppm HPG in the blend (3000/1000), there is a slight extension thickening at high Reynolds numbers.

The results discussed above show that the capability of the PEO to increase the apparent viscosity of the solution at low Reynolds number diminishes as the HPG concentration is increased in the blend. In fact, for a concentration of HPG of 3000 ppm, the presence of the PEO does not alter the resistance coefficient at low Reynolds numbers. This behavior is better appreciated in Fig. 11.37 where the results corresponding to a concentration of 3000 ppm HPG in the blends are shown. Note that, at the limit of low Reynolds number, the resistance coefficient for the 1000 ppm PEO solution is approximately three times larger than the water value (Fig. 11.36), whereas in the 3000/1000 blend, the same PEO concentration leads to no noticeable change in the Λ values at low Reynolds numbers.

The results obtained with HPG/PEO blends indicate that the presence of HPG in the solution has a definite effect on the extension thickening nature of the PEO. A solvent of different viscosity with respect to water would also influence the extension thickening of the PEO. Figure 11.38 shows the resistance coefficient of 1000 ppm PEO in a viscous Newtonian solvent. The solvent (60% water and 40% glycerol v/v) was chosen so that the shear viscosity of the resulting solution was equal to the apparent viscosity of the 1000/1000 blend at low Reynolds number. Notice that the resistance coefficient at low Reynolds numbers is the same for the 1000/1000 blend and the 0/1000 (w/g) solution. In this

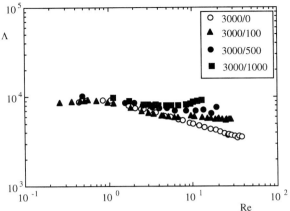

Fig. 11.37. Resistance coefficients for aqueous solutions of HPG/PEO mixtures for high HPG concentrations. Specifications as in Fig. 11.35 [41]

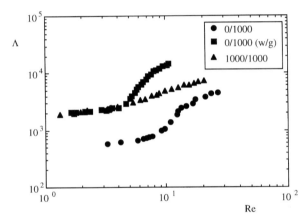

Fig. 11.38. Resistance coefficients for 1000 ppm PEO solutions in various environments. Specifications as in Fig. 11.35, (w/g) refers to a solution for which the solvent is a mixture water/ glycerol 60%/40% v/v [41]

case, the onset Reynolds number is the same for both solutions, but the rate of increase of the extension thickening is very different. This would indicate that the PEO molecules are exposed to solvents with different thermodynamic quality, if one interprets the 1000 ppm HPG solution as being the solvent for PEO in the 1000/1000 blend. However, one must keep in mind that the effect might be a dynamic one in the sense that the formation of a transient molecular network of PEO molecules in the presence of HPG would involve an interaction between both polymers whose nature is not straightforward to characterize.

Gamboa et al. [41] also performed degradation experiments with HPG/PEO blends. These experiments consisted of passing the solution repeatedly through the porous medium at a constant Reynolds number, simultaneously measuring the resistance coefficient for each pass. The results are presented in Fig. 11.39. In this figure, Λ_0 represents the resistance coefficient in the first pass through the porous medium. The HPG solution does not exhibit noticeable degradation.

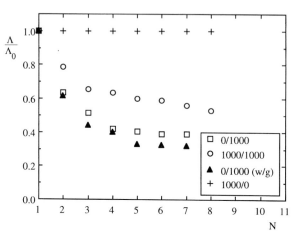

Fig. 11.39. Decrease in resistance coefficient during mechanical degradation of PEO in various environments. The solution in *w/g* was passed at $Re = 10$; the rest were passed at $Re = 16$. Specifications as in Fig. 11.35 and 11.38 [41]

In contrast, the PEO degrades rapidly in the first four passes. An increase in solvent viscosity should lead to an increase in degradation rate, since higher viscosities imply larger molecular stresses [106]. This behavior is confirmed by the results presented in Fig. 11.39: in the water/glycerol mixture, the degradation is faster than in pure water, even though the Reynolds number is lower for the former. On the other hand, the 1000/1000 blend exhibits a lower degradation rate than the PEO solution in pure water, even though its viscosity is appreciably larger. This result indicates that the presence of HPG inhibits the degradation of the PEO.

The results obtained by Kauser et al. [12] with HPAA/PEO mixtures are presented in Figs. 11.40–11.42. Figures 11.40 and 11.41 correspond to solutions

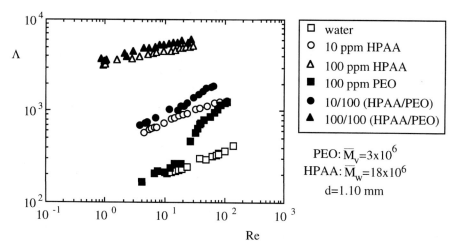

Fig. 11.40. Resistance coefficients for aqueous solutions of HPAA/PEO mixtures for low PEO concentrations. Legend shows HPAA/PEO concentrations in ppm [12]

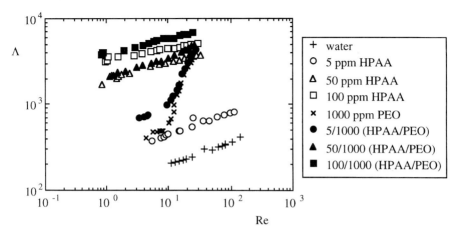

Fig. 11.41. Resistance coefficients for aqueous solutions of HPAA/PEO mixtures for high PEO concentrations. Specifications as in Fig. 11.40. Legend shows HPAA/PEO concentrations in ppm [12]

in deionized water, for which the HPAA behaves as an expanded coil. The trends observed are very similar to those obtained with HPG/PEO mixtures: as the concentration of the semi-rigid polymer increases, the degree of extension thickening of the PEO decreases and the effect even almost disappears. Figure 11.42 presents a comparison between porous media flow (Fig. 11.42 a) and opposed jets flow (Fig. 11.42 b) for the same solutions. As was reported in Sect. 11.3.2, the qualitative trends of the resistance coefficient in the porous medium and the ratio between pressure drop and strain rate in the jets are very similar. In this case, the solution of the mixture exhibits an apparent viscosity that is appreciably higher than those of the solutions of the pure polymers. The solution of the 10/1000 (HPAA/PEO) mixture exhibits a more critical behavior in the opposed jets, as it is the case for the 1000 ppm PEO solution.

In the presence of NaCl (Fig. 11.43) the resistance coefficients are less sensitive to HPAA concentration than without salt. This is a consequence of the fact that, at low Reynolds numbers, the HPAA adopts a coiled conformation. The PEO controls the extension thickening behavior of the mixtures. Notice that the addition of HPAA decreases slightly the onset Reynolds number and makes the extension thickening more gradual.

From the standpoint of the increase in the apparent viscosity of the mixtures, the results presented above indicate that it is more effective to add HPAA to a PEO solution in the absence of NaCl. The results obtained in the opposed jets (Fig. 11.42 b) show that the addition of relatively small amounts of HPAA to the PEO solution induces the formation of transient networks at much lower strain rates. One might conclude that there are definite molecular interactions between the polymers in solution.

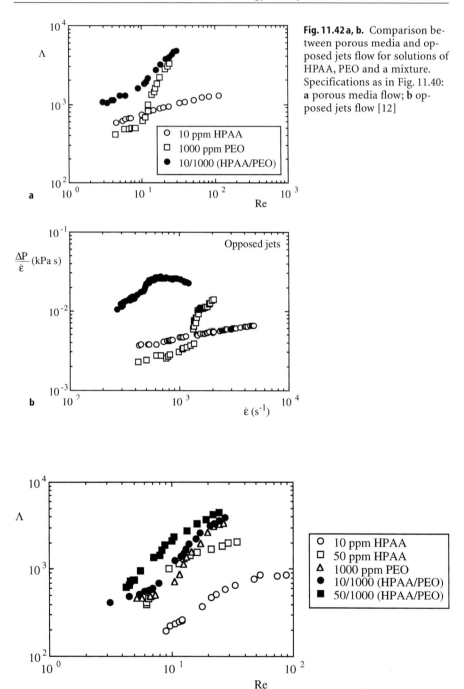

Fig. 11.42 a, b. Comparison between porous media and opposed jets flow for solutions of HPAA, PEO and a mixture. Specifications as in Fig. 11.40: **a** porous media flow; **b** opposed jets flow [12]

Fig. 11.43. Resistance coefficients for solutions of HPAA/PEO mixtures in 0.5 mol/l NaCl. Specifications as in Fig. 11.40 [12]

11.6
Concluding Remarks

Solutions of flexible polymers exhibit extension thickening in flow through porous media. In the range of independent parameters in which this effect is commonly encountered, the extension thickening seems to be a consequence not only of the coil-stretch transition but mostly of the formation of transient entanglement networks of polymer molecules, even at concentrations below the equilibrium coil overlap concentration. This conclusion stems from a variety of facts that cannot be solely explained on the basis of the coil-stretch transition theory. These facts include the criticality of the extension thickening effect, its scaling with concentration and molecular weight, and the effects of flow-induced degradation. Furthermore, we have shown that there exists a remarkable similarity between the extension thickening in porous media flow and results obtained in closely ideal extensional flows where the existence of transient networks can be ascertained by means of optical and mechanical measurements. This similarity indicates that extensional flow at the pore level is the dominating feature of the porous media flow of solutions of flexible polymers.

Solutions of semi-rigid polymers at moderate concentrations exhibit shear thinning in flow through porous media. However, the residual flexibility of the macromolecule allows it to undergo conformational transitions, including the formation of transient entanglement networks, and the induction of a certain degree of extension thickening that might be hidden by shear thinning. This results in a decreased shear thinning effect when the apparent viscosity in porous media flows is compared with the shear viscosity measured in a rotational or capillary viscometer.

The extensive work reported on the flow of polymer solutions through porous media has provided a comprehensive picture of the effect of independent parameters on the onset and degree of extension thickening effects. Most of the information is, however, qualitative in the sense that there are no predictive models but empirical correlations. Further work should be done to establish quantitative links between simple rheological measurements, such as shear and elongational rheometry, and solution behavior in the porous medium. These links should take into account the macroscopic consequence of local shear and elongation rates at the pore level, as well as macromolecular deformation as the polymer undergoes conformation changes and network formation. In this respect, it is expected that further developments in rheo-optical techniques will allow the measurement of local macromolecular dynamics in the pores through, for example, experimental determination of birefringence profiles.

The results presented on cross-linking and solutions of polymer-polymer blends indicate that intermolecular interactions play an important role in the flow resistance of polymer solutions. Further work should be performed in order to quantify these effects.

Acknowledgments. We would like to express our appreciation to J. A. Odell for valuable discussions. Part of the work presented here was supported by grants from the European Union (EC cooperative award No. CI1-CT91-0887) and the Decanato de Investigación y Desarrollo, Universidad Simón Bolívar (S1-CAI-507). We would also like to thank A. C. Gamboa, S. Rodríguez and M. L. Sargenti for their help in preparing this review.

List of Symbols and Abbrevations

A	viscous Ergun constant
	Avogadro's number
B	inertial Ergun constant
C	polymer concentration
C^+	minimum concentration for the formation of transient entanglement networks
$C*$	equilibrium coil overlap concentration
d	particle diameter
d_i	particle diameter of particle size range i
De	Deborah number
De_e	effective Deborah number
De_p	pore Deborah number
I	intensity of optical retardation
k	Boltzmann's constant
k_1	geometric parameter defined in Eq. (11.19)
$\mathbf{K}$	permeability tensor
l	characteristic length of the microscopic scale in porous media flows (pore level)
L	characteristic length of the macroscopic scale in porous media flows
	length over which pressure drop is measured
M	inertial parameter
	polymer molecular weight
$\bar{M}_p$	peak-average molecular weight
$\bar{M}_v$	viscometric average molecular weight
$\bar{M}_w$	weight-average molecular weight
Δn	birefringence
n_i	number of particles of size range i
P	volume-averaged modified pressure
ΔP	pressure drop
$\Delta P'$	corrected pressure drop
r_0	characteristic size of the averaging volume
Re	Reynolds number
$Re*$	Reynolds number in terms of zero shear rate viscosity
Re_o	onset Reynolds number for extension thickening effects
Re_p	pore-scale Reynolds number
t	time
T	temperature

t_c	characteristic time of the flow
$\boldsymbol{v}$	superficial velocity vector
v_p	interstitial velocity of the fluid
X_i	volume fraction of particle size range i

Greek symbols

γ	particle size ratio
$\dot{\gamma}$	shear rate
$\dot{\varepsilon}$	strain rate
$\dot{\varepsilon}_o$	onset strain rate for extension thickening effects
λ	relaxation time of the fluid
Λ	resistance coefficient
Λ^*	resistance coefficient in terms of zero shear rate viscosity
Λ_0	resistance coefficient of polymer solution as Re $\rightarrow$ 0
Λ_s	resistance coefficient of the solvent
Λ_∞	resistance coefficient of polymer solution in the high-Reynolds number plateau
ϱ	density of the fluid
μ	viscosity of the fluid
η	shear viscosity
$[\eta]$	intrinsic viscosity of the polymer
η_0	shear viscosity of low-shear rate Newtonian plateau
η_∞	shear viscosity of high-shear rate Newtonian plateau
η_e^*	reduced effective extensional viscosity
η_e'	effective excess extensional viscosity
η_s	viscosity of the solvent
ϕ	porosity of the porous medium
τ_{11}, τ_{22}	normal stresses
τ_{12}	shear stress

References

1. Jennings RR, Rogers JH, West TJ (1971) J Petroleum Tech March 391
2. Lake LW (1989) Enhanced oil recovery. Prentice Hall, Englewood Cliffs
3. Sorbie KS (1991) Polymer-improved oil recovery. CRC Press, Boca Raton
4. Slater GE, Farouq Ali SM (1968) Prod Monthly 32:22
5. Savins JG (1969) Ind Eng Chem 61:18
6. Gleasure RW, Phillips CR (1990) SPE Reservoir Eng 5:481
7. Whitaker S (1986) Transport in Porous Media 1:3
8. Ergun S (1952) Chem Eng Progress 48:89
9. Macdonald IG, El-Sayed MS, Mow K, Dullien FAL (1979) Ind Eng Chem Fundam 18:199
10. Sargenti ML, Müller AJ, Sáez AE, unpublished results
11. Tatham JP, Carrington S, Odell JA, Gamboa AC, Müller AJ, Sáez AE (1995) J Rheol 39:961
12. Kauser N, Dos Santos L, Delgado M, Müller AJ, Sáez AE (1998) J Appl Polym Sci 72:783
13. Haas R, Kulicke W-M (1985) In: Gampert B (ed) The influence of polymer additives on velocity and temperature fields. Springer, Berlin Heidelberg New York
14. Vorwerk J, Brunn PO (1991) J Non-Newt Fluid Mech 41:119

15. Hill HJ, Brew JR, Claridge EL, Hite JR, Pope GA (1974) paper SPE 4748
16. Sadowski TJ (1963) Non-Newtonian flow through porous media. PhD Thesis, University of Wisconsin, Madison
17. Christopher RH, Middleman S (1965) Ind Eng Chem Fundam 4:422
18. Duda JL, Hong S-A, Klaus EE (1983) Ind Eng Chem Fundam 22:299
19. Chakrabarti S, Seidl B, Vorwerk J, Brunn PO (1991) Rheol Acta 30:114
20. Chauvetau G (1982) J Rheol 26:111
21. Canella WJ, Huh C, Seright RS (1988) paper SPE 18089
22. Sorbie KS, Huang Y (1991) J Colloid Interface Sci 145:74
23. Huang Y, Sorbie KS (1993) paper SPE 25173
24. Durst F, Haas R, Kaczmar BU (1981) J Applied Polym Sci 26:3125
25. Dauben DL, Menzie DE (1967) J Petroleum Tech 19:1065
26. James DF, McLaren DR (1975) J Fluid Mech 70:733
27. Laufer G, Gutfinger C, Abuaf N (1976) Ind Eng Chem Fundam 15:74
28. Elata C, Burger J, Michlin J, Takserman U (1977) Phys Fluids 20:S49
29. Naudascher E, Killen JM (1977) Phys Fluids 20:S280
30. Durst F, Haas R, Naudascher E, Schroeder M (1979) Experimental study of polymer flows through porous media. Report SFB80/E/129, Universität Karlsruhe
31. Käser F, Keller RJ (1980) J Eng Mech Div ASCE EM 3:525
32. Rodríguez S, Romero C, Sargenti ML, Müller AJ, Sáez AE, Odell JA (1993) J Non-Newt Fluid Mech 49:63
33. Pye DJ (1964) Trans SPE 231 (Part I): 911
34. Marshall RJ, Metzner AB (1967) Ind Eng Chem Fundam 6:393
35. Jones WM, Davies OH (1976) J Phys D: Appl Phys 9:753
36. Michele M (1977) Rheol Acta 16:413
37. Kulicke W-M, Haas R (1984) Ind Eng Chem Fundam 23:306
38. Farinato RS, Yen WS (1987) J Appl Polym Sci 33:2353
39. Sáez AE, Müller AJ, Odell JA (1994) Colloid Polym Sci 272:1224
40. Flew S, Sellin RHJ (1993) J Non-Newt Fluid Mech 47:169
41. Gamboa AC, Sáez AE, Müller AJ (1994) Polymer Bull 33:717
42. Sandiford BB (1964) Trans SPE 231 (Part I): 917
43. Jones WM, Maddock JL (1966) Nature 212:388
44. Gogarty WB (1967) Trans SPE 240 (Part I): 151
45. Bird RB, Armstrong RC, Hassager O (1987) Dynamics of polymeric liquids. Volume 1: fluid mechanics, 2nd edn. Wiley-Interscience, New York
46. Franzen P (1979) Rheol Acta 18:392
47. Jones WM, Maddock JL (1969) British J Appl Phys (J Phys D) 2:797
48. Maerker JM (1975) SPE J August: 311
49. De Gennes PG (1974) J Chem Phys 60:5030
50. Hinch EJ (1974) Coll Internationaux C.N.R.S. 233:241
51. Müller AJ, Odell JA, Keller A (1988) J Non-Newt Fluid Mech 30:99
52. Batcheolor GK (1971) J Fluid Mech 46:813
53. Durst F, Haas R (1981) Rheol Acta 20:179
54. Peterlin A (1966) Pure Appl Chem 12:563
55. Warner HJ (1972) Ind Eng Chem Fundam 11:379
56. Bird RB, Armstrong RC, Hassager O (1987) Dynamics of polymeric liquids. Volume 2: kinetic theory, 2nd edn. Wiley-Interscience, New York
57. Haas R, Durst F (1982) Rheol Acta 21:15
58. Hoagland DA, Prud'homme RK (1989) Macromolecules 22:775
59. Deiber JA, Schowalter WR (1981) AIChE J 27:912
60. Heemskerk J, Janssen-Van Rosmalen R, Holtslag RJ, Teeuw D (1984) paper SPE/DOE 12652
61. Gupta RK, Sridhar T (1985) Rheol Acta 24:148
62. Ghoniem SA-A (1985) Rheol Acta 24:588
63. Vorverk J, Brunn PO (1994) J Non-Newt Fluid Mech 51:79

64. Skartsis L, Khomami B, Kardos JL (1992) J Rheol 36:589
65. Dunlap PN, Leal LG (1987) J Non-Newt Fluid Mech 23:5
66. Odell JA, Müller AJ, Keller A (1988) Polymer 29:1179
67. Keller A, Müller AJ, Odell JA (1987) Progress Colloid Polym Sci 75:179
68. Evans AR, Shaqfeh ESG, Frattini PL (1994) J Fluid Mech 281:319
69. Sridhar T (1990) J Non-Newt Fluid Mech 35:85
70. Geffroy E, Leal LG (1990) J Non-Newt Fluid Mech 35:361
71. Dybbs A, Edwards RV (1984) In: Bear J, Corapcioglu Y (eds) Fundamentals of transport phenomena in porous media, Martinus Nijhoff, The Hague, p 199
72. Vossoughi S, Seyer FA (1974) Can J Chem Eng 52:666
73. Moan M, Chauveteau G, Ghoniem S (1979) J Non-Newt Fluid Mech 5:463
74. Ouibrahim A, Fruman DH (1980) J Non-Newt Fluid Mech 7:315
75. James DF, Saringer JH (1980) J Fluid Mech 97:655
76. Haas R, Durst F (1982) Rheol Acta 21:566
77. Chauveteau G, Moan M, Magueur A (1984) J Non-Newt Fluid Mech 16:315
78. Chauveteau G (1986) In: Glass JE (ed) Water-soluble polymers. ACS, Washington (Advances in chemistry series, vol 213)
79. Jones DM, Walters K (1989) Rheol Acta 28:482
80. Chmielewski C, Petty CA, Jayaraman K (1990) J Non-Newt Fluid Mech 35:309
81. Talwar KK, Khomami B (1995) J Non-Newt Fluid Mech 57:177
82. Müller AJ, Medina LI, Pérez-Martín O, Rodríguez S, Romero C, Sargenti ML, Sáez AE (1993) Appl Mech Rev 46:S63
83. Chow A, Keller A, Müller AJ, Odell JA (1988) Macromolecules 21:250
84. Harlen OG, Hinch EJ, Rallison JM (1992) J Non-Newt Fluid Mech 44:229
85. Doi M, Edwards SF (1986) The theory of polymer dynamics. Clarendon Press, Oxford
86. Keller A, Odell JA (1985) Colloid Polym Sci 263:181
87. Nguyen TW, Yu G, Kausch H-H (1995) Macromolecules 28:4851
88. Odell JA, Keller A, Müller AJ (1989) In: Glass JE (ed) Polymers in aqueous media – performance through association. ACS, Washington, p 193
89. Müller AJ, Odell JA, Tatham JP (1990) J Non-Newt Fluid Mech 35:231
90. Chin S, Hoagland DA, Muri JJ, Parkhe AD (1989) Rheol Acta 28:202
91. Dyakonova NE, Odell JA, Brestkin YuV, Lyulin AV, Sáez AE (1996) J Non-Newt Fluid Mech 67:285
92. Rando M, Smitter L, Socías P, Müller AJ, Sáez AE (in preparation)
93. Chauveteau G, Kohler N (1974) paper SPE 4745
94. Müller AJ, Gamboa AC, Sáez AE (1994) In: Proceedings of the 4th Latin-American polymer symposium. Gramado, Brazil, p 372
95. Kulicke W-M, Kniewske R, Klein J (1982) Progress Polym Sci 8:373
96. Haas R, Kulicke W-M (1984) Ind Eng Chem Fundam 23:316
97. Tsiklauri MG (1994) J Eng Phys Thermophys 66:233
98. Smith FW (1970) J Petroleum Tech Feb: 148
99. Maerker JM (1976) SPE J August: 172
100. Morris CW, Jackson KM (1978) paper SPE 7064
101. Seright RS (1980) paper SPE 9297
102. Wellington SL (1983) SPE J December: 901
103. Southwick JG, Manke CW (1986) paper SPE 15652
104. Martin FD (1984) paper SPE/DOE 12651
105. Moreno RA, Müller AJ, Sáez AE (1996) Polym Bull 37:663
106. Odell JA, Müller AJ, Narh KA, Keller A (1990) Macromolecules 23:3092
107. Odell JA, Keller A, Müller AJ (1992) Colloid Polym Sci 270:307
108. Müller AJ, Odell JA, Carrington S (1992) Polymer 33:2598
109. Seright RS, Maerker JM, Holzwarth G (1981) Polym Preprints 22:30
110. Dingilian G, Ruckenstein E (1974) AIChE J 20:1222
111. Malhotra JP, Chaturvedi PN, Singh RP (1988) J Appl Polym Sci 36:837

Subject Index

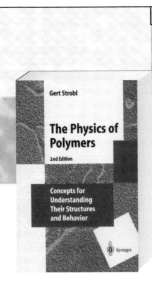

G.R. Strobl

Physics of Polymers

**Concepts for Understanding
Their Structures and Behavior**

2nd corr. ed.1997.XII, 439 pp.
218 figs. 5 tabs.
Softcover DM 58,-*
ISBN 3-540-63203-4

Polymer physics is one of the key lectures not only in polymer science but also in materials science. Strobl presents in his textbook the elements of polymer physics to the necessary extent in a very didactical way. His main focus lays on the concepts of polymer physics, not on theoretical aspects or mere physical methods.

**Please order from:
Springer-Verlag
P.O. Box 14 02 01
D-14302 Berlin, Germany
Fax: +49 30 827 87 301
e-mail: orders@springer.de
or through your bookseller**

Springer

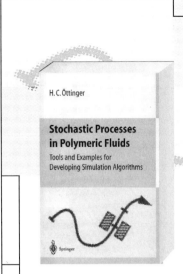

H.C. Öttinger

Stochastic Processes in Polymeric Fluids

Tooks and Examples for Developing Simulation Algorithms

1996. XXIV, 362 pp. 34 figs., 3 tabs. Softcover DM 98,-* ISBN 3-540-58353-X

This book consists of two strongly interweaved parts: the mathematical theory of stochastic processes and its applications to molecular theories of polymeric fluids. The comprehensive mathematical background provided in the first part should be equally useful in many other branches of engineering and the natural sciences. As a benefit from the second part one gains a more direct understanding of polymer dynamics, one can more easily identify exactly solvable models, and one can develop efficient computer simulation algorithms in a straightforward manner. In view of the examples and applications to problems from the front line of science, this volume may be used as a basic textbook or as a reference book.

Program examples written in FORTRAN are available from: ftp.springer.de/pub/chemistry/polysim/.

Please order from:
Springer-Verlag
P.O. Box 14 02 01
D-14302 Berlin, Germany
Fax: +49 30 827 87 301
e-mail: orders@springer.de
or through your bookseller

* This price applies in Germany/Austria/Switzerland and is a recommended retail price. Prices and other details are subject to change without notice. In EU countries the local VAT is effective. d&p · 65046/2 SF · Gha

Computer to Film: Saladruck, Berlin
Binding: H. Stürtz AG, Würzburg